A Short Introduction to String Theory

Suitable for graduate students in physics and mathematics, this book presents a concise and pedagogical introduction to string theory. It focuses on explaining the key concepts of string theory, such as bosonic strings, D-branes, supersymmetry, and superstrings and on clarifying the relationship between particles, fields, and strings without assuming an advanced background in particle theory or quantum field theory, thus making it widely accessible to interested readers from a range of backgrounds. Important ideas underpinning current research, such as partition functions, compactification, gauge symmetries, and T-duality are analysed both from the world-sheet (conformal field theory) and the space-time (effective field theory) perspectives. Ideal for either self-study or a one semester graduate course, *A Short Introduction to String Theory* is an essential resource for students studying string theory, containing examples and homework problems to develop understanding, with fully worked solutions available to instructors.

Thomas Mohaupt obtained his PhD in Theoretical Physics from the University of Münster and is Reader in Theoretical Physics at the University of Liverpool. He has published extensively on string theory, specialising in black hole entropy and supergravity, and has taught various courses on related topics in the UK and Germany. He has been a visiting scholar at Stanford University, a visiting professor at the University of Nancy, and a Senior Fellow at the Erwin Schrödinger International Institute for Mathematics and Physics in Vienna.

A Short Introduction to String Theory

THOMAS MOHAUPT

University of Liverpool

CAMBRIDGE
UNIVERSITY PRESS

CAMBRIDGE
UNIVERSITY PRESS

University Printing House, Cambridge CB2 8BS, United Kingdom

One Liberty Plaza, 20th Floor, New York, NY 10006, USA

477 Williamstown Road, Port Melbourne, VIC 3207, Australia

314–321, 3rd Floor, Plot 3, Splendor Forum, Jasola District Centre, New Delhi – 110025, India

103 Penang Road, #05–06/07, Visioncrest Commercial, Singapore 238467

Cambridge University Press is part of the University of Cambridge.

It furthers the University's mission by disseminating knowledge in the pursuit of education, learning, and research at the highest international levels of excellence.

www.cambridge.org
Information on this title: www.cambridge.org/9781108481380
DOI: 10.1017/9781108611619

First published 2022

A catalogue record for this publication is available from the British Library.

ISBN 978-1-108-48138-0 Hardback

Additional resources for this publication at www.cambridge.org/stringtheory.

This book is dedicated to my parents, Dr Helga and Manfred Mohaupt.

Contents

Preface

This book gives a short introduction into the basic ideas and concepts of string theory. It is aimed, primarily, at PhD students who are commencing their studies and for other readers who would like to approach the topic at a gentler pace than in other textbooks, and with a minimum of prerequisites. The level of this book is in between Zwiebach (2009), which is a pedagogical introduction into string theory for advanced undergraduates, and textbooks such as Polchinski (1998a), Blumenhagen et al. (2013), Becker et al. (2007), and Schomerus (2017), which go quite steeply into the subject and also are quite comprehensive with regard to the topics they cover.

The approach and selection of topics reflect my experience with teaching fledgling PhD students in the United Kingdom for the past 15 years, including a short course for the MAGIC consortium since 2013. I have expanded the treatment of topics where, in my experience, beginners tend to struggle or to get confused. Part I covers the classical theory and quantisation of particles, strings, and, to some extent, fields. Some material that is typically found in introductory quantum field theory texts has been included: first, because it helps to see this material in context, and second, to start developing the effective field theory perspective, which is one of the threads of this book. Parts II and III unfold the world-sheet and the space-time perspectives of string theory. Part II introduces the minimal background in conformal field theory needed to underpin covariant quantisation, and later toroidal compactification. Part III analyses the physical spectrum in the covariant and light-cone approach and includes a mini-review of the representation theory of the Poincaré group in order to provide a solid understanding of why string states correspond to elementary particles. The critical dimension $D = 26$ of the bosonic string and its shift $a = 1$ in ground state energy are derived in the light-cone formulation as a guided exercise. Parts II and III also contain chapters on partition functions which, in particular, introduce the concept of modular invariance. By the end of Part III, the reader should have a solid understanding of the free quantum bosonic string with periodic, Neumann, and Dirichlet boundary conditions. Part IV goes deeper (and steeper) into three topics: interactions, compactifications, and matter (fermions). Besides a general discussion of amplitudes, the closed string four-scalar amplitude is computed and its structure is analysed and compared to point particle amplitudes. Curved backgrounds are discussed briefly from the sigma model and effective field theory perspective, putting emphasis on the fact that scattering amplitudes and curved backgrounds are complementary ways to describe string interactions. A comprehensive chapter on compactification starts with a detailed treatment of the simplest case, compactification on a circle, as this already allows one to uncover most of the special features of string compactifications. Orbifolds are exemplified using the orbicircle and general toroidal compactifications are treated in some detail, including their generic massless spectra, their effective actions, symmetry enhancement, and the Higgs effect.

T-duality is developed for both closed and open strings, including the Buscher rules. We close this chapter with some remarks on more general compactifications, the vacuum selection problem, and the swampland programme. In the last chapter, we explain how matter (fermions) arises through world-sheet supersymmetry and how modular invariances leads one to the existence of 5 superstring theories defined on 10-dimensional Minkowski space. The chapter concludes with a brief overview of the dualities that relates these theories to each other and to 11-dimensional M-theory.

I hope that this book will be useful for self-study and as a companion for an introductory string theory course at postgraduate level. It includes a number of exercises which, besides encouraging active reading, can be used in classes. The price for a detailed and pedagogical treatment in a short volume is the reduction in the number of topics. D-branes, superstrings, and gravitational aspects of string theory are treated relatively briefly, but the basic ideas are introduced and illustrated. Unfortunately, black hole entropy and the AdS/CFT correspondence had to be omitted, as this would have required to abandon the very idea of a 'short introduction'. Also, we have restricted ourselves to old covariant quantisation and light-cone quantisation, which are sufficient for the topics we cover in this short introduction, while leaving out the path integral and BRST formalisms, which are well covered in the more advanced and comprehensive textbooks mentioned above and in the Introduction.

Large parts of this book should be accessible to any reader with a solid background in quantum mechanics and special relativity. I imagine that the typical reader will already have some background in quantum field theory, general relativity, and group theory and will hopefully develop their understanding of these topics in parallel to studying string theory. I have tried throughout to minimise assumptions about the reader's background, thus making the book as accessible and self-contained as possible. Some relevant background material is provided through appendices or 'narrative summaries' within the main text, together with pointers to the literature.

Acknowledgements

This book is based on various long and short courses I gave on string theory over the past 20 years, in particular a full-year course for advanced undergraduates at the University of Jena, introductory courses for fledgling PhD students at the University of Liverpool, and a short course I have been contributing to the programme of the MAGIC consortium since 2013. While this book does not directly follow any of these lectures, it has grown out of this experience, and the questions and comments I have received from the students attending my courses were essential in shaping it. I would like to thank my colleagues and the PhD students in the String and Beyond the Standard Model Phenomenology research cluster at the Department of Mathematical Sciences, University of Liverpool, for providing a stimulating environment, in particular through our weekly group meetings. I am particularly indebted to Gabriel Lopes Cardoso, Harold Erbin, Maxime Médevielle, and Flavio Tonioni for their detailed feedback.

Introduction

String theory is an extension of quantum field theory which includes gravity and provides a framework for constructing models in particle physics and cosmology. It aims to be a complete fundamental theory of space, time, and matter. Given that the particle concept becomes subtle once special relativity and quantum theory merge into quantum field theory, not to speak of curved backgrounds or even full-fledged quantum gravity, the idea 'to replace point particles by one-dimensional extended objects, called strings' may sound somewhat naive. And historically, string theory did not start with this idea, but with the observation that certain amplitudes with desirable properties could be interpreted as scattering amplitudes of strings. The history of string theory has many twists and turns, which tell us a lot about the erratic nature of theory evolution. In the author's view, string theory is an example of a theory which has not been constructed according to a master plan or programme, but has been discovered – or, if you prefer, developed – in continuous reaction to often unexpected new insights and observations. This said, we will avoid historical and philosophical reflections and develop string theory systematically on the basis of quantum theory and special relativity as taught in an undergraduate course. In time, we will make quite a few intriguing observations, which hopefully will convince the reader that this pursuit is worth our time and effort.

While we will present explicit computations, and ask the reader to perform several more as exercises, equal emphasis will be given to ideas and concepts. Some themes are recurrent throughout this book. One is the complementarity between the *space-time* and the *world-sheet perspectives*. Essentially, this is the observation that the mechanics of a one-dimensional continuum can be formulated as a field theory in $1 + 1$ dimensions. In the currently dominant 'first quantised' formulation of string theory, the world-sheet (trajectory, generalised world-line) plays a central role, in contrast to quantum field theory which, in its standard formulation, takes the space-time perspective. As the particle spectrum of a quantised string contains an infinite tower of massive states – at least as long as we neglect the backreaction on space-time which may lead to the generation of black holes – it is natural to restrict oneself to the dynamics of the most relevant, that is the lightest modes, and often just the massless modes. This naturally leads us to consider the *effective field theory* of the light modes, a concept that universally applies whenever we distinguish between a microscopic and potentially fundamental, and a macroscosopic and effective level of description. Another thread is the comparison between particles and strings, that is the question what makes strings special. Here, we will see that world-sheet *modular invariance* acts like a built-in UV cut-off, while *T-duality* (also called space-time modular invariance) introduces a fundamental length scale and forces us to re-consider how we think about space-time geometry. We will also encounter the power of anomalies, which makes string theory highly restricted and potentially unique. In particular, we will see how the Lorentz covariance of string theory in Minkowski space fixes the number of space-time dimensions. While this makes the theory unique in theory (if you pardon the pun), in practice non-uniqueness bites back in the form of the non-uniqueness of solutions: the predicted number of space-time dimensions is larger than the observed one, thus forcing us to compactify the extra dimensions. At some points in the book we will pause to bring in some additional mathematical tools, or to summarise results from quantum field theory to provide background. This is done in a minimalistic but, hopefully, sufficiently self-contained way and references for further study are provided.

The book is divided into four parts. Part I compares the classical theories of relativistic particles and strings and outlines their quantisation. It also reviews how particles – and, by extension, individual excitations of strings – are described using quantum field theory. We explain the role of constraints and show that for strings we have the choice between *periodic, Neumann*, and *Dirichlet boundary conditions*, the latter implying the existence of *D-branes*. At the end of Part I, we formulate the problem of Lorentz covariant quantisation of strings. Part II develops the world-sheet perspective of string theory, that is, two-dimensional *conformal field theory*. This provides us with the tools for formulating the quantum theory in Part III. Part III takes the space-time point of view. We review the representation theory of the *Poincaré group* in order to understand precisely how string states correspond to elementary particles with a specific mass and spin (or the higher-dimensional analogue of spin, conceptualised as the *little group*). We also discuss how string states can be described from an effective field theory perspective, which helps us to make contact with the usual framework of particle theory based on quantum field theory. Besides Lorentz covariant quantisation we discuss *light-cone quantisation*, which allows us to give a complete treatment of how the number of space-time dimensions is fixed. At the end of Part III, we will have a solid understanding of the quantum theory of free bosonic strings. In Part IV, we provide an overview of three further developments of the theory: *interactions, compactifications* and *matter* (fermions). Interactions can be desribed in two ways: as scattering amplitudes in a fixed background and as the interaction of a string with a non-trivial on-shell background of coherent string states. We explain how these viewpoints are related and in which sense the theory is background independent. The next topic is how to reduce the number of visible space-time dimensions by compactification. We focus on flat backgrounds, that is circles, tori, and orbifolds thereof. The spectrum, symmetry enhancement, and symmetry breaking is discussed from the world-sheet and space-time perspectives. The existence of winding modes distinguishes strings from particles, and leads to a new symmetry, T-duality, which relates large and small compactification scales and introduces a fundamental length scale. The plethora of possible compactifications poses a serious challenge to identifying a string theory model of our universe. We briefly discuss the concepts of *landscape* and *swampland*. Compactifications on two-tori are used to introduce *Calabi – Yau manifolds* and *mirror symmetry*. While we use the bosonic string to explain the concepts and properties of string theory, in the last chapter we show how the theory can be generalised to include *fermions*. As this is intimately related to *supersymmetry*, we start by introducing supersymmetry, first using the harmonic oscillator, then as an extension of Poincaré symmetry. The RNS model serves as a simple example of a two-dimensional supersymmetric field theory, explains how space-time fermions arise through world-sheet supersymmetry, and serves as a stepping-stone to Type-II superstring theories. We sketch how modular invariance imposes restrictions which lead one to 5, 10-dimensional supersymmetric string theories. Type-I and heterotic strings are briefly introduced and we conclude with the network of dualities that relates the 5 superstring theories to each other and to 11-dimensional M-theory.

This book will show the influence of the textbooks that I have used over the years to prepare my lectures and I have not tried to be original for originality's sake. When I entered the field, Green et al. (1987) was the only textbook and, while it does not include the new developments since the later 1980s, it is still unsurpassed in what it does. My other formative text was the predecessor of Blumenhagen et al. (2013) which, in its current incarnation, is one of my standard references, together with Polchinski (1998a), Polchinski (1998b), and Becker et al. (2007). There are of course, other excellent textbooks, including Kaku (1988), Kaku (1991), Kiritsis (2007), West (2012), and Schomerus (2017) and readers who have grown out of this book will not have difficulties in finding one that fits their preferences in style and choice of topics and perspective. There also are books which provide in-depth coverage of topics that we treat relatively briefly or have been forced to omit, in particular, Johnson (2003) for D-branes, Ortin (2004) for a space-time and (super-)gravity based approach to string theory, Ibáñez and Uranga (2012) for string phenomenology, and Ammon and Erdmenger (2015) for the AdS/CFT correspondence. Mathematically inclined readers will enjoy the two volumes Deligne et al. (1999) which cover various topics in string theory and quantum field theory.

PART I

FROM PARTICLES TO STRINGS

Classical Relativistic Point Particles

In this chapter, we review classical relativistic point particles and set out our conventions and notation. Readers familiar with this material can skip through the chapter and use it as a reference when needed.

1.1 Minkowski Space

According to Einstein's theory of special relativity, space and time are combined into 'space-time', which is modelled by Minkowski space $\mathbb{M}$.[1] The elements $P, Q, \ldots \in \mathbb{M}$ are called *events*. We leave the dimension D of space-time unspecified. Minkowski space is homogeneous and thus has no preferred origin, which makes it a point space (affine space) rather than a vector space (linear space). However, displacements relating events P, Q are vectors,

$$x = \overrightarrow{PQ} \in \mathbb{R}^D, \tag{1.1}$$

and once we choose a point $O \in \mathbb{M}$ as the origin of our coordinate system there is a one-to-one correspondence between events P and position vectors

$$x_P = \overrightarrow{OP}. \tag{1.2}$$

The components

$$(x^\mu)_{\mu=0,1,\ldots,D-1} = (x^0, \vec{x}) \,, \quad \vec{x} = (x^i)_{i=1,\ldots,D-1} \tag{1.3}$$

of vectors $x \in \mathbb{R}^D$ provide linear coordinates on $\mathbb{M}$. We assume that $x^i = 0$ is the world-line of an inertial (force-free) observer, so that $x^0 = ct$ is proportional to the time t measured in the associated inertial system, while x^i provide linear coordinates on space. We will normally use natural units where we set the speed of light to unity, $c = 1$.[2]

To measure the distance between events, we use the indefinite scalar product

$$x \cdot y = \eta_{\mu\nu} x^\mu y^\nu, \tag{1.4}$$

on the vector space $\mathbb{R}^D$, with Gram matrix

$$\eta = (\eta_{\mu\nu}) = \begin{pmatrix} -1 & \vec{0}^{\,T} \\ \vec{0} & \mathbb{1}_{D-1} \end{pmatrix}. \tag{1.5}$$

---

[1] For brevity's sake we will use 'Minkowksi space' instead of 'Minkowski space-time'.
[2] Conventions and units are reviewed in Appendices A and B, respectively.

Note that we use the *mostly plus convention* for the metric η. Since the metric allows us to identify the vector space $\mathbb{R}^D$ with its dual, vectors have covariant components x_μ as well as contravariant components x^μ, which are related by *raising and lowering indices* $x_\mu = \eta_{\mu\nu}x^\nu$, and $x^\mu = \eta^{\mu\nu}x_\nu$, where Einstein's summation convention is understood. The corresponding line element on Minkowski space is

$$ds^2 = \eta_{\mu\nu}dx^\mu dx^\nu = -dt^2 + d\vec{x}^2. \tag{1.6}$$

The most general class of transformations which preserve this line element (its *isometries*) are the Poincaré transformations

$$x^\mu \rightarrow \Lambda^\mu{}_\nu x^\nu + a^\mu, \quad \cdot \tag{1.7}$$

where $a = (a^\mu) \in \mathbb{R}^D$ and where $\Lambda = (\Lambda^\mu{}_\nu)$ is an invertible $D \times D$ matrix satisfying

$$\Lambda^T \eta \Lambda = \eta. \tag{1.8}$$

The matrices Λ describe *Lorentz transformations*, which are the most general linear transformations preserving the metric. The Lorentz transformations form a Lie group of dimension $\frac{1}{2}D(D-1)$, called the *Lorentz group* $O(1, D-1)$. Elements $\Lambda \in O(1, D-1)$ have determinant $\det \Lambda = \pm 1$, and satisfy $|\Lambda^0{}_0| \geq 1$. The matrices with $\det \Lambda = 1$ form a subgroup $SO(1, D-1)$. This subgroup still has two connected components, since $\Lambda^0{}_0 \geq 1$ or $\Lambda^0{}_0 \leq -1$. The connected component containing the unit matrix $\mathbb{1} \in O(1, D-1)$ is the *connected* or *proper orthochronous Lorentz group* $SO_0(1, D-1)$. The corresponding Lie algebra is $\mathfrak{so}(1, D-1)$.

The Lorentz group and translation group combine into the *Poincaré group*, or inhomogeneous Lorentz group, $IO(1, D-1)$, which is a Lie group of dimension $\frac{1}{2}D(D+1)$. Since Lorentz transformations and translations do not commute, the Poincaré group is not a direct product. The composition law

$$(\Lambda, a) \circ (\Lambda', a') = (\Lambda\Lambda', a + \Lambda a') \tag{1.9}$$

shows that the Lorentz group operates on the translation group by the fundamental or vector representation. Therefore, the Poincaré group is the *semi-direct product* of the Lorentz and translation group,

$$IO(1, D-1) = O(1, D-1) \ltimes \mathbb{R}^D. \tag{1.10}$$

Since the Minkowski metric (1.6) is defined by an indefinite scalar product, the square-distance or square-norm $x^2 = x \cdot x$ can be positive, zero, or negative. For terminological simplicity, we will henceforth refer to $x^2 = x \cdot x$ as the *norm* or *length* of x, omitting the qualifier 'square'. This convention will be applied whenever we deal with an indefinite scalar product.

Vectors are classified as *time-like*, *light-like* (also called *null*), or *space-like* according to their norm:

$$x \text{ time-like} \quad \Leftrightarrow \quad x \cdot x < 0,$$
$$x \text{ light-like} \quad \Leftrightarrow \quad x \cdot x = 0,$$
$$x \text{ space-like} \quad \Leftrightarrow \quad x \cdot x > 0.$$

Since signals can only travel with speed $v \leq 1 (= c)$, this encodes information about the causal relations between events. Two events P, Q are called time-like, light-like, or space-like relative to each other, if the displacement vector $x = \overrightarrow{PQ}$ is time-like, light-like, or space-like, respectively. Only non-space-like events can be causally related, and their causal order is invariant under orthochronous Poincaré transformations, which exclude the time inversion $T : t \to -t, \vec{x} \to \vec{x}$. Since some particle interactions are not invariant under the space inversion $P : t \to t, \vec{x} \to -\vec{x}$, the symmetry group relevant for particle physics is the proper orthochronous Poincaré group $SO_0(1, D-1) \ltimes \mathbb{R}^D$. This is the connected component of the unit element of the full Poincaré group, which has three further connected components which contain T, P, and their product TP.

1.2 Particles

The fundamental constituents of matter are usually modelled as particles, that is, as objects that are localised and can be characterised by a small number of parameters, such as mass, spin, and charges. While some particles are bound states of others, the standard model of particle physics is based on a list of particles, assumed to be elementary in the sense that they do not have constituents and, therefore, no internal excitations. In classical mechanics, such particles are modelled as mathematical points. The motion of such a point particle, or *particle* for short, is described by a parametrised curve called the *world-line*. If we restrict ourselves to inertial frames, it is natural to choose the coordinate time t as the curve parameter. Then, the world-line of a particle is a parametrised curve

$$C : I \to \mathbb{M} : t \mapsto x(t) = (x^\mu(t)) = (t, \vec{x}(t)), \tag{1.11}$$

where $I \subset \mathbb{R}$ is the time interval for which the particle is observed. $I = \mathbb{R}$ is included as a limiting case.

The *velocity* of a particle relative to an inertial frame is

$$\vec{v} = \frac{d\vec{x}}{dt}, \tag{1.12}$$

and $v = \sqrt{\vec{v} \cdot \vec{v}} \geq 0$ is the speed. Since t and $\vec{v}$ are not covariant quantities (Lorentz tensors), it is useful to formulate relativistic mechanics using the Lorentz vector x^μ and its derivatives with respect to a curve parameter which is a Lorentz scalar. This works differently for massive and for massless particles.

The inertial *mass m* of a particle measures its resistance against a change of velocity. Massive particles, $m > 0$, propagate with velocities $v < 1$ and have time-like world-lines, that is world-lines where the tangent vector is time-like everywhere. Massless particles, $m = 0$, propagate with velocity $v = 1$ and have light-like world-lines. Poincaré symmetry also admits *tachyons*, that is, particles with negative mass-squared, $m^2 < 0$, which propagate with velocity $v > 1$ and have space-like world-lines. Such tachyons are discarded because they would allow a-causal effects, such as sending signals backwards in time. In quantum field theory, tachyons are re-interpreted as indicating instabilities resulting from expanding a theory around a local maximum of the potential. This is a physical effect and does not involve particles propagating with superluminal speed (see Section 7.7).

For massive particles, we can use the *proper time* τ as a curve parameter. Infinitesimally, the relation between proper time and coordinate time is

$$-d\tau^2 = [-dt^2 + d\vec{x}^2]_{|C} = \left[-1 + \left(\frac{d\vec{x}}{dt}\right)^2\right] dt^2 \Rightarrow d\tau = dt\sqrt{1 - \vec{v}^2}. \qquad (1.13)$$

Since τ is, by construction, Lorentz invariant, the *relativistic velocity*

$$\dot{x}^\mu = \frac{dx^\mu}{d\tau} = \left(\frac{dt}{d\tau}, \frac{d\vec{x}}{dt}\frac{dt}{d\tau}\right) = \frac{1}{\sqrt{1 - \vec{v}^2}}(1, \vec{v}) \qquad (1.14)$$

is a Lorentz vector. Moreover, it is a time-like *unit tangent vector* to the world-line, since $\dot{x}^2 = \dot{x}^\mu \dot{x}_\mu = -1$. The norm is in particular constant, which makes τ an *affine curve parameter*. The name 'affine' curve parameter reflects the fact that such curve parameters are unique up to affine transformations, $\tau \mapsto a\tau + b$, $a, b \in \mathbb{R}$, $a \neq 0$.

By further differentiation, we obtain the *relativistic acceleration*,

$$a^\mu = \ddot{x}^\mu. \qquad (1.15)$$

Newton's first law states that force-free particles are unaccelerated relative to inertial frames.

The *relativistic momentum* of a particle is

$$p^\mu = m\dot{x}^\mu = (p^0, \vec{p}) = \left(\frac{m}{\sqrt{1 - \vec{v}^2}}, \frac{m\vec{v}}{\sqrt{1 - \vec{v}^2}}\right). \qquad (1.16)$$

The component $p^0 = E$ is the total energy of the particle. The norm of p^μ is minus its mass squared

$$p^\mu p_\mu = -m^2 = -E^2 + \vec{p}^2. \qquad (1.17)$$

Note the minus sign which is due to us using the mostly plus convention for the metric.

Force-free particles propagate with constant velocity, which means that their world-lines are straight lines. The relativistic version of *Newton's second law* states that motion under a force is determined by the equation

$$\frac{dp^\mu}{d\tau} = m\frac{d^2x^\mu}{d\tau^2} = f^\mu, \qquad (1.18)$$

where the Lorentz vector f^μ is the relativistic force. Note that we assume that the mass m is constant, which is satisfied for stable elementary particles but may not hold in other applications of relativistic mechanics (e.g., for the motion of a rocket which expels fuel).

1.3　A Non-covariant Action Principle for Relativistic Particles

The equations of motion of all fundamental physical theories can be obtained from variational principles. In this approach, a theory is defined by specifying its *action* which is a functional on the configuration space. The equations of motion are the Euler–Lagrange equations obtained by imposing that the action is invariant under infinitesimal variations of the path, with the initial and final position kept fixed.

For a point particle, the configuration space is parametrised by its position $\vec{x}$ and velocity $\vec{v}$. The action functional takes the form

$$S[\vec{x}] = \int dt\, L(\vec{x}(t), \vec{v}(t)). \tag{1.19}$$

In principle, the Lagrangian L can have an explicit dependence on time, corresponding to a time-dependent potential or external field. In fundamental theories, we assume the invariance of the field equations under time-translations, which forbids an explicit time dependence of L.

The action for a free, massive, relativistic particle is proportional to the proper time along the world-line, and given by minus the product of its mass and the proper time:

$$S = -m \int dt \sqrt{1 - \vec{v}^2}. \tag{1.20}$$

The minus sign has been introduced so that L has the conventional form $L = T - V$ where T is the part quadratic in time derivatives, that is, the *kinetic energy*. The remaining part V is the *potential energy*. We work in units where the speed of light and the reduced Planck constant have been set to unity, $c = 1$, $\hbar = 1$. In such *natural units* the action S is dimensionless. To verify that the action principle reproduces the equation of motion (1.18), we consider the motion $\vec{x}(t)$ of a particle between the initial postion $\vec{x}_1 = \vec{x}(t_1)$ and the final position $\vec{x}_2 = \vec{x}(t_2)$. Then, we compute the first order variation of the action under infinitesimal variations $\vec{x} \to \vec{x} + \delta\vec{x}$, which are arbitrary, except for the boundary conditions $\delta\vec{x}(t_i) = 0, i = 1, 2$ (see Figure 1.1). To compare the initial and deformed path we Taylor expand in $\delta\vec{x}(t)$:

$$S[\vec{x}(t) + \delta\vec{x}(t)] = S[\vec{x}(t), \vec{v}(t)] + \delta S[\vec{x}(t), \vec{v}(t)] + \cdots \tag{1.21}$$

where the omitted terms are of quadratic and higher order in $\delta\vec{x}(t)$. The equations of motion are found by imposing that the first variation δS vanishes.

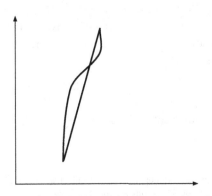

Fig. 1.1   The action principle selects the paths for which the action is stationary under variation. The endpoints are kept fixed.

Practical manipulations are most easily performed using the following observations:

1. The variation δ *acts like a derivative*. For example, for a function $f(\vec{x})$ we have the chain rule

$$\delta f = \partial_i f \delta x^i, \tag{1.22}$$

as is easily verified by Taylor expanding $f(\vec{x} + \delta\vec{x})$. Similarly, sum, constant factor, product, and quotient apply, for example, $\delta(fg) = \delta f g + f \delta g$.

2. $\vec{v} = \frac{d\vec{x}}{dt}$ is not an independent quantity, and therefore

$$\delta\vec{v} = \delta\frac{d\vec{x}}{dt} = \frac{d}{dt}\delta\vec{x}. \tag{1.23}$$

3. To find δS, we need to collect all terms proportional to $\delta\vec{x}$. Derivatives acting on $\delta\vec{x}$ have to be removed through integration by parts which creates boundary terms. If such boundary terms are not automatically zero, we must impose that they vanish which restricts the class of configurations which qualify as solutions.

Solving the variational problem for the action (1.20), we obtain the equation of motion

$$\frac{d}{dt}\frac{m\vec{v}}{\sqrt{1 - \vec{v}^2}} = \frac{d}{dt}\vec{p} = \vec{0}, \tag{1.24}$$

which is equivalent to (1.18) in the absence of forces, $f^\mu = 0$.

Remark: When performing the variation without specifying the Lagrangian L, one obtains the Euler–Lagrange equations

$$\frac{\partial L}{\partial x^i} - \frac{d}{dt}\frac{\partial L}{\partial v^i} = 0. \tag{1.25}$$

For $L = -m\sqrt{1 - \vec{v}^2}$ this is easily seen to give (1.24).[3]

Exercise 1.3.1 Verify that the variation of (1.20) takes the form

$$\delta S = -\int_{t_1}^{t_2}\left(\frac{d}{dt}\frac{mv_i}{\sqrt{1 - \vec{v}^2}}\right)\delta x^i dt + \frac{mv_i}{\sqrt{1 - \vec{v}^2}}\delta x^i\bigg|_{t_1}^{t_2}. \tag{1.26}$$

Does the boundary term impose any conditions on the dynamics?

Exercise 1.3.2 Verify that (1.24) is equivalent to (1.18) with $f^\mu = 0$. Then, extend this to the case where a force term is present. This requires one to know, or to derive, the relation between the relativistic force f^μ and the non-relativistic expression $\vec{F}$. Show that the non-covariant version of Newton's second law,

$$\frac{d\vec{p}}{dt} = \vec{F}, \quad \text{where} \quad \vec{p} = \frac{m\vec{v}}{\sqrt{1 - v^2}} \tag{1.27}$$

is equivalent to (1.18).

---

[3] In my opinion, it is more natural, convenient, and insightful to obtain the equations of motion for a given concrete theory by varying the corresponding action, as done above, instead of plugging the Lagrangian into the Euler–Lagrange equations. This procedure reminds one that there may be boundary terms that one has to worry about, as we will see when replacing particles by strings.

1.4 Canonical Momenta and Hamiltonian

We now turn to the Hamiltonian description of the relativistic particle. In the Lagrangian formalism we use the configuration space variables $(\vec{x}, \vec{v}) = (x^i, v^i)$. In the Hamiltonian formalism, the velocity $\vec{v}$ is replaced by the canonical momentum

$$\pi^i := \frac{\partial L}{\partial v_i}. \tag{1.28}$$

For the Lagrangian $L = -m\sqrt{1 - \vec{v}^2}$, the canonical momentum agrees with the kinetic momentum, $\vec{\pi} = \vec{p} = (1 - \vec{v}^2)^{-1/2} m\vec{v}$. However, conceptually canonical and kinetic momentum are different quantities. A standard example where the two quantities are not equal is a charged particle in a magnetic field (see Section 13.6, i.p. formula (13.169)).

The Hamiltonian $H(\vec{x}, \vec{\pi})$ is obtained from the Lagrangian $L(\vec{x}, \vec{v})$ by a Legendre transformation:

$$H(\vec{x}, \vec{\pi}) = \vec{\pi} \cdot \vec{v} - L(\vec{x}, \vec{v}(\vec{x}, \vec{\pi})). \tag{1.29}$$

For $L = -m\sqrt{1 - \vec{v}^2}$ the Hamiltonian is equal to the total energy:

$$H = \vec{\pi} \cdot \vec{v} - L = \vec{p} \cdot \vec{v} - L = \frac{m}{\sqrt{1 - \vec{v}^2}} = p^0 = E. \tag{1.30}$$

Describing relativistic particles using the action (1.20) has the following disadvantages:

- We can describe massive particles, but photons, gluons, and the hypothetical gravitons underlying gravity are believed to be massless. How can we describe massless particles?
- The independent variables $\vec{x}, \vec{v}$ are not Lorentz vectors. Therefore, our formalism lacks manifest Lorentz covariance. How can we formulate an action principle that is Lorentz covariant?
- We have picked a particular curve parameter for the world-line, namely the inertial time with respect to a Lorentz frame. While this is a natural choice, 'physics', that is, observational data, cannot depend on how we label points on the world-line. How can we formulate an action principle that is manifestly covariant with respect to reparametrisations of the world-line?

We will answer these questions in reverse order.

1.5 Length, Proper Time, and Reparametrisations

To prepare for the following, we first discuss general curve parameters and reparametrisations. Consider a smooth parametrised curve in Minkowski space,

$$C : I \ni \sigma \longrightarrow x^\mu(\sigma) \in \mathbb{M}, \tag{1.31}$$

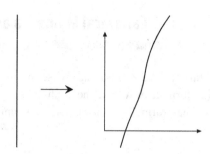

Fig. 1.2 The world-line of a particle is described by a parametrised curve, that is, by a map from a parameter interval into space-time. Physical quantities do not depend on the parametrisation.

where σ is an arbitrary curve parameter, taking values in an interval $I \subset \mathbb{R}$ (Figure 1.2).

We can *reparametrise* the curve by introducing a new curve parameter $\tilde{\sigma} \in \tilde{I}$ which is related to σ by an invertible map

$$\sigma \to \tilde{\sigma}(\sigma) , \quad \text{where} \quad \frac{d\tilde{\sigma}}{d\sigma} \neq 0. \tag{1.32}$$

While $C : I \to \mathbb{M}$ and $\tilde{C} : \tilde{I} \to \mathbb{M}$ are different maps, they have the same image in $\mathbb{M}$ and we regard them as different descriptions (parametrisations) of the same curve. The quantity $d\tilde{\sigma}/d\sigma$ is the *Jacobian* of this reparametrisation.

Often, one imposes the stronger condition

$$\frac{d\tilde{\sigma}}{d\sigma} > 0, \tag{1.33}$$

which means that the orientation (direction) of the curve is preserved.

The tangent vector field of the curve is

$$x'^{\mu} := \frac{dx^{\mu}}{d\sigma}. \tag{1.34}$$

A curve $C : I \to \mathbb{M}$ is called space-like, light-like, or space-like if its tangent vector field is space-like, light-like, or space-like, respectively, for all $\sigma \in I$. This property is reparametrisation invariant.

For a space-like curve, $I = [\sigma_1, \sigma_2] \to \mathbb{M}$, the *length* (or 'proper length') is defined as

$$L = \int_{\sigma_1}^{\sigma_2} d\sigma \sqrt{\eta_{\mu\nu} \frac{dx^{\mu}}{d\sigma} \frac{dx^{\nu}}{d\sigma}}. \tag{1.35}$$

For a time-like curve, we can define a 'length' by

$$\tau(\sigma_1, \sigma_2) = \int_{\sigma_1}^{\sigma_2} d\sigma \sqrt{-\eta_{\mu\nu} \frac{dx^{\mu}}{d\sigma} \frac{dx^{\nu}}{d\sigma}}, \tag{1.36}$$

and this quantity is precisely the proper time for a particle that has this curve as its world-line. We note that the proper length and proper time are distinguished affine curve parameters, characterised by the tangent vector field having unit norm. For light-like curves there is no analogous quantity, but we will see that there still is a distinguished class of affine curve parameters for the world-lines of massless particles.

Exercise 1.5.1   Verify that the length (1.35) of a space-like curve is reparametrisation invariant. Why does this not depend on whether the reparametrisation preserves the orientation of the curve?

Exercise 1.5.2   Show that the tangent vector field $\frac{dx^\mu}{d\tau}$ for the curve parameter τ defined by (1.36) has norm $\dot{x}^2 = -1$, thus verifying that τ is the proper time.

1.6   A Covariant Action for Massive Relativistic Particles

Using the concepts of the previous section, we introduce the following action:

$$S[x] = -m \int d\sigma \sqrt{-\eta_{\mu\nu} \frac{dx^\mu}{d\sigma} \frac{dx^\nu}{d\sigma}}. \tag{1.37}$$

Up to the constant factor $-m$, the action is the proper time for the motion of the particle along the world-line. We use an arbitrary curve parameter σ, and configuration space variables $(x, x') = (x^\mu, x'^\mu)$, which transform covariantly under Lorentz transformations. The action (1.37) has the following symmetries (invariances):

- The action is invariant under reparametrisations $\sigma \to \tilde{\sigma}(\sigma)$ of the world-line.
- The action is invariant under Poincaré transformations of space-time.

To verify that the new action (1.37) leads to the same field equations as (1.20), we perform the variation $x^\mu \to x^\mu + \delta x^\mu$ and obtain:

$$\frac{\delta S}{\delta x^\mu} = 0 \Leftrightarrow \frac{d}{d\sigma} \left(\frac{m\, x'^\mu}{\sqrt{-x' \cdot x'}} \right) = 0. \tag{1.38}$$

To get the physical interpretation, we choose the curve parameter σ to be the proper time τ:

$$\frac{d}{d\tau} \left(m \frac{dx^\mu}{d\tau} \right) = m\ddot{x}^\mu = 0, \tag{1.39}$$

where a 'dot' denotes the derivative with respect to proper time. This is indeed (1.18) with $f^\mu = 0$.

The general solution of this equation, which describes the motion of a free massive particle in Minkowski space is the straight world-line

$$x^\mu(\tau) = x^\mu(0) + \dot{x}^\mu(0)\tau. \tag{1.40}$$

Remark: *Reparametrisations vs Diffeomorphisms.* Reparametrisation invariance is also referred to as diffeomorphism invariance. We use the term reparametrisation, rather than diffeomorphism, to emphasise that we interpret the map $\sigma \mapsto \tilde{\sigma}$ *passively*, that is, as a change parametrisation. In contrast, an *active* transformation maps a given point to another point. The expressions for passive and active transformation agree up to an overall minus sign, as we will see in later examples, in particular, in Exercise 5.2.2.

Remark: *'Local' vs 'global' in mathematical and physical terminology.* In mathematics, 'local' refers to statements which hold on open neighbourhoods around each point, whereas 'global' refers to statements holding for the whole space. In contrast, physicists call symmetries 'global' or 'rigid' if the transformation parameters are independent of space-time, and 'local' if the transformation parameters are functions on space-time. In the case of the point particle action, Poincaré transformations are global symmetries, while reparametrisations are local. I will try to reduce the risk of confusion by saying 'rigid symmetry' rather than 'global symmetry', and when a symmetry is referred to as local, it is meant in the physicist's sense. Also, it is common for physicists to talk about statements which are true locally (in the mathematician's sense) but not necessarily true globally, using 'global terminology'.[4]

1.7 Particle Interactions

So far we have considered free particles. Interactions can be introduced by adding terms which couple the particle to external fields. The most important examples are the following:

- If the force f^μ has a potential, $f_\mu = -\partial_\mu V(x)$, then the equation of motion (1.18) follows from the action

$$S = -m \int \sqrt{-\dot{x}^2} d\tau - \int V(x(\tau)) d\tau. \tag{1.41}$$

- If f^μ is the Lorentz force acting on a particle with charge q, that is $f^\mu = qF^{\mu\nu}\dot{x}_\nu$, then the action is

$$S = -m \int \sqrt{-\dot{x}^2} d\tau + q \int A_\mu dx^\mu. \tag{1.42}$$

In the second term, the vector potential A_μ is integrated along the world-line of the particle

$$\int A_\mu dx^\mu = \int A_\mu(x(\tau)) \frac{dx^\mu}{d\tau} d\tau. \tag{1.43}$$

The resulting equation of motion is

$$m\ddot{x}_\mu = qF_{\mu\nu}\dot{x}^\nu, \tag{1.44}$$

where $F_{\mu\nu} = \partial_\mu A_\nu - \partial_\nu A_\mu$ is the field strength tensor. Equation (1.44) is the manifestly covariant version of the Lorentz force

$$\frac{d\vec{p}}{dt} = q\left(\vec{E} + \vec{v} \times \vec{B}\right). \tag{1.45}$$

---

[4] An example where the distinction between local and global aspects is relevant will be given later in Section 4.2 when we discuss the actions of the conformal Lie algebra and of the conformal group on the string world-sheet.

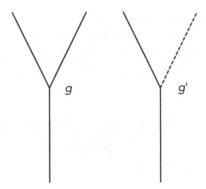

Particle interactions can be described by the splitting and joining of world-lines. Note that every type of vertex will in general have its own independent coupling constant.

- The coupling to gravity can be obtained by replacing the Minkowski metric $\eta_{\mu\nu}$ by a general pseudo-Riemannian metric $g_{\mu\nu}(x)$:

$$S = -m \int d\tau \sqrt{-g_{\mu\nu}(x)\dot{x}^\mu \dot{x}^\nu}. \tag{1.46}$$

The resulting equation of motion is the geodesic equation

$$\ddot{x}^\mu + \Gamma^\mu_{\ \nu\rho}\dot{x}^\nu \dot{x}^\rho = 0, \tag{1.47}$$

with affine curve parameter τ.

Exercise 1.7.1 Derive the Euler–Lagrange equations for the actions (1.41), (1.44), and (1.46). How do the equations of motion look like for a general (non-affine) curve parameter?

Remark: *Coupling to fields vs splitting word-lines.* In all three examples the interaction is introduced by coupling the particle to a background field.[5] A different way to describe particle interactions is to allow the word-lines of particles to split and join (see Figure 1.3). Note that the resulting particle trajectories are not one-dimensional manifolds, but graphs. Each type of interaction between particles has its own type of vertex, and the associated coupling constants will, in general, be independent.

In a quantum theory both descriptions are related, since particles and fields can be viewed as different types of excitations of *quantum fields*. Particles correspond to eigenstates of the particle number operator, whereas fields correspond to *coherent states* which are eigenstates of the particle annihilation operator. The standard formalism used in theoretical particle physics is obtained by quantising classical field theories, rather than classical particles. However, the Feynman rules are naturally interpreted in terms of a particle picture. For example, the left vertex in Figure 1.3

---

[5] *Background field* means that the field is 'given', and while it may solve some field equations of its own, the *backreaction* of the particle on the field is neglected. The full coupled theory of particles and fields would require us to consider both the field equations for the particle and for the field. In electrodynamics this would be the combined system of Lorentz forces and Maxwell equations, with the particle providing a source for the Maxwell field.

can be interpreted as representing interactions in a scalar ϕ^3 theory. Quantum field theory can be reformulated using the *word-line formalism*, which builds upon the classical particle picture. In string theory, the *world-sheet formalism* is the standard formulation, while *string field theory* is less well-developed.

1.8 Canonical Momenta and Hamiltonian for the Covariant Action

From the action (1.37)

$$S = \int L d\sigma = -m \int d\sigma \sqrt{-x'^2}, \tag{1.48}$$

we obtain the following canonical momentum vector:

$$\pi^\mu = \frac{\partial L}{\partial x'_\mu} = m \frac{x'^\mu}{\sqrt{-x'^2}} = m\dot{x}^\mu. \tag{1.49}$$

A new feature compared to the action (1.20) is that the components of the canonical momentum are not independent, but subject to the *constraint*

$$\pi^\mu \pi_\mu = -m^2. \tag{1.50}$$

Since canonical and kinetic momenta agree, $\pi^\mu = p^\mu$, we can interpret the constraint as the mass shell condition $p^2 = -m^2$. The Hamiltonian associated to (1.37) is

$$H = \pi^\mu \dot{x}_\mu - L = 0. \tag{1.51}$$

Thus the Hamiltonian is not equal to the total energy, but rather vanishes. Since $H \propto p^2 + m^2$, the Hamiltonian does not vanish identically, but only for the subset of configurations which satisfy the mass shell condition. Thus, $H = 0$ is a *constraint* which needs to be imposed on top of the dynamical field equations. This is sometimes denoted $H \simeq 0$, and one says that the Hamiltonian is *weakly zero*. This type of constraint arises when mechanical or field theoretical systems are formulated in a manifestly Lorentz covariant or manifestly reparametrisation invariant way. We will not need to go deeply into *constrained dynamcis*, because the constraints we will encounter can be imposed as *initial conditions*. We will demonstrate this explicitly for strings later, see Section 2.2.5.

Remark: *Hamiltonian and time evolution.* For those readers who are familiar with the formulation of mechanics using Poisson brackets, we add that while the Hamiltonian is weakly zero, it still generates the infinitesimal time evolution of physical quantities. Similarly, in the quantum version of the theory, the infinitesimal time evolution of an operator in the Heisenberg picture is still given by its commutator with the Hamiltonian. For the point particle, the only constraint is the vanishing of the Hamiltonian.

By accepting that constraints are the prize to pay for a covariant formalism, we can describe relativistic massive particles in a Lorentz covariant and reparametrisation invariant way. But we still need to find a way to include massless particles.

1.9 A Covariant Action for Massless and Massive Particles

To include massless particles, we use a trick which works by introducing an auxiliary field $e(\sigma)$ on the word-line. We require that $e\,d\sigma$ is reparamentrisation invariant:

$$e\,d\sigma = \tilde{e}\,d\tilde{\sigma}. \tag{1.52}$$

This implies that e transforms inversely to a coordinate differential:

$$\tilde{e}(\tilde{\sigma}) = e(\sigma)\frac{d\sigma}{d\tilde{\sigma}}. \tag{1.53}$$

We also require that e does not have any zeros, which makes $e\,d\sigma$ a one-dimensional volume element. We take $e > 0$ for definiteness, and this condition is preserved under reparametrisations which respect the orientation of the world-line.

Using the invariant line element, we write down the following action:

$$S[x, e] = \frac{1}{2}\int e\,d\sigma \left(\frac{1}{e^2}\left(\frac{dx^\mu}{d\sigma}\right)^2 - K\right), \tag{1.54}$$

where K is a real constant.

The action (1.54) has the following symmetries:

- $S[x, e]$ is invariant under reparametrisations $\sigma \to \tilde{\sigma}$.
- $S[x, e]$ is invariant under Poincaré transformations $x^\mu \to \Lambda^\mu_{\ \nu}x^\nu + a^\mu$.

The action depends on the fields $x = (x^\mu)$ and e. Performing the variations $x \to x + \delta x$ and $e \to e + \delta e$, respectively, we obtain the following equations of motion.

$$\frac{d}{d\sigma}\left(\frac{x'^{\,\mu}}{e}\right) = 0, \tag{1.55}$$

$$x'^{\,2} + e^2K = 0. \tag{1.56}$$

Exercise 1.9.1 Derive the equations of motion (1.55) and (1.56) by variation of the action (1.54).

The equation of motion for e is algebraic, and tells us that for $K > 0$ the solution is a time-like curve, while for $K = 0$ it is light-like and for $K < 0$ space-like. To show that the time-like case brings us back to (1.37), we set $K = m^2$ and solve for the auxiliary field e:

$$e = \frac{\sqrt{-x'^{\,2}}}{m}, \tag{1.57}$$

where we have used that $e > 0$. Substituting the solution for e into (1.54) we recover the action (1.37) for a massive particle of mass m.

The advantage of (1.54) is that it includes the case of massless particles as well. Let us consider $K = m^2 \geq 0$. Instead of solving for e we now fix it by *imposing a gauge*, which, in this case means that we pick a particular parametrisation of the world-line.

- For $m^2 > 0$, we impose the gauge

$$e = \frac{1}{m}.$$

$\qquad$ (1.58)

The equations of motion become:

$$\ddot{x}^\mu = 0 \,, \quad \dot{x}^2 = -1.$$

$\qquad$ (1.59)

The second equation tells us that this gauge is equivalent to choosing the proper time τ as curve parameter.

- For $m^2 = 0$, we impose the gauge

$$e = 1.$$

$\qquad$ (1.60)

The equations of motion become:

$$\ddot{x}^\mu = 0 \,, \quad \dot{x}^2 = 0.$$

$\qquad$ (1.61)

The second equation tells us that the world-line is light-like, as expected for a massless particle. In this case there is no proper time, but choosing $e = 1$ still corresponds to choosing a distinguished curve parameter. Observe that the first, dynamical equation of motion only simplifies to $\ddot{x}^\mu = 0$ if we choose e to be constant. Conversely, the equation $\ddot{x}^\mu = 0$ is only invariant under *affine* reparametrisations $\sigma \mapsto a\sigma + b$, $a \neq 0$ of the world-line. Imposing $e = 1$ (or any other constant value) corresponds to choosing an *affine curve parameter*. Since for light-like curves the concept of length or proper time does not exist, choosing an affine curve parameter serves as a substitute. We can fix the affine parameter up to an additive constant by imposing that $p^\mu = \dot{x}^\mu$, where p^μ is the momentum of the massless particle. While a relativistic velocity cannot be defined for massless particles, the relativistic momentum, which is related to measurable quantities, is well defined.

Remark: Readers who are familiar with field theory will observe that the first term of the action (1.54) looks like the action for a one-dimensional free massless scalar field. Readers familiar with general relativity will recognise that the one-dimensional invariant volume element $e\,d\sigma$ is analogous to the D-dimensional invariant volume element $\sqrt{|g|}d^D x$ appearing in the Einstein–Hilbert action, and in actions describing the coupling of matter to gravity. Since $g = \det(g_{\mu\nu})$ is the determinant of the metric, we can interpret e as the square root of the determinant of an intrinsic one-dimensional metric on the world-line. A one-dimensional metric has only one component, given by $e^2(\sigma)$. Therefore, the first term of (1.54) is the action for a one-dimensional free massless scalar field coupled to one-dimensional gravity.

Exercise 1.9.2 Demonstrate that the gauge choices (1.58), (1.60) are possible by showing that you can change the parametrisation of the world-line accordingly.

Remark: The action (1.54) is not only useful in physics, but also in geometry, because it allows one to formulate a variational principle which treats null (light-like) curves on the same footing as non-null (time-like and space-like) curves. Actions of the type (1.54) and their higher-dimensional generalisations are known as *sigma models* in physics. Geometrically, (1.54) is an *energy functional* (sometimes called the Dirichlet energy functional) for maps between two Riemannian spaces:[6] the world-line and Minkowski space. Maps which extremise the energy functional are called *harmonic maps*, and in the specific case of a one-dimensional action functional, *geodesic curves*. Note that we have seen above that the solutions for the Minkowski space are straight lines in affine parametrisation, and therefore geodesics.

1.10 Literature

Some remarks in this chapter assume a basic knowledge of general relativity and quantum field theory, which can be found in any standard textbook. *Coherent states* are frequently used in quantum optics, but do not belong to the standard canon of quantum field theory as used by particle theorists. See, for example, Duncan (2012) who also discusses the subtleties of the particle concept. The *world-line formalism* is reviewed in Schubert (2001). We also assume some background in *Lie groups and Lie algebras*. While the material included in particle theory and quantum field theory textbooks should be sufficient to follow the text, we mention some books for further reading: Gilmore (1974), Cahn (1984), Cornwell (1997), Fuchs and Schweigert (1997), Ramond (2010), cover group theory and its applications to particle theory. Sexl and Urbantke (2001) give a detailed discussion of the Lorentz and Poincaré group in the context of special relativity, while Woit (2017) is a comprehensive and pedagogical introduction into group theory as used in quantum theory. Mathematical texts are Humphreys (1972), Fulton and Harris (1991), Bump (2004), and the classic, Weyl (1939).

The study of *constrained dynamics* is a research subject in its own right, starting with Dirac (1964). Classical textbooks on the subject are Sudarshan and Mukunda (1974) and Sundermeyer (1982). Modern approaches to the quantisation of 'general gauge theories', including string theory, use the BRS and BV formalisms which are covered in detail in the monograph Henneaux and Teitelboim (1992), see also the review article Marnelius (1982).

---

[6] For terminological simplicity we will refer to real manifolds which are equipped with a symmetric and non-degenerate rank two co-tensor field as Riemannanian spaces or Riemannian manifolds. In the literarature one often uses the terms pseudo-Riemannian or semi-Riemannian to indicate that non-definite signatures are admitted.

Classical Relativistic Strings

In this chapter, we introduce classical relativistic strings as generalisations of classical relativistic particles.

2.1 The Nambu–Goto Action

2.1.1 Action, Equations of Motion, and Bounday Conditions

If we replace point particles by strings, that is by one-dimensional extended objects, then the world-line is replaced by a *world-sheet* Σ, which is a surface embedded into Minkowksi space:[1]

$$X : \Sigma \ni P \longrightarrow X(P) \in \mathbb{M}. \tag{2.1}$$

On space-time we choose linear coordinates associated to a Lorentz frame, denoted $X = (X^\mu)$, where $\mu = 0, 1, \ldots, D-1$. On the world-sheet we choose local coordinates $\sigma = (\sigma^0, \sigma^1) = (\sigma^\alpha)$. Depending on the global structure of the world-sheet, it may or may not be possible to cover Σ with a single coordinate system. We will have to verify that physical quantites, such as the action and the equations of motion are covariant with respect to reparametrisations. For the time being we will only consider *free strings* which do not split or join. In this case, the world-sheet Σ has the topology of a strip (open strings; see Figure 2.1) or of a cylinder (closed strings).

At each point of Σ we can choose a time-like tangent vector ('the direction towards the future') and a space-like tangent vector ('the direction along the string'). We adopt the convention that the coordinate σ^0 is time-like (that is, the corresponding tangent vector $\partial_0 = \frac{\partial}{\partial \sigma^0}$ is a time-like vector), while the coordinate σ^1 is space-like:[2]

$$\dot{X}^2 \leq 0 , \quad (X')^2 > 0. \tag{2.2}$$

Here we use the following notation for tangent vectors:

$$\dot{X} = (\partial_0 X^\mu) = \left(\frac{\partial X^\mu}{\partial \sigma^0} \right) , \quad X' = (\partial_1 X^\mu) = \left(\frac{\partial X^\mu}{\partial \sigma^1} \right). \tag{2.3}$$

We also make conventional choices for the range of the world-sheet coordinates. The space-like coordinate takes values $\sigma^1 \in [0, \pi]$, whereas the time-like coordinate

[1] To be precise, the map $\Sigma \to \mathbb{M}$ only needs to be an *immersion*, that is its differential must be invertible. This means that we allow strings to self-intersect. Note that such self-intersections do not give rise to interactions, which are instead encoded in the topology of Σ. An immersion becomes an embedding if we make its domain sufficiently small, and therefore we will use the more intuitive term embedding.

[2] Here we use the interpretation of vector fields as differential operators, $v(x) = v^\mu(x)\partial_\mu$. We will see later that the tangent vector $\dot{X}$ can become light-like at special points.

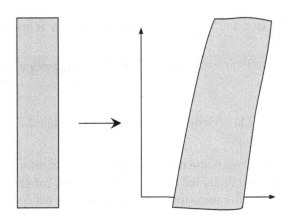

Fig. 2.1   The world-sheet of a string is described by an embedded surface. For a free open string the world-sheet has the topology of a strip.

takes values $\sigma^0 \in [\sigma^0_{(1)}, \sigma^0_{(2)}] \subset \mathbb{R}$. The limiting case $\sigma^0 \in \mathbb{R}$ is allowed and describes the asymptotic time evolution of a string from the infinite past to the infinite future.[3]

The *Nambu–Goto action* is the direct generalisation of (1.37), and thus proportional to the area of the world-sheet Σ, measured with the metric induced on Σ by the Minkowski metric:

$$S_{\text{NG}}[X] = -TA(\Sigma) = -T \int_\Sigma d^2 A. \tag{2.4}$$

The constant T must have the dimension (length)$^{-2}$ or energy/length, in units where $\hbar = 1$, $c = 1$. It is, therefore, called the *string tension*. This quantity generalises the mass of a point particle and is the only dimensionful parameter entering into the definition of the theory.[4]

The Minkowski metric $\eta_{\mu\nu}$ induces a metric $g_{\alpha\beta}$ on Σ by *pullback:*

$$g_{\alpha\beta} = \partial_\alpha X^\mu \partial_\beta X^\nu \eta_{\mu\nu}, \tag{2.5}$$

which is used to define the invariant area element

$$d^2 A = d^2 \sigma \sqrt{|\det(g_{\alpha\beta})|} \tag{2.6}$$

on Σ. Note that $\det(g_{\alpha\beta}) < 0$ since Σ has one time-like and one space-like direction, which is why we have to take the absolut value. Combining the above formulae, the Nambu–Goto action is

$$S = -T \int d^2 \sigma \sqrt{\left|\det\left(\partial_\alpha X^\mu \partial_\beta X^\nu \eta_{\mu\nu}\right)\right|}. \tag{2.7}$$

---

[3] The coordinate σ^0 can be thought of as *world-sheet time*. Note that it is in general different from Minkowski time $X^0 = t$.

[4] Note that while T is a constant, the length of a string cannot be constant, because special relativity does not allow rigid bodies, as they would allow to transfer energy with infinite speed. The length of a relativistic string is dynamical, and its total energy depends on its state of motion. We will see later how the constant T enters into the formula for the total energy of a string.

The area, and, hence, the Nambu–Goto action, is invariant under reparametrisations of Σ,

$$\sigma^\alpha \to \tilde\sigma^\alpha(\sigma^0, \sigma^1)\,, \quad \text{where} \quad \det\left(\frac{\partial\tilde\sigma^\alpha}{\partial\sigma^\beta}\right) \neq 0. \tag{2.8}$$

Exercise 2.1.1   Verify that the Nambu–Goto action (2.7) is invariant under reparametrisations.

The action is also invariant under Poincaré transformations of $\mathbb{M}$, which is manifest since the Lorentz indices are contracted in the correct way to obtain a Lorentz scalar. To perform computations, it is useful to write out the action more explicitly:

$$S_{\mathrm{NG}} = \int d^2\sigma \mathcal{L} = -T \int d^2\sigma \sqrt{(\dot X \cdot X')^2 - \dot X^2 (X')^2}. \tag{2.9}$$

Here $\mathcal{L}$ is the Lagrangian density, or Langrangian for short.

The *world-sheet momentum densities* are defined as

$$P^\alpha_\mu = \frac{\partial\mathcal{L}}{\partial\partial_\alpha X^\mu}.$$

Evaluating the components explicitly, we find:

$$\Pi_\mu := P^0_\mu := \frac{\partial\mathcal{L}}{\partial\dot X^\mu} \;=\; -T\frac{(X')^2\dot X_\mu - (\dot X \cdot X')X'_\mu}{\sqrt{(\dot X \cdot X')^2 - \dot X^2(X')^2}},$$

$$P^1_\mu := \frac{\partial\mathcal{L}}{\partial X'_\mu} \;=\; T\frac{\dot X^2 X'_\mu - (\dot X \cdot X')\dot X_\mu}{\sqrt{(\dot X \cdot X')^2 - \dot X^2(X')^2}}. \tag{2.10}$$

Note that $P^0_\mu = \Pi_\mu$ is the canonical momentum (density). While these expressions look complicated, a more intuitive form is obtained if we assume that we can choose coordinates such that $X' \cdot \dot X = 0$, $\dot X^2 = -1$, $(X')^2 = 1$:[5]

$$(P^\alpha_\mu) = T(\dot X_\mu\,, -X'_\mu) = T(-\partial^0 X_\mu\,, -\partial^1 X_\mu). \tag{2.11}$$

This momentum density generalises the momentum $p^\mu = m\dot x^\mu$ along the world-line of a relativistic particle.

The equations of motion are found by imposing that the action (2.9) is invariant under variations $X \to X + \delta X$, subject to the condition that the initial and final positions of the string are kept fixed: $\delta X(\sigma^0 = \sigma^0_{(1)}) = 0$, $\delta X(\sigma^0 = \sigma^0_{(2)}) = 0$. When carrying out the variation, formulae take a more compact form when using the expressions (2.10):

$$\delta S = \int d^2\sigma \left(P^0_\mu \delta\dot X^\mu + P^1_\mu \delta X'^\mu\right).$$

We need to perform an integration by parts which creates two boundary terms:

$$\delta S = \int_0^\pi d\sigma^1 \left[P^0_\mu \delta X^\mu\right]_{\sigma^0_{(1)}}^{\sigma^0_{(2)}} + \int_{\sigma^0_{(1)}}^{\sigma^0_{(2)}} d\sigma^0 \left[P^1_\mu \delta X^\mu\right]_{\sigma^1=0}^{\sigma^1=\pi} - \int d^2\sigma\, \partial_\alpha P^\alpha_\mu \delta X^\mu.$$

---

[5] See Zwiebach (2009) for a detailed account of how to explicitly construct various types of useful coordinates systems on the world-sheet.

The first boundary term vanishes because the initial and final position of the string are kept fixed. However, the second boundary term does not vanish automatically, and to have a consistent variational principle we have to impose that

$$\delta S = \int_{\sigma_{(1)}^0}^{\sigma_{(2)}^0} d\sigma^0 \left[P_\mu^1 \delta X^\mu \right]_{\sigma^1=0}^{\sigma^1=\pi} \overset{!}{=} 0. \tag{2.12}$$

This condition tells us that we must impose boundary conditions, for which we have the following options:

1. *Periodic* boundary conditions:

$$X(\sigma^1) = X(\sigma^1 + \pi). \tag{2.13}$$

This corresponds to *closed strings*, where the world-sheet can only have space-like boundaries corresponding to the initial and final configuration.

2. *Neumann* boundary conditions:

$$P_\mu^1 \big|_{\sigma^1=0,\pi} = \frac{\partial \mathcal{L}}{\partial X_\mu'} \bigg|_{\sigma^1=0,\pi} = 0. \tag{2.14}$$

Since P_μ^1, evaluated at $\sigma^1 = 0, \pi$, is the component of the world-sheet momentum density normal to the boundary, Neumann boundary conditions imply momentum conservation at the ends of the string. They describe open strings whose ends can move freely. As we will see in Exercise 2.2.10, the ends of an open string always move with the speed of light.

3. *Dirichlet* boundary conditions. For space-like directions $i = 1, \ldots, D - 1$ we can impose

$$P_i^0 \big|_{\sigma^1=0,\pi} = \frac{\partial \mathcal{L}}{\partial \dot{X}_i} \bigg|_{\sigma^1=0,\pi} = 0.$$

Since this implies that the tangential component of the world-sheet momentum vanishes at the boundary, it correponds to keeping the ends of the string fixed in the i-th direction:

$$X^i(\sigma^1 = 0) = x_o^i, \quad X^i(\sigma^1 = \pi) = x_1^i. \tag{2.15}$$

In this case, translation invariance in Minkowski space is broken by the boundary conditions, and momentum is not conserved at the ends of the string. To restore momentum conservation one must couple open strings with Dirichlet boundary conditions to new types of dynamical objects, called D-*branes*.

2.1.2 D-branes

Let us have a closer look into Dirichlet boundary conditions and D-branes. For concreteness we consider open strings ending on a p-dimensional D-brane, or D-p-brane for short, which is located at $x^{p+1} = x_0^{p+1}, x^{p+2} = x_0^{p+2}, \ldots, x^{D-1} = x_0^{D-1}$. This means that we impose Neumann boundary conditions for the p space-like coordinates $x^1, \ldots, x^p$, and for time x^0. These are the direction parallel to the $(p+1)$-dimensional *world-volume* of the D-p-brane. Dirichlet boundary conditions are imposed for the transverse coordinates $x^{p+1}, \ldots, x^{D-1}$. Thus, the ends of open strings with such boundary conditions are located on the same D-p-brane (see Figure 2.2).

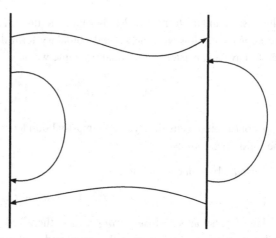

Fig. 2.2   A configuration with two parallel D-branes. There are four types of open strings, classified by their start and endpoints. Note that since strings are oriented, as indicated by the arrowheads, start and endpoint can be distinguished.

For $p = D - 1$, we have Neumann boundary conditions in all directions and the D-$(D - 1)$-brane is *space-filling*. The other extreme is the D-0-brane or D-particle where Dirichlet boundary conditions are imposed in all spatial directions. D-1-branes are strings, which are called D-strings to distinguish them from the fundamental strings defined by the Nambu–Goto action. D-2-branes are membranes, and D-p-branes with $p > 2$ are higher-dimensional versions of membranes.

As a generalisation we can consider configurations with more than one D-brane. The simplest configuration is two parallel D-p-branes located at different positions in the $(p + 1)$-direction, as in Figure 2.2:

$$x_{(1)}^{p+1} \neq x_{(2)}^{p+1} , \quad x_{(1)}^{p+k} = x_{(2)}^{p+k}, \; k = 2, \ldots k = D - (p + 1).$$

As a further generalisation, we can consider configurations involving any number of D-branes, and with different values for p. This is certainly possible as far as imposing boundary conditions is concerned, but we need to treat the D-branes as dynamical objects if we want to preserve momentum conservation. This raises difficult dynamical question, since now D-branes can move and collide. Moreover, 'string theory' now seems to be a theory of strings *and* D-branes. One way of thinking about D-branes is as *solitons*, understood in a suitably relaxed sense. In field theory the term soliton refers to solutions of the field equations that behave like particles, which means that they are and remain localised, have finite total energy, and are regular. Solitons break translations invariance, which a D-branes also does transverse to its world-volume directions. Since D-branes with $p > 0$ are infinitely extended, we need to replace the condition of finite energy by finite energy per world-volume, but this is natural since D-branes are translation invariant along their world-volume. Therefore we could compactify these directions (impose a periodic identification) and obtain a finite mass point-like object in the remaining non-compact directions.[6] In the supersymmetric Type-II string theories one can show explicitly that there are,

---

[6] We also need an appropriate interpretation of what 'regular' means, but we will not enter into the details.

for specific values of p, static solutions of the low-energy effective field theory,[7] called supergravity p-branes, which correspond to D-p-branes. We will come back to the D-p-branes of Type-II superstring theories in Chapter 14. We remark that D-branes are a major tool in string theory based model building for particle physics. The advantage of this approach is that one can control the particle spectrum through the choice of the D-brane configuration.

Let us also consider briefly whether Dirichlet boundary conditions in the time direction make sense. Such configurations correspond to event-like branes which exist only for a given moment of time. One way to make sense of this is to perform an analytic continuation (Wick rotation) in time and to consider the *Euclidean formulation* of the theory. Then time is on the same footing as space, and Euclidean D-branes are interpreted as *instantons*, again with a suitably relaxed interpretation of the term. In field theory, instantons are finite action solutions to the Euclidean equations of motion and therefore correspond to saddle points of the Euclidean functional integral. When expanding around saddle points, they contribute subleading terms to observables, which are suppressed by a factor $e^{-S_{\text{inst}}}$, where S_{inst} is the action of the instanton field configuration. For gauge theories (and also for string theory), the coupling g enters into this weight factor as e^{-1/g^2}, which is smooth but not analytic at $g = 0$. Therefore, such contributions are not visible in perturbation theory (which is a power series in g around $g = 0$), they are *non-perturbative*.

The D-(-1)-brane, where Dirichlet boundary conditions are imposed in all directions, is called the D-instanton, and open string correlation functions computed with such boundary conditions show indeed the exponential suppression expected for an instanton correction. In the Type-IIB superstring theory one can find a corresponding Euclidean solution to the low-energy effective field theory. Euclidean D-p-branes with $p \geq 0$ give rise to instanton corrections if all their world-volume directions are along compact dimensions, so that the instanton action is finite. We will see later that the consistency of the quantum theory requires $D > 4$ so that compactification of the extra direction is required. Euclidean D-p-branes are therefore relevant for computing instanton corrections in string theory.

Dirichlet boundary conditions in the time direction have also been considered in Lorentz signature. As we will see later, there is a symmetry in string theory, called T-duality, which for open strings exchanges Dirichlet and Neumann boundary conditions (see Section 13.6). If one allows T-duality transformations along time-like directions, this introduces branes with Dirichlet boundary conditions in time. The corresponding solutions of the low-energy effective field theory are time-dependent and can be interpreted as cosmological solutions. Such branes are called E-branes (since they have a Euclidean world-volume, however, the same term is sometimes used for D-branes embedded into a Euclidean space-time).

2.1.3 Constraints

Once we have chosen boundary conditions, the vanishing of the remaining, non-boundary terms in the variation of the action implies the following equations of motion:

---

[7] The concept of a low-energy effective field theory will be introduced in Section 8.3 and further developed in Part IV.

$$\partial_\alpha P_\mu^\alpha = 0. \tag{2.16}$$

While, given (2.10), this looks very complicated, it becomes the two-dimensional wave equation upon choosing coordinates where $\dot{X} \cdot X' = 0$, $\dot{X}^2 = -1$, $X'^2 = 1$. Instead of showing how such a coordinate system can be constructed, we will derive this result in a different way using the Polyakov action in Section 2.2.

The canonical momenta are not independent, but subject to two independent constraints, which generalise the mass shell condition of a particle:

$$\Pi^\mu X'_\mu = 0 \,, \quad \Pi^2 + T^2 (X')^2 = 0. \tag{2.17}$$

The canonical Hamiltonian (density) is obtained from the Lagrangian by a Legendre transformation:

$$\mathcal{H}_{\text{can}} = \dot{X}\Pi - \mathcal{L} = 0. \tag{2.18}$$

As for the relativistic particle, the Hamiltonian is not equal to the energy (density), and it is (weakly) zero.

Exercise 2.1.2   Verify the relations (2.17) and (2.18).

2.2 The Polyakov Action

2.2.1 Action, Symmetries, Equations of Motion

The Polyakov action is related to the Nambu–Goto action in the same way as for point particles the action (1.54) is related to (1.37). That is, we replace the area by the corresponding energy functional, or, in physical terms, sigma model. This requires us to introduce an intrinsic metric $h_{\alpha\beta}(\sigma)$ on the world-sheet Σ. The Polyakov action is

$$S_\text{P}[X, h] = -\frac{T}{2} \int d^2\sigma \sqrt{h} h^{\alpha\beta} \partial_\alpha X^\mu \partial_\beta X^\nu \eta_{\mu\nu}, \tag{2.19}$$

where $h = -\det(h_{\alpha\beta}) = |\det(h_{\alpha\beta})|$. The intrinsic metric $h_{\alpha\beta}$ is a priori unrelated to the induced metric $g_{\alpha\beta}$, but we require that $h_{\alpha\beta}$ has the same signature $(-+)$ as $g_{\alpha\beta}$. The embedding $X : (\Sigma, h_{\alpha\beta}) \to (\mathbb{M}, \eta_{\mu\nu})$ is now a map between two Riemannian manifolds.

The Polyakov action is invariant under Poincaré transformations on space-time, and has the following local symmetries with respect to the world-sheet Σ:

1. *Reparametrisations* $\sigma \to \tilde{\sigma}(\sigma)$, which act by

$$\tilde{X}^\mu(\tilde{\sigma}) = X^\mu(\sigma) \,, \quad \tilde{h}_{\alpha\beta}(\tilde{\sigma}) = \frac{\partial \sigma^\gamma}{\partial \tilde{\sigma}^\alpha} \frac{\partial \sigma^\delta}{\partial \tilde{\sigma}^\beta} h_{\gamma\delta}(\sigma). \tag{2.20}$$

2. *Weyl transformations*:

$$h_{\alpha\beta}(\sigma) \to e^{2\Lambda(\sigma)} h_{\alpha\beta}(\sigma). \tag{2.21}$$

Remarks:

1. Weyl transformations do not act on the coordinates and are therefore different from reparametrisations. Mathematicians usually call them conformal transformations, because they change the metric but preserve the *conformal structure* of $(\Sigma, h_{\alpha\beta})$. A conformal structure on a manifold Σ is an equivalence class $[h_{\alpha\beta}]$ of metrics which differ by multiplication by a local scale factor which is positive everywhere, $h_{\alpha\beta}(\sigma) \cong e^{2\Omega(\sigma)} h_{\alpha\beta}(\sigma)$.

2. The invariance of the action under Weyl transformation is a special property of strings. Higher-dimensional versions of the Polyakov action are not Weyl invariant.[8]

3. Combining Weyl with reparametrisation invariance, one has three local transformations which can be used to gauge-fix the metric $h_{\alpha\beta}$ completely. Thus, $h_{\alpha\beta}$ does not introduce new local degrees of freedom, as required for an auxiliary field.

The equations of motion are obtained by imposing stationarity with respect to the variations $X^\mu \to X^\mu + \delta X^\mu$ and $h_{\alpha\beta} \to h_{\alpha\beta} + \delta h_{\alpha\beta}$:

$$\frac{1}{\sqrt{h}} \partial_\alpha \left(\sqrt{h} h^{\alpha\beta} \partial_\beta X^\mu \right) = 0, \tag{2.22}$$

$$\partial_\alpha X^\mu \partial_\beta X_\mu - \frac{1}{2} h_{\alpha\beta} h^{\gamma\delta} \partial_\gamma X^\mu \partial_\delta X_\mu = 0. \tag{2.23}$$

Exercise 2.2.1 Show that (2.22) and (2.23) are the Euler–Lagrange equations of the action (2.19). For reference, the variation of $\sqrt{h} = \sqrt{|\det h_{\alpha\beta}|}$ is

$$\delta\sqrt{h} = \frac{1}{2}\sqrt{h} h^{\alpha\beta} \delta h_{\alpha\beta} = -\frac{1}{2}\sqrt{h} h_{\alpha\beta} \delta h^{\alpha\beta}. \tag{2.24}$$

As for the Nambu–Goto action, we have to make sure that boundary terms vanish, which gives us the choice between periodic, Neumann and Dirichlet boundary conditions.

The X-equation (2.22) is the two-dimensional wave equation on the Riemannian manifold $(\Sigma, h_{\alpha\beta})$, which can be rewritten in various ways:

$$\Box X^\mu = 0 \Leftrightarrow \nabla_\alpha \nabla^\alpha X^\mu = 0 \Leftrightarrow \nabla_\alpha \partial^\alpha X^\mu = 0, \tag{2.25}$$

where ∇_α is the covariant derivative with respect to the world-sheet metric $h_{\alpha\beta}$. The h-equation (2.23) is algebraic and implies that

$$\det(g_{\alpha\beta}) = \frac{1}{4} \det(h_{\alpha\beta}) (h^{\gamma\delta} g_{\gamma\delta})^2. \tag{2.26}$$

When imposing the h-equation, the Polyakov action becomes the Nambu–Goto action.

Exercise 2.2.2 Show that (2.23) implies that $h_{\alpha\beta}$ and $g_{\alpha\beta}$ are related by (2.26). Use this result to show that upon imposing (2.23) the Polyakov becomes the Nambu–Goto action.

---

[8] One can construct Weyl-invariant actions for higher-dimensional objects by introducing additional auxiliary fields, see, e.g., Nieto (2001).

2.2.2 Interpretation as a Two-Dimensional Field Theory

The advantage of the Polyakov action is that it does not involve a square root and takes the form of a standard two-dimensional field theory action for free massless scalar fields. This allows us to take an alternative point of view and to interpret Σ as a two-dimensional space-time, populated by D scalar fields $X = (X^\mu)$, which take values in the *target space* $\mathbb{M}$. We can then use methods, results and intuition from field theory. We will call this point of view the *world-sheet perspective* in contrast to the *space-time perspective* where (2.19) is interpreted in terms of a string in Minkowksi space.

When studying a field theory on a Riemannian manifold, one defines the energy momentum tensor of an action by its variation with respect to the metric.[9]

The energy-momentum tensor of the Polyakov action is:

$$T_{\alpha\beta} := -\frac{4}{T}\frac{1}{\sqrt{h}}\frac{\delta S_P}{\delta h^{\alpha\beta}} = 2\partial_\alpha X^\mu \partial_\beta X_\mu - h_{\alpha\beta}h^{\gamma\delta}\partial_\gamma X^\mu \partial_\delta X_\mu, \qquad (2.27)$$

where the overall normalisation is conventional.

Expressing the h-equation of motion in terms of $T_{\alpha\beta}$ gives

$$T_{\alpha\beta} = 0. \qquad (2.28)$$

This is an algebraic equation for the auxiliary field $h_{\alpha\beta}$, which has to be imposed as a constraint on solutions of the dynamical X-equation. Note that (2.28) resembles the Einstein equations, which in general dimension n take the form $R_{\mu\nu} - \frac{1}{2}Rg_{\mu\nu} = \kappa^2 T_{\mu\nu}$, where $g_{\mu\nu}$ is the metric, $R_{\mu\nu}$ the Ricci tensor, R the Ricci scalar and κ the gravitational coupling constant. The Einstein–Hilbert action, which generates the l.h.s. of this equation is $S_{EH} = \frac{1}{2\kappa^2}\int d^n x\sqrt{|g|}R$. Could we modify the Polyakov action by adding a two-dimensional Einstein–Hilbert term? The answer is that adding an Einstein–Hilbert term does not change the local dynamics, because the variation of the two-dimensional Einstein–Hilbert action is a total derivative, and therefore does not contribute to the equations of motion. In other words, $T_{\alpha\beta} = 0$ is already the two-dimensional Einstein equation. The story is more interesting if we consider the world-sheet Σ globally. The integral of the Einstein–Hilbert action gives a number, which is invariant under variations of the metric. This number turns out to be a *topological invariant* of Σ, its *Euler number*. We will come back to this in Section 12.2.1.

Let us now investigate the properties of the energy momentum tensor. On a flat world-sheet $T_{\alpha\beta}$ is conserved (has a vanishing divergence). By the equivalence principle[10] we expect that $T_{\alpha\beta}$ is covariantly conserved on a general world-sheet $(\Sigma, h_{\alpha\beta})$:

$$\nabla^\alpha T_{\alpha\beta} = 0. \qquad (2.29)$$

---

[9] This can also be applied to field theories on a flat space-time, by introducing a background metric that can then be varied. An alternative definition of the energy momentum tensor uses the Noether theorem. The resulting energy momentum tensor is in general not symmetric, and may differ from the one obtained by variation of the metric by a total derivative.

[10] Einstein's equivalence principle states that physics in a freely falling frame under gravity is point-wise the same as in special relativity. Geometrically this is the statement that by introducing Riemann normal coordinates around a point P, we can arrange that the covariant and partial derivative agree at P.

This conservation equations only holds on-shell, that is for field configurations which satisfy the equations of motion. The energy-momentum tensor is traceless:

$$T^\alpha{}_\alpha = h^{\alpha\beta} T_{\alpha\beta} = 0, \tag{2.30}$$

and this holds off-shell, that is without using the equations of motion. Note that the trace of a tensor is defined using contraction with the metric, and is therefore in general different from the trace of the matrix $T_{\alpha\beta}$. Since $T_{\alpha\beta}$ is symmetric and traceless, it has two independent components.

Exercise 2.2.3 Show that $T_{\alpha\beta}$ is covariantly conserved on-shell, that is modulo the equations of motion.

Exercise 2.2.4 Show that $T_{\alpha\beta}$ is traceless, without using the equations of motion.

2.2.3 The Conformal Gauge

One way to use symmetries is to impose *gauge conditions* which bring expressions to a standard form. The Polyakov action has three local symmetries: the reparametrisations of two coordinates and Weyl transformations. Since the metric $h_{\alpha\beta}$ has three independent components, counting suggests that we can fix its form completely, in particular that we can impose the so-called *conformal gauge* where it takes the form of the standard two-dimensional Minkowski metric:

$$h_{\alpha\beta} \overset{!}{=} \eta_{\alpha\beta} = \begin{pmatrix} -1 & 0 \\ 0 & 1 \end{pmatrix}. \tag{2.31}$$

This is indeed true locally, because any two-dimensional Riemannian metric is conformally flat and can be written locally as the product of the standard flat metric with a conformal factor. Since we can apply Weyl transformations in addition to reparametrisations, the conformal factor can be removed, leaving us with the standard flat metric:

$$h_{\alpha\beta} \to e^{2\Omega(\sigma)} \eta_{\alpha\beta} \to \eta_{\alpha\beta}. \tag{2.32}$$

Note that on general world-sheets the reparametrisations needed for the first step only exist locally, but not globally. Thus global gauge fixing is a more complicated problem, on which we will comment in Section 12.1.1.

We now work out various useful formulas for the Polyakov string in the conformal gauge. Some care is required when imposing gauge conditions on the action itself rather than on the equations of motion. In the present case, substituting the gauge condition (2.31) into the Polyakov action gives

$$S_P = -\frac{T}{2} \int d^2\sigma \, \eta^{\alpha\beta} \partial_\alpha X^\mu \partial_\beta X_\mu. \tag{2.33}$$

Variation of this action with respect to X gives indeed the correct equation of motion, namely the gauge fixed version of the previous X-equation:

$$\Box X^\mu = -(\partial_0^2 - \partial_1^2) X^\mu = 0. \tag{2.34}$$

This is the standard ('flat') two-dimensional wave equation, which is known to have the general solution

$$X^\mu(\sigma) = X_L^\mu(\sigma^0 + \sigma^1) + X_R^\mu(\sigma^0 - \sigma^1), \tag{2.35}$$

describing decoupled left- and right-moving waves.

However, not all solutions of the two-dimensional wave equation are solutions of string theory. First of all, we have to impose boundary conditions. In the conformal gauge, the consistent boundary conditions take the following form:

$$X^\mu(\sigma^1 + \pi) = X^\mu(\sigma), \quad \text{(periodic)}, \tag{2.36}$$

$$X'_\mu\big|_{\sigma^1=0,\pi} = 0, \quad \text{(Neumann)}, \tag{2.37}$$

$$\dot{X}_\mu\big|_{\sigma^1=0,\pi} = 0, \quad \text{(Dirichlet)}. \tag{2.38}$$

Morever, the equations coming from the h-variation of the Polyakov action must now be added by hand:

$$T_{\alpha\beta} = 0. \tag{2.39}$$

The energy momentum tensor is traceless

$$\text{Trace}(T) = T^\alpha{}_\alpha = \eta^{\alpha\beta} T_{\alpha\beta} = -T_{00} + T_{11} = 0, \tag{2.40}$$

and since this holds off-shell we only have two non-trivial constraints:

$$T_{01} = T_{10} = 2\dot{X}X' = 0, \quad T_{00} = T_{11} = \dot{X}^2 + X'^2 = 0. \tag{2.41}$$

In the Hamiltonian formulation of the Nambu–Goto action, constraints arose from relations between the canonical momenta. For the Polyakov action, the canonical momenta are

$$\Pi^\mu = \frac{\partial \mathcal{L}_P}{\partial \dot{X}_\mu} = T\dot{X}^\mu, \tag{2.42}$$

and the canonical Hamiltonian is

$$H_{\text{can}} = \int_0^\pi d\sigma^1 \left(\dot{X}\Pi - \mathcal{L}_P \right) = \frac{T}{2} \int_0^\pi d\sigma^1 \left(\dot{X}^2 + X'^2 \right). \tag{2.43}$$

Thus, $T_{00} = T_{01} = 0$ implies that the Hamiltonian vanishes on shell.

Exercise 2.2.5 Compute the world-sheet momentum densities $P_\mu^\alpha = \partial\mathcal{L}/\partial(\partial_\alpha X^\mu)$ of the Polyakov action in the conformal gauge. Using that $\Pi_\mu = P_\mu^0$ is the canonical momentum, show that the constraints (2.41) are equivalent to the constraints (2.17).

We have thus verified that as long as we impose the gauge fixed h-equation by hand, we can obtain the gauge-fixed versions of all other equations by variation of the gauge-fixed action. In general, one needs to be careful when imposing conditions directly on the action instead of on the field equations.

2.2.4 Light-Cone Coordinates

Equation (2.35) suggests to introduce *light-cone coordinates* (also called *null coordinates*):

$$\sigma^\pm := \sigma^0 \pm \sigma^1. \tag{2.44}$$

We adopt a convention where we write σ^a, with $a = +,-$ for light-cone coordinates and σ^α, with $\alpha = 0, 1$ for non-null coordinates.

To relate quantities in both types of coordinate systems, we compute the Jacobian of the coordinate transformation and its inverse:

$$(J_\alpha{}^a) = \frac{D(\sigma^+, \sigma^-)}{D(\sigma^0, \sigma^1)} = \begin{pmatrix} 1 & 1 \\ 1 & -1 \end{pmatrix}, \quad (J_a{}^\alpha) = \frac{D(\sigma^0, \sigma^1)}{D(\sigma^+, \sigma^-)} = \frac{1}{2}\begin{pmatrix} 1 & 1 \\ 1 & -1 \end{pmatrix}. \quad (2.45)$$

Tensor indices are converted using these Jacobians. In particular, the light-cone expressions for coordinate differentials and derivatives are

$$d\sigma^\pm = d\sigma^0 \pm d\sigma^1, \quad \partial_\pm = \frac{1}{2}(\partial_0 \pm \partial_1). \quad (2.46)$$

For reference, we note that the standard Minkowski metric $\eta_{\alpha\beta}$ takes the following form in light-cone coordinates:

$$(\eta_{ab}) = -\frac{1}{2}\begin{pmatrix} 0 & 1 \\ 1 & 0 \end{pmatrix}, \quad (\eta^{ab}) = -2\begin{pmatrix} 0 & 1 \\ 1 & 0 \end{pmatrix}. \quad (2.47)$$

The light-cone components of the energy-momentum tensor are

$$T_{++} = \frac{1}{2}(T_{00} + T_{01}), \quad T_{--} = \frac{1}{2}(T_{00} - T_{01}), \quad T_{+-} = 0 = T_{-+}. \quad (2.48)$$

Note that the trace, evaluated in light-cone coordinates, is

$$\text{Trace}(T) = \eta^{ab}T_{ab} = 2\eta^{+-}T_{+-} = -4T_{+-}. \quad (2.49)$$

Thus tracelessness means $T_{+-} = 0$, and the two independent components are T_{++} and T_{--}.

We can write the gauge-fixed action in light-cone coordinates:

$$S_P = -\frac{T}{2}\int d^2\sigma \eta^{\alpha\beta}\partial_\alpha X^\mu \partial_\beta X_\mu$$
$$= \frac{T}{2}\int d^2\sigma = (\dot{X}^2 - X'^2) = 2T\int d^2\sigma \partial_+ X^\mu \partial_- X_\mu. \quad (2.50)$$

For reference, the equations of motion take the form

$$\Box X^\mu = -(\partial_0^2 - \partial_1^2)X^\mu = -4\partial_+\partial_- X^\mu = 0. \quad (2.51)$$

Thus, in light-cone coordinates it is obvious that the general solution decomposes into independent left- and right-moving waves with arbitrary profile:

$$X^\mu(\sigma) = X_L^\mu(\sigma^0 + \sigma^1) + X_R^\mu(\sigma^0 - \sigma^1). \quad (2.52)$$

We also give the constraints in light-cone coordinates:

$$T_{++} = 2\partial_+ X^\mu \partial_+ X_\mu = 0 \quad \Leftrightarrow \quad \dot{X}_L^2 = 0,$$
$$T_{--} = 2\partial_- X^\mu \partial_- X_\mu = 0 \quad \Leftrightarrow \quad \dot{X}_R^2 = 0. \quad (2.53)$$

2.2.5 From Symmetries to Conservation Laws

One of the most central insights in physics is that symmetries, to be precise, rigid symmetries, lead to conservation laws. The formal statement of this relation is the famous first *Noether theorem*, which shows that given an action with a rigid symmetry, we can construct a *conserved current*, from which in turn a *conserved charge* is obtained by integration of the current over a space-like hypersurface.

We will not formulate the Noether theorem in generality, but give a self-contained treatment of world-sheet conformal and space-time Poincaré symmetry.

2.2.5.1 Conformal Transformations on Σ

When imposing the conformal gauge $h_{\alpha\beta} \overset{!}{=} \eta_{\alpha\beta}$ we made use of both reparametrisations and Weyl transformations. However, we have not completely gauge-fixed these symmetries, because some reparametrisations only change the metric by a conformal factor, and can therefore be undone by a compensating Weyl transformation:

$$\eta_{\alpha\beta} \to e^{2\Omega(\sigma)}\eta_{\alpha\beta} \to \eta_{\alpha\beta}.$$

We will refer to reparametrisations which have this restricted form as *conformal transformations*, and to the gauge-fixed theory as *conformally invariant*. In Section 4.2, we will see that conformal transformations take the following form in light-cone coordinates:

$$\sigma^+ \to \tilde{\sigma}^+(\sigma^+) , \quad \sigma^- \to \tilde{\sigma}^-(\sigma^-).$$

That is, conformal transformations are precisely those reparametrisations which do not mix the light-cone coordinates.

We have already verified that the energy momentum tensor is conserved, $\partial^\alpha T_{\alpha\beta} = 0$. In light-cone coordinates, this statement reads

$$\partial_- T_{++} = 0 , \quad \partial_+ T_{--} = 0,$$

which implies that each independent component only depends on one of the light-cone coordinates,

$$T_{++} = T_{++}(\sigma^+) , \quad T_{--} = T_{--}(\sigma^-).$$

Therefore, our conserved current $T_{\alpha\beta}$ decomposes into two *chiral* conserved currents. The decoupling of left- and right-moving quantities, which we have already seen in other expressions before, is characteristic for massless two-dimensional theories, and hence for strings. Given this chiral decomposition, it is sufficient to look at one chiral current, say T_{++}.

Since we are interested in string theory rather than in two-dimensional scalar field theory on global Minkowski space, we need to impose boundary conditions. Let us choose periodic boundary conditions for definiteness. This implies that T_{++} is periodic, $T_{++}(\sigma^+ + \pi) = T_{++}(\sigma^+)$. We observe that we can create infinitely many conserved chiral currents by multiplying T_{++} with an arbitrary (smooth) periodic function $f(\sigma^+)$ since

$$\partial_- \left(f(\sigma^+)T_{++}\right) = 0. \tag{2.54}$$

We claim that the corresponding conserved charge is:

$$L_f = T \int_0^\pi d\sigma^1 f(\sigma^+)T_{++}. \tag{2.55}$$

Exercise 2.2.6 Show that L_f as defined in (2.55) is a conserved charge, i.e. verify that

$$\frac{d}{d\sigma^0}L_f = 0.$$

Since $f(\sigma^+)$ is periodic, we can expand it in a Fourier series. The Fourier basis $\{e^{2im\sigma^+} \mid m \in \mathbb{Z}\}$ then provides us with a basis $\{\tilde{L}_m \mid m \in \mathbb{Z}\}$ for the conserved charges, where

$$\tilde{L}_m = \frac{T}{4} \int_0^\pi d\sigma^1 e^{2im\sigma^1} T_{++}. \tag{2.56}$$

Exercise 2.2.7   Why can we write $e^{2im\sigma^1}$ instead of $e^{2im\sigma^+}$ in formula (2.56)?

By repeating the same steps for T_{--} we obtain a second infinite set of conserved charges,

$$L_m = \frac{T}{4} \int_0^\pi d\sigma^1 e^{-2im\sigma^1} T_{--}. \tag{2.57}$$

Using the conserved charges, we can rewrite the constraints $T_{++} = 0 = T_{--}$ as

$$L_m = 0 = \tilde{L}_m. \tag{2.58}$$

Since $L_m, \tilde{L}_m$ are conserved charges, we have now justified our previous claim that we can impose the constraints as initial conditions.

   If one repeats this analysis for open strings, one finds that since left- and right-moving waves couple through the boundary conditions, there is only one infinite set of conserved charges:

$$L_m = \frac{T}{2} \int_0^\pi d\sigma^1 \left(e^{im\sigma^1} T_{++} + e^{-im\sigma^1} T_{--} \right). \tag{2.59}$$

Exercise 2.2.8   Derive (2.59) for Neumann boundary conditions. One helpful observation is that one can formally combine left- and right-moving quantities into quantities which are periodic on the doubled interval $\sigma^1 \in [-\pi, \pi]$.

2.2.5.2 Poincaré Transformations on $\mathbb{M}$ – Momentum and Angular Momentum of the String

The Polyakov action is invariant under Poincaré transformations of M:

$$X^\mu \to \Lambda^\mu{}_\nu X^\nu + a^\mu. \tag{2.60}$$

To derive the associated conserved currents, we will use the *Noether method*, which underlies the proof of the general Noether theorem. Instead of the general case, let us consider space-time translations $X^\mu \to X^\mu + a^\mu$ and space-time Lorentz transformations $X^\mu \to \Lambda^\mu{}_\nu X^\nu$. From the world-sheet point of view, these are inner symmetries. The conserved current can be derived by considering what happens if we promote rigid symmetry transformations to local transformations where the transformation parameters become functions of the world-sheet coordinates. This procedure is often referred to as *gauging* the rigid symmetry. The infinitesimal form of a gauged translation is $\delta X^\mu = \epsilon^\mu(\sigma)$, and for a gauged Lorentz transformation

$\delta X^\mu = \omega^\mu{}_\nu(\sigma) X^\nu$. These are no longer symmetries of the action, but we know that (i) the action becomes invariant if we restrict them to rigid transformations, and (ii) the action is invariant under all variations if we impose the equations of motion.

Exercise 2.2.9 Write the variations of the gauge-fixed action (2.33) under infinitesimal gauged translations $\delta X^\mu = \epsilon^\mu(\sigma)$ and Lorentz transformations $\delta X^\mu = \omega^\mu{}_\nu(\sigma) X^\nu$ in the form

$$\delta_\epsilon S = \int d^2\sigma (\partial^\alpha J_\alpha^\mu) \epsilon_\mu(\sigma) , \quad \delta_\omega S = \int d^2\sigma (\partial^\alpha J_\alpha^{\mu\nu}) \frac{1}{2} \omega_{\nu\mu}(\sigma). \quad (2.61)$$

Read off the Noether currents J_α^μ and $J_\alpha^{\mu\nu}$ for translations and Lorentz transformations, and explain why these currents are conserved on-shell on the world-sheet.[11] Note that $\omega_{\mu\nu} = -\omega_{\nu\mu}$. Why did we include a factor $\frac{1}{2}$ in the expression for $\delta_\omega S$? Why will this method produce a conserved current for any rigid symmetry?

The associated conserved charges are obtained by integration of the time-like component of the current along any space-like hypersurface $\sigma^0 = \text{const.}$ in Σ:

$$P^\mu = \int_0^\pi d\sigma^1 J_0^\mu = T \int_0^\pi d\sigma^1 \dot{X}^\mu,$$

$$J^{\mu\nu} = \int_0^\pi d\sigma^1 J_0^{\mu\nu} = T \int_0^\pi d\sigma^1 \left(X^\mu \dot{X}^\nu - X^\nu \dot{X}^\mu \right). \quad (2.62)$$

The quantities P_μ and $J_{\mu\nu}$ are the total relativistic momentum and angular momentum of the string.

2.2.6 Explicit Solutions – Periodic Boundary Conditions

So far we have extracted information without solving the equations of motion explicitly. We now turn to this problem, which requires to select solutions of the two-dimensional wave equation (2.52) which satisfy the boundary conditions.

We start with periodic boundary conditions. The most general solution of the two-dimensional wave equation which is periodic in σ^1 can be parametrised in the form

$$X^\mu(\sigma) = a^\mu + b^\mu \sigma^0 + \sum_{n \neq 0} c_n^\mu e^{-2in\sigma^-} + \sum_{n \neq 0} d_n^\mu e^{-2in\sigma^+}, \quad (2.63)$$

where $a^\mu, b^\mu \in \mathbb{R}$ and $(c_n^\mu)^* = c_{-n}^\mu$ and $(d_n^\mu)^* = d_{-n}^\mu$, since X^μ is real. The term linear in σ^0 is allowed by the boundary conditions and solves the wave equation.

The conventional parametrisation of the solution used in string theory looks somewhat different from (2.63):

$$X^\mu(\sigma) = x^\mu + L_s^2 p^\mu \sigma^0 + \frac{i}{2} L_s \sum_{n \neq 0} \frac{1}{n} \alpha_n^\mu e^{-2in\sigma^-} + \frac{i}{2} L_s \sum_{n \neq 0} \frac{1}{n} \tilde{\alpha}_n^\mu e^{-2in\sigma^+}, \quad (2.64)$$

---

[11] While this can be verified explicitly by using the equations of motion, try to give a general argument.

where $x^\mu, p^\mu \in \mathbb{R}$ and $(\alpha_n^\mu)^* = \alpha_{-n}^\mu$ and $(\tilde{\alpha}_n^\mu)^* = \tilde{\alpha}_{-n}^\mu$. The *string length* L_S is defined as[12]

$$L_S = \frac{1}{\sqrt{\pi T}}.$$

Note that the coordinates σ^α and the coefficients $\alpha_n^\mu, \tilde{\alpha}_n^\mu$ are dimensionless, while p^μ has the dimension of an inverse length, given that $c = 1$ and $\hbar = 1$. To see explicitly that X^μ splits into left- and right-moving parts, as in (2.52), note that $p^\mu \sigma^0 = \frac{1}{2} p^\mu (\sigma^+ + \sigma^-)$.

While looking more complicated than (2.63), equation (2.64) is better adapted to the physical interpretation of the coefficients. To see this, we compute the total momentum:

$$P^\mu = T \int_0^\pi d\sigma^1 \dot{X}^\mu = p^\mu.$$

Thus, the coefficient p^μ of the term proportional to σ^1 is equal to the total momentum. Next, we compute the motion of the centre of mass:

$$x_{CM}^\mu = \frac{1}{\pi} \int_0^\pi d\sigma^1 X^\mu(\sigma) = x^\mu + p^\mu \sigma^0.$$

For time-like p^μ we can match this with the world-line of a massive relativistic particle,

$$x^\mu(\tau) = x^\mu(0) + \frac{dx^\mu}{d\tau}(0)\tau,$$

and conclude that the centre of mass of a string behaves like a relativistic particle and moves, in the absence of forces, on a straight line in Minkowski space. Since $p^\mu = m\dot{x}^\mu$, world-sheet time σ^0 and proper time τ are related by $\sigma^0 = m^{-1}\tau$. The mass of the string is given by $P^\mu P_\mu = p^\mu p_\mu = -m^2$, and we will work out explicit expressions later. Thus the motion of a string decomposes into two parts: a *zero mode part* corresponding to the centre of mass motion, and the remaining terms, which describe left- and right-moving waves. As required for dimensional reasons, the string tension T appears explicitly in (2.64). It is convenient to use so-called string units where one sets

$$L_S = 1 \Leftrightarrow T = \frac{1}{\pi} \quad \text{in addition to} \quad c = 1, \hbar = 1. \tag{2.65}$$

This is an example of a system of units where all quantities are measured in multiples of fundamental constants. Another example are Planck units where $G_N = 1$, $\hbar = 1$, $c = 1$. Units are discussed in Appendix B.

To impose the constraints, we now evaluate the conserved charges $L_m, \tilde{L}_m$ for our solution. Since they are conserved, we can evaluate them for any world-sheet time σ^0, the most convenient choice being $\sigma^0 = 0$. Using $T_{\pm\pm} = 2(\partial_\pm X)^2$ we find:

---

[12] Note that in later chapters it will be convenient to use a rescaled string length, denoted l_S, which is related to L_S by $l_S = \sqrt{2} L_S$.

$$L_m := \frac{T}{4} \int_0^\pi d\sigma^1 \, e^{-2im\sigma^1} T_{--} = \frac{\pi T}{2} \sum_{n=-\infty}^{\infty} \alpha_{m-n} \cdot \alpha_n, \qquad (2.66)$$

$$\tilde{L}_m := \frac{T}{4} \int_0^\pi d\sigma^1 \, e^{2im\sigma^1} T_{++} = \frac{\pi T}{2} \sum_{n=-\infty}^{\infty} \tilde{\alpha}_{m-n} \cdot \tilde{\alpha}_n. \qquad (2.67)$$

While x^μ does not appear in $T_{\pm\pm}$, the momentum p^μ is present and has been included in the sum by defining

$$\alpha_0 = \tilde{\alpha}_0 = \frac{1}{\sqrt{4\pi T}} p \overset{\pi T = 1}{=} \frac{1}{2} p. \qquad (2.68)$$

The constraints $T_{\pm\pm} = 0$ imply

$$L_m = \tilde{L}_m = 0. \qquad (2.69)$$

The *canonical Hamiltonian* is[13]

$$H = \int_0^\pi d\sigma^1 \left(\dot{X}\Pi - \mathcal{L} \right) = \frac{T}{2} \int_0^\pi (\dot{X}^2 + (X')^2)^2 = L_0 + \tilde{L}_0. \qquad (2.70)$$

The Hamiltonian constraint $H = 0$ allows us to express the mass of a state in terms of its Fourier coefficients:

$$H = L_0 + \tilde{L}_0 = \frac{\pi T}{2} \sum_{n=-\infty}^{\infty} (\alpha_{-n} \cdot \alpha_n + \tilde{\alpha}_{-n} \cdot \tilde{\alpha}_n) = \frac{p^2}{4} + \pi T (N + \tilde{N}) = 0, \qquad (2.71)$$

where we defined the total occupation numbers

$$N = \sum_{n=1}^{\infty} \alpha_{-n} \cdot \alpha_n \,, \quad \tilde{N} = \sum_{n=1}^{\infty} \tilde{\alpha}_{-n} \cdot \tilde{\alpha}_n.$$

This provides us with the *mass shell condition*:

$$M^2 = -p^2 = 4\pi T (N + \tilde{N}). \qquad (2.72)$$

Since $L_0 = \tilde{L}_0$ we have the additional constraint

$$N = \tilde{N}.$$

This is called *level matching*, because it implies that left- and right-moving modes contribute equally to the mass. The physical interpretation of the other constraints $L_m = 0 = \tilde{L}_m$, $m \neq 0$ will be discussed in Chapters 7 and 9.

2.2.7 Explicit Solutions – Neumann Boundary Conditions

We now turn to open strings. The solution of the two-dimensional wave equation with Neumann boundary conditions is

---

[13] In Chapter 4, we will see explicitly that $L_0 + \tilde{L}_0$ generates translations of σ^0. For this reason it is also called the *world-sheet Hamiltonian*.

$$X^{\mu}(\sigma) = x^{\mu} + L_S^2 p^{\mu} \sigma^0 + i L_S \sum_{n \neq 0} \frac{1}{n} \alpha_n^{\mu} e^{-in\sigma^0} \cos(n\sigma^1) \,, \tag{2.73}$$

where $x^{\mu}, p^{\mu} \in \mathbb{R}$ and $(\alpha_m^{\mu})^* = \alpha_{-m}^{\mu}$. There is only one set of Fourier coefficients, since left- and right-moving waves couple through the boundary conditions and combine into standing waves. We can, of course, re-write (2.73) in the form $X = X_L(\sigma^+) + X_R(\sigma^-)$:

$$X_{L/R}^{\mu}(\sigma^{\pm}) = \frac{1}{2}x^{\mu} + L_S^2 p_{L/R}^{\mu} \sigma^{\pm} + L_S \sum_{n \neq 0} \frac{1}{n} \alpha_{n\,(L/R)}^{\mu} e^{-in\sigma^{\pm}} \,, \tag{2.74}$$

subject to

$$p_L^{\mu} = p_R^{\mu} = \frac{1}{2}p^{\mu} \,, \quad \alpha_{n(L)}^{\mu} = \alpha_{n(R)}^{\mu}. \tag{2.75}$$

Exercise 2.2.10   Show that the ends of an open string must move with the speed of light.

We mentioned before that with Neumann boundary conditions there is only one set of conserved charges L_m. Here is their explicit form in terms of Fourier coefficients:

$$L_m = \frac{T}{2} \int_0^{\pi} d\sigma^1 \left(e^{im\sigma^1} T_{++} + e^{-im\sigma^1} T_{--} \right) = \frac{T}{16} \int_{-\pi}^{\pi} e^{im\sigma^1} \left(\dot{X} + X' \right)^2$$

$$= \frac{1}{2} \pi T \sum_n \alpha_{m-n} \alpha_n, \tag{2.76}$$

where we defined $\alpha_0 = p$. The canonical Hamiltonian is $H = L_0$. As for closed strings, the Hamiltonian constraint $H = L_0 = 0$ is the *mass shell condition*:

$$M^2 = -p^2 = 2\pi T N. \tag{2.77}$$

2.2.8   Explicit Solutions – Dirichlet Boundary Conditions

We now turn to the case where the coordinates X^i, $i = p+1, \ldots, D-1$ are subject to Dirichlet boundary conditions. Concretely, we impose that the ends of open strings are constrained to live on two parallel D-p-branes:

$$X^i(\sigma^0, \sigma^1 = 0) = x_0^i \,, \quad X^i(\sigma^0, \sigma^1 = \pi) = x_1^i. \tag{2.78}$$

The most general solution of the two-dimensional wave equation with these boundary conditions is

$$X^i(\sigma) = x_0^i + 2w^i \sigma^1 - L_S \sum_{n \neq 0} \frac{1}{n} \alpha_n^i e^{-in\sigma^0} \sin(n\sigma^1) \,, \quad \text{where} \quad w^i = \frac{1}{2\pi}(x_1^i - x_0^i), \tag{2.79}$$

and where $(\alpha_n^i)^* = \alpha_{-n}^i$. Observe that while Neumann boundary conditions allow a term linear in σ^0, Dirichlet boundary conditions allow a term linear in σ^1, and in fact require it if the D-branes do not coincide. Left- and right-moving waves still combine into standing waves, but momentum is no longer conserved at the ends, as $\partial_1 X^i(\sigma^0, \sigma^1 = 0, \pi) \neq 0$. We can decompose the solution explicitly into a left- and a right-moving part, $X^i = X_L^i(\sigma^+) + X_R^i(\sigma^-)$ with

$$X_{L/R}^i(\sigma^\pm) = \frac{x_0^i}{2} + p_{L/R}^i \sigma^\pm + \frac{i}{2} L_S \sum_{\neq 0} \frac{1}{n} \alpha_{n(L/R)}^i e^{-in\sigma^\pm},$$

with

$$p_L^i = -p_R^i = w^i , \quad \alpha_{n(L)}^i = -\alpha_{n(R)}^i.$$

The conserved charges L_m take the same form (2.76) as for Neumann boundary conditions, but with α_0^i now defined as

$$\alpha_0^i = 2\sqrt{\pi T} p_L^i = 2\sqrt{\pi T} w^i = \sqrt{\frac{T}{\pi}}(x_1^i - x_0^i).$$

To understand the effect of Dirichlet boundary conditions on the mass of a string, consider the case where we split the string coordinates as $X^\mu = (X^m, X^i)$, with X^m, $m = 0, 1, \dots, p$ satisfying Neumann boundary conditions and X^i, $i = p+1, \dots, D-1$ satisfying Dirichlet boundary conditions (2.78). The mass is obtained by solving the constraint $L_0 = 0$ for the mass M. With our choice of boundary conditions, we find that

$$L_0 = \frac{1}{2}\alpha_0^2 + N = \frac{1}{2\pi T}p^2 + \frac{T}{2\pi}|x_1 - x_0|^2 + N, \quad N = \sum_{n>0} \eta_{\mu\nu} \alpha_n^\mu \alpha_n^\nu. \qquad (2.80)$$

Solving the constraint $L_0 = 0$ for the mass, we obtain the mass formula

$$M^2 = T^2 |x_1 - x_0|^2 + 2\pi T N. \qquad (2.81)$$

The first term corresponds to the energy of a string which is stretched between the two D-branes in a configuration of minimal length. Observe that this is consistent with interpreting T as the string tension.

Exercise 2.2.11   Fill in the details of the derivation of the mass formula (2.81).

2.2.9 Non-oriented Strings

All solutions we have obtained for open and closed strings can be built from solutions which are either even or odd with respect to *world-sheet parity*

$$\Omega : \quad \sigma^1 \mapsto \pi - \sigma^1 \cong -\sigma^1 \quad \mathrm{mod}\ \pi . \qquad (2.82)$$

As indicated, Ω can be interpreted as a *reflection up to boundary conditions* of the spatial world-sheet coordinate σ^1. Since $\Omega^2 = \mathbb{1}$, the eigenvalues of this transformation are ± 1, and we refer to the corresponding eigenstates as even and odd, respectively. For closed strings world-sheet parity interchanges left- and right-moving excitations, whereas for open strings the mode number decides which excitations are even and odd:

$$\Omega : \begin{cases} \alpha_m^\mu \leftrightarrow \tilde{\alpha}_m^\mu, & \text{(closed)}, \\ \alpha_n^\mu \mapsto (-1)^n \alpha_n^\mu, & \text{(Neumann)}, \\ \alpha_n^\mu \mapsto (-1)^{n+1} \alpha_n^\mu, & \text{(Dirichlet)}. \end{cases}$$

Non-oriented strings theories are defined by restricting to configurations which are even. Since in such a theory all configurations are invariant under world-sheet parity, one looses the information about the orientation of the string. In other words, one cannot distinguish between the ends of the string and the two sides of the world-sheet. Non-oriented closed strings are symmetric under exchange of left- and right-moving waves, while for non-oriented open strings, depending on the boundary conditions, either the odd- or the even-numbered modes cannot be excited. While this reduces the number of independent excitations, it increases the number of world-sheet topologies, because we now have to admit non-orientable surfaces, such as the Moebius strip (for non-oriented open strings) and the Klein bottle (for non-oriented closed strings).

2.2.10 Literature

Zwiebach (2009) covers many aspects of this chapter in more detail, in particular how to construct various types of adapted coordinate systems. Readers who would like to see a more detailed treatment of *D-branes* can find this in the textbooks Polchinski (1998a), Blumenhagen et al. (2013), and in the monograph Johnson (2003). Textbooks which cover *solitons* and *instantons* in detail are Rajaraman (1982) and Weinberg (2012), see also the classical Erice lectures Coleman (1985). Branes which are localised in time are a less explored topic, see, for example, Hull (1998), Hull (2001), Gutperle and Strominger (2002). At some points we have assumed that the reader is familiar with covariant derivatives as used in general relativity (the Levi–Civita connection). This material can be found in any textbook on general relativity, and of course in books on differential geometry. A good reference for *differential geometry* as used in theoretical physics is Frankel (2004). More on the Noether theorem can be found in textbooks on mechanics, classical, and quantum field theory.

3 Quantised Relativistic Particles and Strings

In this chapter, we introduce the covariant quantisation of relativistic particles and strings, and compare this to the quantisation of (scalar) fields. The detailed analysis of the Hilbert space of states for relativistic strings will be performed in Part III, after we have developed various useful tools from two-dimensional conformal field theory in Part II.

3.1 Quantised Relativistic Particles

The usual heuristic approach to *quantisation* is to promote the canonical coordinates and canonical momenta of a classical theory to self-adjoint operators acting on a separable Hilbert space $\mathcal{H}$, and to impose the canoncial commutation relations. The canonical commutation relations can be motivated through replacing the *Poisson brackets* $\{\cdot, \cdot\}$ of the classical theory by quantum commutators $[\cdot, \cdot]$, using the formal substitution rule $\{\cdot, \cdot\} \rightarrow -i[\cdot, \cdot]$. We will not evaluate the Poisson brackets of the classical theory, but directly postulate the canonical commutation relations.

In the case of a free non-relativistic particle with Cartesian coordinates x^i and momenta p^j, the canonical commutation relations are

$$[x^i, p^j] = i\delta^{ij}, \tag{3.1}$$

where we have set $\hbar = 1$, and where the unit operator on the Hilbert space $\mathcal{H}$ is understood on the right-hand side. We will procede formally and ignore the technical complications caused by the fact that x^i, p^j are unbounded operators on an infinite dimensional Hilbert space.

For a relativistic particle the natural generalisation of (3.1) is

$$[x^\mu, p^\nu] = i\eta^{\mu\nu}. \tag{3.2}$$

However, we know that the components of the relativistic momentum are subject to the mass shell condition $p^2 + m^2 = 0$. One option is to solve this constraint in the classical theory, and then to quantise the theory using only gauge-inequivalent quantities. A specific version of this procedure is the so-called light cone quantisation, which will be discussed later (see Chapter 10). Any such scheme has the disadvantage that Lorentz invariance is no longer manifest. Here we will follow the complementary, covariant approach, where canonical commutation relations are imposed on Lorentz covariant quantities, while the constraint $p^2 + m^2 = 0$ is imposed afterwards and selects a subspace of *physical states*.

We start by constructing a representation space $\mathcal{F}$ for the commutations relations (3.2), which we call the *Fock space*. This is done by postulating the existence of a distinguished state, the *vacuum* or *ground state* $|0\rangle$, which is translation invariant:

$$p^\nu|0\rangle = 0. \tag{3.3}$$

The Fock space $\mathcal{F}$ is generated by applying operators built out of the canonically conjugate operator x^μ. We assume that p^ν has a complete set of eigenstates, so that $\mathcal{F}$ is spanned by momentum eigenstates $|k\rangle$,

$$p^\nu|k\rangle = k^\nu|k\rangle. \tag{3.4}$$

The scalar product between momentum eigenstates is defined using the Dirac delta function:

$$\langle k|k'\rangle = \delta^D(k - k'). \tag{3.5}$$

Momentum eigenstates are not normalisable. To obtain normalisable states we form superpositions ('wave packages') of the form

$$|\Phi\rangle = \int d^D k\, \tilde\Phi(k)|k\rangle \tag{3.6}$$

and require them to be square-integrable:

$$|\langle\Phi|\Phi\rangle| < \infty. \tag{3.7}$$

Since the scalar product between normalisable states is

$$\langle\Phi|\Phi'\rangle = \int d^D k d^D k'\, \tilde\Phi^*(k)\tilde\Phi(k')\langle k|k'\rangle = \int d^D k\, \tilde\Phi^*(k)\tilde\Phi(k), \tag{3.8}$$

the resulting Hilbert space $\mathcal{H} \subset \mathcal{F}$ is isomorphic to $L^2(\mathbb{R}^D)$, the space of square-integrable functions in D variables. This is, however, not the Hilbert space of physical states, because we still have to impose the constraint $p^2 + m^2 = 0$. Therefore, we define the subspace of physical states by $\mathcal{F}_{\text{phys}} \subset \mathcal{F}$, where

$$\mathcal{F}_{\text{phys}} = \{|\Phi\rangle \in \mathcal{F}\,|\,(p^2 + m^2)|\Phi\rangle = 0\}. \tag{3.9}$$

The physical Hilbert space $\mathcal{H}_{\text{phys}} \subset \mathcal{F}_{\text{phys}}$ is the subspace of normalisable physical states.

States satisfying the mass shell condition can be parametrised as

$$|\Phi\rangle = \int d^D k\, \delta(k^2 + m^2)\tilde\phi(k)|k\rangle, \tag{3.10}$$

where the δ-function forces the momenta to be zero away from the mass shell $k^2 + m^2 = 0$. The mass shell condition has two solutions for the energy k^0:

$$k^0 = \pm\sqrt{\vec{k}^2 + m^2}. \tag{3.11}$$

Restricting to positive energies $k^0 > 0$, and using that[1]

$$\theta(k^0)\delta(k^2 + m^2) = \frac{1}{2|k^0|}\delta\left(k^0 - \sqrt{\vec{k}^2 + m^2}\right), \tag{3.12}$$

---

[1] Here we use the following identity for δ-functions, which can be verified by applying the chain rule when integrating against test functions: if a function $g(x)$ has zeros at $x_1, x_2, \ldots, x_N$, such that $g'(x_i) \neq 0$ for $i = 1, \ldots, N$, then $\delta(g(x)) = \sum_{i=1}^N \frac{1}{|g'(x_i)|}\delta(x - x_i)$.

we obtain

$$|\Phi\rangle = \int \frac{d^{D-1}\vec{k}}{2\omega_{\vec{k}}} \tilde{\phi}(k)|k\rangle, \qquad (3.13)$$

where $\omega_{\vec{k}} = \sqrt{\vec{k}^2 + m^2}$, and where k is restricted to values on the hyperboloid $k^2 + m^2 = 0, k^0 > 0$.

The scalar product between two normalisable states $|\Phi_i\rangle \in \mathcal{H}_{\text{phys}}$, $i = 1, 2$ which satisfy the mass shell condition is

$$\langle \Phi_1 | \Phi_2 \rangle = \int \frac{d^{D-1}\vec{k}}{2\omega_{\vec{k}}} \tilde{\phi}_1^*(k) \tilde{\phi}_2(k). \qquad (3.14)$$

The resulting Hilbert space $\mathcal{H}_{\text{phys}}$ is isomorphic to $L^2(\mathbb{R}^{D-1}, d\mu)$, where

$$d\mu = \frac{d^{D-1}\vec{k}}{2\omega_{\vec{k}}} \qquad (3.15)$$

is the Lorentz invariant measure on the mass hyperboloid $k^2 + m^2 = 0, k^0 > 0$.

As an extension, one can also admit negative energies $k^0 < 0$. The corresponding states are not interpreted as states of negative energy, but as positive energy modes of an *antiparticle* which is distinct from the particle. The particle-antiparticle Hilbert space consists of two orthogonal copies of $L^2(\mathbb{R}^{D-1}, d\mu)$. Particles and antiparticles carry opposite charges under all global symmetries. Therefore it is generic that particles and antiparticles are distinct. Examples of neutral particles are photons and the hypothetical gravitons.

Exercise 3.1.1 Show that the position space representation

$$\Phi(x) = \langle x | \Phi \rangle, \qquad (3.16)$$

of a physical state $|\Phi\rangle \in \mathcal{H}_{\text{phys}}$ satisfies the Klein–Gordon equation:

$$(-\Box + m^2)\Phi(x) = 0, \qquad (3.17)$$

where $\Box = \partial_\mu \partial^\mu = -\partial_0^2 + \Delta$ is the wave operator.

Remark: For solutions of the Klein–Gordon equation one can define a time-independent scalar product using the conserved current

$$j^\mu = i(\Phi^* \partial^\mu \Phi - \partial^\mu \Phi^* \Phi), \qquad (3.18)$$

which has the interpretation of a *charge density*. Note that the current is only non-vanishing for complex-valued $\Phi(x)$, and that the associated scalar product

$$\langle \Phi_1 | \Phi_2 \rangle = i \int_{x^0 = \text{const}} d^{D-1}\vec{x}\, j^0$$

$$= i \int_{x^0 = \text{const}} d^{D-1}\vec{x}\, \left(\Phi_1^*(x) \partial^0 \Phi_2(x) - (\partial^0 \Phi_1(x)) \Phi_2^*(x) \right)$$

has a relative sign between positive and negative energy solutions, which makes it indefinite.

Remark: Intepreting $\langle x | \Phi \rangle$ as a wave function with the properties familiar from the Schrödinger formulation of quantum mechanics is problematic. One can show that in contrast to non-relativistic theories there are no arbitrarily sharply localised 'position eigenstates' in the Hilbert space of a relativistic quantum theory. The standard conclusion is that relativistic quantum theories need to be formulated as multiparticle theories. For this purpose *field quantisation*, which often somewhat misleadingly is called second quantisation is a more convenient framework.

3.2 Field Quantisation and Quantum Field Theory

As Exercise 3.1.1 has shown, the Klein–Gordon equation can be viewed as the implementation of the constraint $p^2 + m^2 = 0$ which selects the physical states in the Fock space of a relativistic particle. With this interpretation the Klein–Gordon equation is analogous to the Schrödinger equation, in the sense that it is a condition imposed on quantum states. There is an alternative interpretation of the Klein–Gordon equation as a *classical field equation*, analogous to the Maxwell equations. We will now quantise the classical complex Klein–Gordon field and compare the result to the quantisation of a relativistic particle. The Klein–Gordon equation

$$(-\Box + m^2)\Phi(x) = 0 \tag{3.19}$$

follows from the action

$$S[\Phi] = \int d^D x \left(-\partial_\mu \Phi \partial^\mu \Phi^* - m^2 \Phi \Phi^*\right). \tag{3.20}$$

The canonical momentum of this field theory is

$$\Pi(x) = \frac{\partial L}{\partial \partial_0 \Phi} = \partial_0 \Phi^*. \tag{3.21}$$

The Fourier representation of the general solution can be parametrised in the following form:

$$\Phi(x) = \int d^D k \left(\theta(k^0)\delta(k^2 + m^2)\tilde{\phi}_+(k) + \theta(-k^0)\delta(k^2 + m^2)\tilde{\phi}_-^*(k)\right). \tag{3.22}$$

Here we used the step functions $\theta(\pm k^0)$ to separate the two components of the hyperboloid $k^2 + m^2 = 0$. The δ-function can be used to carry out the k^0-integration:

$$\Phi(x) = \frac{1}{(2\pi)^{(D-1)/2}} \int \frac{d^{D-1}\vec{k}}{2\omega_{\vec{k}}} \left(\phi_+(k)e^{ikx} + \phi_-^*(k)e^{-ikx}\right)_{k^0 = \omega_{\vec{k}}}. \tag{3.23}$$

Here $\phi_\pm(k)$ are rescaled versions of $\tilde{\phi}_\pm(k)$. We now quantise the complex Klein–Gordon field by declaring $\Phi(x)$ to be an operator[2] satisfying the canonical commutation relation

$$[\Phi(x), \Pi(y)]_{x^0 = y^0} = i\delta^{D-1}(\vec{x} - \vec{y}). \tag{3.24}$$

Since $\Phi(x)$ depends on time x^0, we are in the Heisenberg picture, where operators depend on time while states are time-independent. We only need to specify the commutator at equal times, because the commutator at other times is fixed by time evolution. The spatial coordinate is treated as a continuous index labelling degrees

---

[2] Technically, this is an operator-valued distribution, which becomes an operator by application to test functions. We proceed formally.

of freedom located at different points. The operator $\Phi(x)$ can be represented as a Fourier integral, with the Fourier coefficients $\phi_\pm(k)$ promoted to operators. Complex conjugated quantities are now interpreted has Hermitian conjugate operators, and denoted $\Phi^\dagger(x)$, $\phi_\pm^\dagger(k)$. The Fourier modes satisfy the relations:[3]

$$[\phi_\pm(\vec{k}), \phi_\pm^\dagger(\vec{k}')] = 2\omega_{\vec{k}}\delta^{D-1}(\vec{k} - \vec{k}') \;, \quad [\phi_\pm(\vec{k}), \phi_\mp^\dagger(\vec{k}')] = 0. \tag{3.25}$$

Now we can construct a Fock space based on a ground state $|0\rangle$ defined by the properties

$$\phi_\pm(\vec{k})|0\rangle = 0 \;, \quad \langle 0|0\rangle = 1. \tag{3.26}$$

The states

$$|\vec{k}\rangle_\pm = \phi_\pm^\dagger(\vec{k})|0\rangle \tag{3.27}$$

are the momentum eigenstates for two particles, which are related by $\Phi \to \Phi^\dagger$, and which are interpreted as a particle and its antiparticle. The mutual scalar products between such states are

$$_\pm\langle\vec{k}|\vec{k}'\rangle_\pm = \langle 0|\phi_\pm(\vec{k})\phi_\pm^\dagger(\vec{k}')|0\rangle = \langle 0|[\phi_\pm(\vec{k})\phi_\pm^\dagger(\vec{k}')]|0\rangle = 2\omega_{\vec{k}}\delta^{D-1}(\vec{k} - \vec{k}') \tag{3.28}$$

and

$$_\mp\langle\vec{k}|\vec{k}'\rangle_\pm = 0. \tag{3.29}$$

Thus particle and antiparticle states are orthogonal. By taking square-integrable superpositions we obtain two orthogonal copies of the Hilbert space $L^2(\mathbb{R}^{D-1}, d\mu)$. This is the same Hilbert space that we obtained above by quantising the relativistic particle, provided that we include both sheets of the mass hyperboloid. The advantage of field quantisation is that it is straightforward to describe multiparticle states. In particular, by repeated application of creation operators $\phi_\pm^\dagger(\vec{k})$ we obtain the multiparticle momentum eigenstates

$$\phi_+^\dagger(\vec{k}_1)\phi_+^\dagger(\vec{k}_2)\cdots\phi_-^\dagger(\vec{k}_1')\phi_-^\dagger(\vec{k}_2')\cdots|0\rangle. \tag{3.30}$$

Since creation operators commute among themselves, the multiparticle Hilbert space obtained by taking square integrable superpositions of momentum eigenstates has the form

$$\mathcal{H}_{\text{phys}}^{\text{multiparticle}} = \mathbb{C} \oplus \mathcal{H}_{\text{phys}}^+ \oplus \bigvee^2 \mathcal{H}_{\text{phys}}^+ \oplus \cdots \oplus \mathcal{H}_{\text{phys}}^- \oplus \bigvee^2 \mathcal{H}_{\text{phys}}^- \oplus \cdots, \tag{3.31}$$

where $\mathbb{C}$ is the zero-particle sector spanned by $|0\rangle$, where $\mathcal{H}_{\text{phys}}^\pm \simeq L^2(\mathbb{R}^{D-1}, d\mu)$ are the one-particle Hilbert spaces for particle and antiparticle, and where $\bigvee^k \mathcal{H}_{\text{phys}}^\pm$ denotes the symmetrised k-th tensor power of $\mathcal{H}_{\text{phys}}^\pm$.

The commutation relations (3.25) resemble those of harmonic oscillators labeled by the momentum $\vec{k}$. We can make this relation manifest by using the rescaled operators

$$a(\vec{k}) = (2\omega_{\vec{k}})^{-1/2}\phi_+(k) \;, \quad b(\vec{k}) = (2\omega_{\vec{k}})^{-1/2}\phi_-(k) \;, \tag{3.32}$$

and their Hermitian conjugates

$$a^\dagger(\vec{k}) = (2\omega_{\vec{k}})^{-1/2}\phi_+^\dagger(k) \;, \quad b^\dagger(\vec{k}) = (2\omega_{\vec{k}})^{-1/2}\phi_-^\dagger(k). \tag{3.33}$$

---

[3] The relation between (3.24) and (3.25) works like in the similar example in Exercise 3.3.1.

These satisfy standard harmonic oscillator, though with a 'continuous index $\vec{k}$':

$$[a(k), a^\dagger(\vec{k})] = \delta^{D-1}(\vec{k} - \vec{k}') \,, \quad [b(k), b^\dagger(\vec{k})] = \delta^{D-1}(\vec{k} - \vec{k}'), \tag{3.34}$$

and the mode expansion of the field operator takes the form

$$\Phi(x) = \frac{1}{(2\pi)^{(D-1)/2}} \int \frac{d^{D-1}\vec{k}}{\sqrt{2\omega_{\vec{k}}}} \left(a(k)e^{ikx} + b^\dagger(k)e^{-ikx} \right)_{k^0 = \omega_{\vec{k}}}. \tag{3.35}$$

Note that while

$$d\mu = \frac{d^{D-1}k}{2\omega_{\vec{k}}} \quad \text{and} \quad 2\omega_{\vec{k}}\delta^{D-1}(\vec{k} - \vec{k}') \tag{3.36}$$

are a Lorentz invariant measure and a Lorentz invariant δ-function on the mass shell $k^2 + m^2 = 0$, the expressions

$$\frac{d^{D-1}x}{\sqrt{2\omega_{\vec{k}}}} \quad \text{and} \quad \delta^{D-1}(\vec{k} - \vec{k}') \tag{3.37}$$

are not Lorentz invariant.[4]

Instead of a classical complex field, we could have considered a classical real field by setting $\Phi(x) = \Phi^\dagger(x)$. This implies $\phi_+(k) = \phi_-(k)$ and $a(\vec{k}) = b(\vec{k})$, and thus identifies particle and antiparticle. Note that this identification is only possible if the theory does not contain expressions which involve the phase of the scalar field Φ. For example, the minimal coupling of a scalar field to an electromagnetic field requires the invariance under local phase transformations, and since particle and antiparticle carry opposite charges they cannot be identified with one another.

In this section we have seen that the advantage of field quantisation ('second quantisation') over particle quantisation ('first quantisation') is that it directly leads to a multiparticle theory, which is what one needs in a relativistic setting where particles can be created and annihilated. Therefore quantum field theory is the standard formulation of quantised relativistic systems based on point particles, although methods based on the 'first quantised' approach, also called *world-line formalism* have applications. The extension of standard quantum field theory to strings is called *string field theory*. String field theory in its current form is quite complicated and has limited practical use, while the 'first quantised' world-sheet formalism is well developed.

3.3 Quantised Relativistic Strings

In the final section of this chapter, we will outline the problem of quantising relativistic strings, and thus motivate why we need to develop certain tools in Part II before being able to formulate a quantum theory of strings in Part III. The classical solutions found in the previous chapter have shown that the degrees of freedom

---

[4] The Lorentz invariance of $d\mu$ is clear because we have obtained this integration measure by localising the volume element $d^D k$ on the mass shell using $\delta(k^2 + m^2)$. Since $\delta^D(k - \vec{k}') = \delta(k^0 - k'^0)\delta^{D-1}(\vec{k} - \vec{k}') = \delta(k \cdot k - k' \cdot k')2|k^0|\delta^{D-1}(\vec{k} - \vec{k}')$, it follows that $2\omega_{\vec{k}}\delta^{D-1}(\vec{k} - \vec{k}')$ is Lorentz invariant.

of a free relativistic string in Minkowski space combine those of a relativistic particle with an infinite set of harmonic oscillators. This gives us a clear idea of how the Hilbert space should look like, and we could postulate canonical relations for x^μ, p^ν and the Fourier coefficients α_m^μ, $\tilde{\alpha}_m^\mu$. To be more systematic we start by imposing canonical commutation on the string coordinates $X^\mu(\sigma^0, \sigma^1)$ and the canonical momenta $\Pi^\nu(\sigma^0, \sigma^1)$. We work with the Polyakov action in the conformal gauge, and can interprete X^μ either as embedding coordinates for a string in space-time $\mathbb{M}$, or as a set of scalar fields on the world-sheet Σ. In the conformal gauge $\Pi^\mu = T\dot{X}^\mu$. Both X^μ and Π^ν are time-dependent operators, and thus we are in the Heisenberg picture of quantum mechanics. Canonical commutators are imposed at equal world-sheet times, and σ^1 is treated as a continuous index, similar to the discrete index μ. For concreteness we consider periodic boundary conditions. Then the canonical commutation relations are:

$$[X^\mu(\sigma^0, \sigma^1), \Pi^\nu(\sigma'^0, \sigma'^1)]_{\sigma^0 = \sigma'^0} = i\eta^{\mu\nu}\delta_\pi(\sigma^1 - \sigma'^1), \qquad (3.38)$$

where

$$\delta_\pi(\sigma^1) = \frac{1}{\pi}\sum_{k=-\infty}^{\infty} e^{-2ik\sigma^1} = \delta_\pi(\sigma^1 + \pi) \qquad (3.39)$$

is the periodic δ-function with period π (see Appendix C). One can view this as the canonical commutation relations of a two-dimensional field theory on a world-sheet Σ, which has the topology of a cylinder.

Exercise 3.3.1 Show that the commutation relations

$$[x^\mu, p^\nu] = i\eta^{\mu\nu}, \quad [\alpha_m^\mu, \alpha_n^\nu] = m\eta^{\mu\nu}\delta_{m+n,0}, \quad [\tilde{\alpha}_m^\mu, \tilde{\alpha}_n^\nu] = m\eta^{\mu\nu}\delta_{m+n,0}$$
$$(3.40)$$

imply the canonical commutation relations (3.38), including $[X^\mu, X^\nu]_{\sigma^1 = \sigma'^1} = 0 = [\Pi^\mu, \Pi^\nu]_{\sigma^1 = \sigma'^1}$, which are 'understood', but have not been listed explicitly.

Since X^μ is Hermitian, it also follows that

$$(x^\mu)^\dagger = x^\mu, \quad (p^\mu)^\dagger = p^\mu, \quad (\alpha_m^\mu)^\dagger = \alpha_{-m}^\mu, \quad (\tilde{\alpha}_m^\mu)^\dagger = \tilde{\alpha}_{-m}^\mu. \qquad (3.41)$$

Comparing the relations for α_m^μ to the standard relations between the creation and annihilation operators of a harmonic oscillator

$$[a, a^\dagger] = 1, \qquad (3.42)$$

we see that we indeed get an infinite set of harmonic oscillators. More precisely, setting

$$\alpha_m^\mu = \begin{cases} \sqrt{m}a_m^\mu & \text{for } m > 0, \\ \sqrt{-m}(a_{-m}^\mu)^\dagger & \text{for } m < 0, \end{cases} \qquad (3.43)$$

we obtain

$$[a_m^\mu, (a_n^\nu)^\dagger] = \eta^{\mu\nu}\delta_{m,n}. \qquad (3.44)$$

While these are standard harmonic oscillator relations for $\mu, \nu \neq 0$, we obtain an additional minus sign in the relations for $\mu = \nu = 0$.

To explore this further, let us build a Fock space $\mathcal{F}_{osc}$ by starting with a ground state $|0\rangle$, defined by

$$\alpha_n^\nu |0\rangle = 0, \quad n > 0. \tag{3.45}$$

If we include the zero mode part corresponding to x^μ, p^ν we should of course add the condition $p^\nu |0\rangle = 0$. But we already know how to deal with this 'zero mode part' of the Fock space and concentrate on the 'oscillator part' $\mathcal{F}_{osc}$ for the time being.

Oscillator eigenstates are generated by applying creation operators α_{-m}^μ, $m > 0$ to the ground state. We take the ground state to be normalised as $\langle 0|0\rangle = 1$. Now consider scalar products of the form

$$\left(\alpha_{-m}^\mu |0\rangle \, , \alpha_{-n}^\nu |0\rangle \right) = \langle 0|\alpha_m^\mu \alpha_{-n}^\nu |0\rangle = \langle 0|[\alpha_m^\mu \, , \alpha_{-n}^\nu]|0\rangle = \eta^{\mu\nu} \delta_{m,n}. \tag{3.46}$$

Thus the natural, relativistically covariant scalar product on $\mathcal{F}_{osc}$ is indefinite, while the space of states in a quantum theory must have a positive definite scalar product. However, we still need to impose the constraints $L_m = 0$ and $\tilde{L}_m = 0$. Since solving them before quantisation obscures relativistic covariance, we will impose them on the states of the quantum theory, and thus define the subspace of physical states $\mathcal{F}_{phys} \subset \mathcal{F}$ by requiring that the matrix elements of $L_m, \tilde{L}_m$ vanish between physical states $|\Phi\rangle, |\Phi'\rangle$:

$$\langle \Phi | L_m | \Phi' \rangle = 0 \, , \quad \langle \Phi | \tilde{L}_m | \Phi' \rangle = 0. \tag{3.47}$$

As we will see later, these conditions will need to be modified for $m = 0$. The space $\mathcal{F}_{phys}$ is not yet the physical Hilbert space. In particular, it is not positive definite but at best *positive semi-definite*:

$$\langle \Phi | \Phi \rangle \geq 0. \tag{3.48}$$

One can show that there are always non-trivial *null states* in $\mathcal{F}_{phys}$, that is non-trivial physical states $|\psi\rangle \neq 0$ with zero norm $\langle \psi | \psi \rangle = 0$. The existence of these states is related to the residual symmetry under conformal transformations, which has not been fixed by imposing the conformal gauge. Physical states which differ by null states are related by conformal transformations, and should thus be identified with one another. Writing $|\Phi\rangle \sim |\Phi'\rangle$ when two physical states differ by a null state, the candidate for the physical Hilbert space is

$$\mathcal{H} = \mathcal{F}_{phys}/ \sim . \tag{3.49}$$

In Part II we will develop various tools from conformal field theory which will help us to better understand the structure of the Hilbert space. The space-time interpretation of the resulting physical states will be the subject of Part III.

Continuing our preview of things to come we remark that further conditions need to be imposed to guarantee that $\mathcal{F}_{phys}$ is positive semi-definite. The so-called *no-ghost theorem* shows the absence of negative norm states for $D \leq 26$, while for $D > 26$ negative norm states always exist. This implies that strings in Minkowksi space can only be quantised consistently if $D \leq 26$. Moreover, once string interactions are included, negative norm states, which naively have been projected out by imposing the constraints, can reoccur as intermediate states in loop diagrams. Since for consistency they then have to be allowed as asymptotic states as well, the *unitarity* of the quantum theory is lost. The only case where negative norm states can be decoupled consistently for strings in a Minkowski background is in $D = 26$ dimensions. Since, at least at length scales accessible to current experiments, we live in a space-time of dimension 4, this raises the question how to account for the the extra dimensions. We will come back to this question in Part IV of the book.

3.4 Literature on Quantum Field Theory

There are many good textbooks on quantum field theory all of which will cover the quantisation of the scalar field. The one actually used for reference when writing this chapter is Peskin and Schroeder (1996). Readers with a deeper interest in conceptual aspects of quantum field theory are referred to Duncan (2012). As most textbooks we have proceeded formally and ignored mathematical (functional analytic) details. For a more mathematical treatment, see for example, Streater and Wightman (1964), Glimm and Jaffe (1981), and Haag (1996). For literature on *constrained quantisation*, see Section 1.10.

PART II

THE WORLD-SHEET PERSPECTIVE

4 The Free Massless Scalar Field on the Complex Plane

Part II of this book is devoted to developing our understanding of the world-sheet aspects of string theory. In this chapter, we will take a bottom-up approach and explore string theory with a one-dimensional target space, formulated as the theory of a two-dimensional massless scalar field on the complex plane. This will allow us to explore the role played by conformal symmetry in an explicit example. In the next chapter, we will take the opposite, top-down approach starting by postulating conformal symmetry, and then exploring its consequences. In Chapter 6, the final chapter of Part II, we will discuss the important topic of partition functions.

4.1 The Cylinder and the Plane

For simplicity and concreteness, we will consider closed string theory with a one-dimensional target space. The world-sheet is taken to be a cylinder, $\Sigma \cong I \times S^1$, which may be finite, semi-infinite or infinite. Compared to Chapter 3, we make the following modifications:

1. We perform a Wick rotation, changing the world-sheet signature from Lorentzian to Euclidean.
2. We perform a conformal map from the cylinder to the complex plane.

The Wick rotation is a standard operation in quantum mechanics and quantum field theory, which allows one to map problems of quantum physics to problems in statistical physics. We note in passing that the global continuation between spaces with Lorentz signature and Euclidean signature is only possible for special cases, which include the spaces relevant for this chapter, namely the world-sheet cylinder and a one-dimensional target space, or, more generally, D-dimensional Minkowski space. When working with the Polyakov action on general world-sheets, one needs to define the theory in Euclidean world-sheet signature, since surfaces in general do not admit a globally defined Lorentzian metric. When working within the path integral approach, one also performs a Wick rotation of the time coordinate X^0.[1] To define scattering amplitudes in Minkowski space one uses that relativistic S-matrices are analytic in the momenta, which allows one to construct amplitudes in the Euclidean regime and then to evaluate them for physical, Lorentzian momenta.

---

[1] Defining path integrals in Lorentzian signature is possible in principle, and requires to add a small imaginary part to X^0 to make the path integral convergent, analogous to the '$i\epsilon$'-prescription used for propagators. Lorentzian path integrals have the advantage of avoiding the issue of global continuation between signatures, but due to their oscillatory behaviour they are much less well understood than Euclidean path integrals. See, e.g., Duncan (2012).

Starting with a world-sheet $\Sigma \cong I \times S^1$, with coordinates σ^0, σ^1, we analytically continue the world-sheet time by setting $\sigma^0 = -i\sigma^2$ and taking σ^2 to be real. On the Euclidean cylinder we define the complex coordinate $w = \sigma^2 + i\sigma^1$. Then we map the cylinder to the complex plane by the conformal map

$$w = \sigma^2 + i\sigma^1 \mapsto z = e^{2w} = e^{2(\sigma^2 + i\sigma^1)}. \tag{4.1}$$

On the complex plane, Euclidean time translations $\sigma^2 \mapsto \sigma^2 + a$ become *dilatations* $z \mapsto e^{2a}z$, time ordering becomes *radial ordering*, and spatial translations $\sigma^1 \mapsto \sigma^1 + \theta$ become *rotations*, $z \mapsto e^{2i\theta}z$. The infinite cylinder $\mathbb{R} \times S^1$ is mapped to the complex plane with the origin removed, $\mathbb{C}^* = \mathbb{C} \backslash \{0\}$, while a finite cylinder gets mapped to an annulus, $a < |z| < b$. On the complex plane it is convenient to work with *complex partial derivatives*

$$\partial_z = \frac{1}{2}(\partial_x - i\partial_y), \quad \partial_{\bar{z}} = \frac{1}{2}(\partial_x + i\partial_y), \tag{4.2}$$

where x, y are the real coordinates given by $z = x + iy$. Complex partial derivatives are analogous to the light-cone derivatives $\partial_\pm = \frac{1}{2}(\partial_0 \pm \partial_1)$ which we use in Minkowski signature. Tensor fields on $\mathbb{C}$ can be transformed from real coordinates (x, y) to the complex coordinate z by applying the Jacobian $D(z, \bar{z})/D(x, y)$ and its inverse. With our conventions, light-cone indices $(-, +)$ correspond to holomorphic and antiholomorphic indices $(z, \bar{z})$.

To see how this works, we start with the Polyakov action in the conformal gauge, reduced to a single string coordinate X:

$$S = -\frac{1}{2\pi} \int_\Sigma d^2\sigma \, \partial_\alpha X \partial^\alpha X = \frac{2}{\pi} \int_\Sigma d^2\sigma \, \partial_+ X \partial_- X. \tag{4.3}$$

Under a Wick rotation combined with the map (4.1) this becomes

$$S_E = \frac{2}{\pi} \int_\Sigma d^2w \, \partial_w X \partial_{\bar{w}} X = \frac{2}{\pi} \int_\mathbb{C} d^2z \, \partial_z X \partial_{\bar{z}} X, \tag{4.4}$$

where $d^2z = dxdy = \frac{i}{2}dzd\bar{z}$, and where the Euclidean action is related to the Wick rotated action by $S_E = -iS$, so that it is real and non-negative. We will drop the subscript 'E' on the Euclidean action in the following.

The Euclidean equation of motion is the Laplace equation,

$$\Delta X = 0 \Leftrightarrow \partial_z \partial_{\bar{z}} X = 0. \tag{4.5}$$

The components of the energy-momentum tensor are:

$$T_{zz} = -2\partial_z X \partial_z X, \quad T_{\bar{z}\bar{z}} = -2\partial_{\bar{z}} X \partial_{\bar{z}} X, \quad T_{z\bar{z}} = 0. \tag{4.6}$$

Energy-momentum conservation

$$\partial_{\bar{z}} T_{zz} = 0, \quad \partial_z T_{\bar{z}\bar{z}} = 0, \tag{4.7}$$

implies that T_{zz} is a holomorphic function, while $T_{\bar{z}\bar{z}}$ is antiholomorphic. We set

$$T(z) := T_{zz}, \quad \bar{T}(\bar{z}) := T_{\bar{z}\bar{z}}. \tag{4.8}$$

In Minkowski signature, left- and right-moving fields are independent. In the Euclidean formalism this reflects itself in the freedom to treat holomorphic and anti-holomorphic fields as independent. In particular, one can treat z and $\bar{z}$ as independent complex variables. Then Minkowski space and Euclidean space are embedded into $\mathbb{C}^2$ as real subspaces. Moreover, in this and in the following chapters we will see that one can define chiral conformal field theories which are based exclusively on holomorphic (or antiholomorphic) fields. This allows one to construct hybrid theories where the holomorphic and antiholomorphic parts are different. One example is the heterotic string which we will briefly discuss in Section 14.7. Finally, to keep Hermitian conjugation aligned with its action in Minkowski signature, 'Hermitian conjugation' in Euclidean signature needs to be defined differently from what one might expect naively. This will be explained in Section 5.2.2.

4.2 Infinitesimal and Finite Conformal Transformations

In the conformal gauge, we can still make reparametrisations which only change the metric by a conformal factor:

$$h_{\alpha\beta} \mapsto e^{2\Lambda(x)} h_{\alpha\beta}, \tag{4.9}$$

where $(x^\alpha) = (x^1, x^2) = (x, y)$ are local real coordinates on the world-sheet. Let us consider infinitesimal reparametrisations

$$x^\alpha \mapsto x^\alpha + \epsilon \xi^\alpha(x^1, x^2), \tag{4.10}$$

where ξ^α is a local vector field on the world-sheet, which generates the reparametrisation. The induced change of the metric is, to order ϵ:

$$\delta h_{\alpha\beta} = -\left(\xi^\gamma \partial_\gamma h_{\alpha\beta} + \partial_\alpha \xi^\gamma h_{\gamma\beta} + \partial_\beta \xi^\gamma h_{\alpha\gamma} \right) \epsilon. \tag{4.11}$$

Now we impose the conformal gauge $h_{\alpha\beta} = \delta_{\alpha\beta}$, and look for reparametrisations which only change the metric by a conformal factor. To first order in ϵ, this implies

$$2\Lambda\delta_{\alpha\beta} = -\left(\partial_\alpha \xi^\gamma \delta_{\gamma\beta} + \partial_\beta \xi^\gamma \delta_{\alpha\gamma} \right). \tag{4.12}$$

Switching to the local complex coordinate $z = x + iy$, and rewriting the above equation in terms of holomorphic and antiholomorphic indices we find[2]

$$\partial_z \xi^{\bar{z}} = 0, \quad \partial_{\bar{z}} \xi^z = 0. \tag{4.13}$$

This shows that our freedom amounts to performing holomorphic reparametrisations $z \mapsto f(z)$, where f is holomorphic. Infinitesimal transformations take the form $z \mapsto z + \epsilon \zeta^z(z)$, where $\xi^z(z)$ is a holomorphic function. This statement is well known from the theory of complex functions: a map from the complex plane to itself is conformal (preserves angles) if and only if it is holomorphic. Vector fields ξ which generate infinitesimal conformal transformations are called *conformal Killing vector fields*. We can parametrise the space of conformal Killing vector fields by expanding $\xi^z(z)$ in a power series:

$$\delta z = \epsilon \xi^z(z) = \epsilon \sum_{n=-\infty}^{\infty} a_n z^n, \tag{4.14}$$

---

[2] The corresponding formula in Minkowski signature is $\partial_\pm \xi^\mp = 0$ which implies that conformal transformations take the form $\sigma^+ \mapsto \tilde{\sigma}^+(\sigma^+)$, $\sigma^- \mapsto \tilde{\sigma}^-(\sigma^-)$ in light-cone coordinates.

where $a_n \in \mathbb{C}$ are coefficients. Note that $a_n \neq 0$ for $n < 0$ is allowed, that is we do not require that ξ^z is holomorphic at $z = 0$. Vector fields can be interpreted as differential operators acting on functions. The vector fields

$$l_m = -z^{m+1}\partial_z, \quad m \in \mathbb{Z}, \tag{4.15}$$

form a basis for the infinitesimal conformal transformations

$$z \to z + \delta_m z, \quad \delta_m z = \epsilon l_m z = -\epsilon z^{m+1}. \tag{4.16}$$

The infinitesimal conformal transformations form an infinite dimensional Lie algebra, with commutation relations between the generators l_m given by

$$[l_m, l_n] = (m - n)l_{m+n}. \tag{4.17}$$

This Lie algebra is known as the *Witt algebra*. It is a so-called $\mathbb{Z}$-graded Lie algebra, that is, it decomposes as a direct sum, $V = \bigoplus_{m \in \mathbb{Z}} V_m$, such that $[V_m, V_n] \subset V_{m+n}$.

Not all infinitesimal conformal transformations give rise to well defined finite conformal transformations. While all vector fields l_m are regular on the punctured complex plane $\mathbb{C}^* = \mathbb{C}\backslash\{0\}$, only those with $m > -1$ are regular at $z = 0$ and generate transformations which are regular on the complex plane $\mathbb{C}$. This is equivalent to the statement that a function is holomorphic on the complex plane if its series expansion only contains non-negative powers of z. As we will see later, the worldsheet used in string perturbation theory is the Riemann sphere $\hat{\mathbb{C}}$, which is obtained by adding an additional point '∞' to the complex plane. The Riemann sphere and the complex plane can be related by stereographic projection, with the south pole corresponding to the origin $0 \in \mathbb{C}$ and the north pole corresponding to ∞. To have conformal transformations which are well defined on $\hat{\mathbb{C}}$, we need to require that the generating vector fields are regular at ∞. The only conformal Killing vector fields which are regular on the Riemann sphere are $l_{-1} = -\partial_z$ (infinitesimal translation), $l_0 = -z\partial_z$ (infinitesimal dilatation), and $l_1 = -z^2\partial_z$ (infinitesimal special conformal transformations). They generate the three-dimensional Lie subalgebra $\mathfrak{sl}(2, \mathbb{C})$ of the Witt algebra. The corresponding group of finite conformal transformations on the Riemann sphere $\hat{\mathbb{C}}$ is $\mathrm{SL}(2, \mathbb{C})$ and acts by fractional linear transformations

$$z \to \frac{az + b}{cz + d}, \quad \begin{pmatrix} a & b \\ c & d \end{pmatrix} \in \mathrm{SL}(2, \mathbb{C}). \tag{4.18}$$

Since $\pm M \in \mathrm{SL}(2, \mathbb{C})$ have the same action on $\hat{\mathbb{C}}$, the automorphism group of the Riemann sphere is isomorphic to $\mathrm{PSL}(2, \mathbb{C}) = \mathrm{SL}(2, \mathbb{C})/\{\pm\mathbb{1}\}$. This group is called the *Möbius group*. For terminological convenience we will also refer to $\mathrm{SL}(2, \mathbb{C})$ as the Möbius group.[3]

Exercise 4.2.1 Investigate which of the vector fields $l_m = -z^{m+1}\partial_z$ are holomorphic at $z = \infty$. (You need to make a coordinate transformation to a coordinate system which is well defined at $z = \infty$.) Hence, show that the only conformal Killing vector fields which are holomorphic on the Riemann sphere are l_{-1}, l_0, l_1.

---

[3] Some authors consider the action of $\mathrm{GL}(2, \mathbb{C})$ on the Riemann sphere. Since $\mathrm{PGL}(2, \mathbb{C}) = \mathrm{GL}(2, \mathbb{C})/\mathbb{C}^* \cong \mathrm{SL}(2, \mathbb{C})/\{\pm\mathbb{1}\}$, the Möbius group is isomorphic to $\mathrm{PGL}(2, \mathbb{C})$.

4.3 From Commutators to Operator Product Expansions

We now turn to a formulation of conformal field theories which works with operator product expansions (OPEs) instead of commutators. OPEs are a technique in quantum field theory which is not restricted to theories which are two-dimensional or conformal, though in this case they are particularly powerful. The idea of an OPE is to express the product of two operators $A(x)$, $B(y)$, located at nearby positions, by a sum over local operators, multiplied by functions which are analytic for $x \neq y$ but possibly singular at $x = y$. If one chooses a basis $O_i(x)$ of local operators, all OPEs can be derived from the mutual OPEs of the basis operators,

$$O_i(x)O_j(y) = \sum_k C_{ij}^k(x - y)O_k(y). \tag{4.19}$$

In this section we use the free massless scalar field $X(z, \bar{z})$ to illustrate how commutators translate into OPEs. The mode expansion of $X(z, \bar{z})$ in the complex plane is

$$X(z, \bar{z}) = x - i\frac{p}{4}\log|z|^2 + \frac{i}{2}\sum_{n \neq 0}\alpha_n z^{-n} + \frac{i}{2}\sum_n \tilde{\alpha}_n \bar{z}^{-n}. \tag{4.20}$$

We have seen that imposing standard equal time commutation relations in Minkowski signature implies

$$[x, p] = i, \quad [\alpha_m, \alpha_n] = m\delta_{m+n,0}, \quad [\tilde{\alpha}_m, \tilde{\alpha}_n] = m\delta_{m+n,0}. \tag{4.21}$$

We can decompose $X(z, \bar{z})$ into a holomorphic and an antiholomorphic part:[4]

$$X(z, \bar{z}) = \frac{1}{2}\left(x - ip_L\log z + i\sum_{n \neq 0}\frac{1}{n}\alpha_n z^{-n}\right) + \frac{1}{2}\left(x - ip_R\log \bar{z} + i\sum_{n \neq 0}\frac{1}{n}\tilde{\alpha}_n \bar{z}^{-n}\right), \tag{4.22}$$

where the left- and right-moving momanta are defined as $p_L = p_R = \frac{p}{2}$. Typical quantities that we want to compute in a quantum field theory are vacuum expectation values of products of field operators, such as the two-point function $\langle X(z, \bar{z})X(w, \bar{w})\rangle$. The field $X(z, \bar{z})$ contains logarithmic terms, and, therefore, it is not single-valued. We will come back to the role of $X(z, \bar{z})$ below. To illustrate the OPE formalism, we consider the correlation function of the holomorphic derivative

$$\phi_z = 2i\partial_z(X(z, \bar{z})) = \sum_n \alpha_n z^{-n-1}, \tag{4.23}$$

where we have set $\alpha_0 := p_L$:

$$G_2(z, w) := \langle 0|\phi_z\phi_w|0\rangle = \sum_{m,n} z^{-m-1}w^{-n-1}\langle 0|\alpha_m\alpha_n|0\rangle. \tag{4.24}$$

Using that $\alpha_m|0\rangle = 0$ for $m \geq 0$, which implies $\langle 0|\alpha_m = 0$ for $m \leq 0$, we can restrict the sums, and then replace the product $\alpha_m\alpha_n$ by a commutator:

$$G_2(z, w) = \sum_{m>0}\sum_{n<0} z^{-m-1}w^{-n-1}\langle 0|\alpha_m\alpha_n|0\rangle = \sum_{m>0}\sum_{n<0} z^{-m-1}w^{-n-1}\langle 0|[\alpha_m\alpha_n]|0\rangle. \tag{4.25}$$

---

[4] One may wonder in which sense $X(z, \bar{z})$ is still a 'real' field, like its Minkowskian counterpart. See Section 5.2.2.

Using the commutation relations (4.21), we can perform the sum over n:

$$G_2(z, w) = \sum_{m>0} m z^{-m-1} w^{m-1} = \frac{1}{z^2} \sum_{m>0} m \left(\frac{w}{z}\right)^{m-1}. \tag{4.26}$$

Recognising this series as the derivative of the geometric series, we obtain

$$G_2(z, w) = \langle 0 | \phi_z \phi_w | 0 \rangle = \frac{1}{(z-w)^2}. \tag{4.27}$$

We now make several observations, which represent general properties of conformal field theories. The series only converges for $|z| > |w|$, which illustrates that in the Euclidean formalism only radially ordered operator products are well defined. However, the correlation function $(z-w)^{-2}$ is holomorphic except for $z = w$ where it has a second order pole. Therefore the vacuum correlation function can be extended to a meromorphic function by analytical continuation. Observe that in contrast to Lorentzian correlation functions the resulting Euclidean correlation function is symmetric under the exchange of z and w. This is a general property of Euclidean correlation functions of bosonic operators. The Euclidean correlators of fermionic fields involve odd powers of $z - w$ and are thus antisymmetric. The operators and their products are formal power series with coefficients that are operators. However, we have seen that upon evaluation between ground states we obtain a convergent series which defines a meromorphic function: This illustrates that Euclidean operator product expansions converge *on the vacuum*.

Exercise 4.3.1   Show that under Wick rotation the time evolution operator e^{iHt}, where H is the Hamiltonian becomes unbounded.[5] What does this imply for the analytic continuation of correlation functions? Show that the continuation of time ordered correlation functions is well defined.

Let us now reverse the procedure, that is we start with the OPE for ϕ_z and want to recover the commutation relations. We postulate that the complete OPE takes the form

$$\phi_z \phi_w = \frac{1}{(z-w)^2} + : \phi_z \phi_w :, \tag{4.28}$$

where the operator $: \phi_z \phi_w :$, called the *normal ordered product*, is regular for $z \to w$. In other words, we impose that the product $\phi_z \phi_w$ has precisely the singular part which we have found before, and we define the normal ordered product $: \phi_z \phi_w :$ by subtracting the singular part of the OPE.

Exercise 4.3.2   Show that the definition of the normal ordering provided by (4.28) is equivalent to the standard definition of normal ordering in quantum field theory, which is to order each expression such that the annihilation operators are to the right of the creation operators.[6]

---

[5] The norm of an operator with a complete spectrum of eigenstates is given by its largest eigenvalue. For reference, the norm of an operator $A : V \to W$ between two normed vector spaces is $||A|| = \inf \{c \geq 0 | \ ||Av|| < c||v|| , \ \forall v \in V\}$ where we used the norms on V and W. Operators where $||A|| < \infty$ are called bounded operators. For the Hamiltonian H, we make the standard assumption that its spectrum is bounded from below by 0 (corresponding to the ground state), while it is unbounded from above.

[6] Note that this exercise only shows that the two definitions agree for a particular example. In general, they can (and will) differ by finite terms.

Individual coefficients can be extracted out of the mode expansion (4.23) by applying the residue theorem. For a Laurent series

$$f(z) = \sum_m a_m z^m, \tag{4.29}$$

the formula is

$$a_m = \frac{1}{2\pi i} \oint_{z=0} dz\, f(z) z^m, \tag{4.30}$$

where we indicated that the integration contour encloses $z = 0$, but no other poles of $f(z)$. Applying this formally to the field (4.23), where the coefficients are operators rather than numbers, we can write

$$\alpha_m = \frac{1}{2\pi i} \oint_{z=0} dz\, \phi_z z^m. \tag{4.31}$$

Given the OPE (4.28), we would like to compute the commutator

$$[\alpha_m, \alpha_n] = \left[\oint_{z=0} \frac{dz}{2\pi i} \phi_z z^m, \oint_{w=0} \frac{dw}{2\pi i} \phi_w w^n \right]. \tag{4.32}$$

The operator product $\phi_z \phi_w$ is only defined if z and w are radially ordered, while the commutator forces us to take the difference between $\phi_z \phi_w$ and $\phi_w \phi_z$. If we perform the z integration first, then in the first term, where $|z| > |w|$, the contour encloses $z = w$ and $z = 0$, while in the second term, where $|w| > |z|$, the contour encloses $z = 0$ but not $z = w$. Since the commutator is the difference between these two terms, we end up with a contour which only encloses the point $z = w$ (see Figure 4.1).

Thus the choice of the integration contour defines what it means to take an equal time commutator in the complex plane:

$$[\alpha_m, \alpha_n] = \oint_{w=0} \frac{dw}{2\pi i} \oint_{z=w} \frac{dz}{2\pi i} \phi_z \phi_w z^m w^n. \tag{4.33}$$

In the following, Euclidean operator products, like $\phi_z \phi_w$, are always understood to be radially ordered. We insert the OPE (4.28) into the commutator:

$$[\alpha_m, \alpha_n] = \oint_{w=0} \frac{dw}{2\pi i} \oint_{z=w} \frac{dz}{2\pi i} \left(\frac{1}{(z-w)^2} + :\phi_z \phi_w: \right) z^m w^n. \tag{4.34}$$

The z-integral reads out the coefficients of first order poles in $z - w$. The normal ordered product $:\phi_z \phi_w:$ is regular for $z \to w$ and, therefore, it does not contribute. To see that the other term contributes, we rewrite z^m using the binomial formula:

$$z^m = [w + (z-w)]^m = w^m + m w^{m-1}(z-w) + \cdots, \tag{4.35}$$

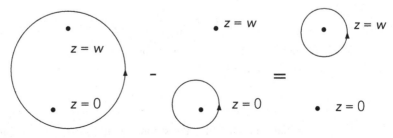

Fig. 4.1 Commutators are encoded by the choice of integration contour when working with OPEs.

where the omitted terms are of higher order in $z - w$. Thus

$$[\alpha_m, \alpha_n] = \oint_{w=0} \frac{dw}{2\pi i} m w^{n+m-1} = m\delta_{m+n,0}, \tag{4.36}$$

and we have recovered the commuation relation (4.21) from the OPE (4.28).

We now turn our attention to the energy momentum tensor. In the classical theory, its holomorphic part is

$$T(z) = -2\partial_z X \partial_z X = \frac{1}{2}\phi_z \phi_z. \tag{4.37}$$

In the quantum theory, this is at first not well defined, since the product $\phi_z \phi_w$ is singular in the limit $w \to z$. Therefore, we define $T(z)$ as the normal ordered product,

$$T(z) = \frac{1}{2} : \phi_z \phi_z := \frac{1}{2} \lim_{w \to z} \left(\phi_z \phi_w - \underbrace{\phi_z \phi_w} \right), \tag{4.38}$$

where the *contraction* $\underbrace{\phi_z \phi_w}$ is the singular part of the OPE,

$$\phi_z \phi_w = \underbrace{\phi_z \phi_w} + : \phi_z \phi_w : . \tag{4.39}$$

To evaluate products between composite operators we need the following version of Wick's theorem:

Wick's theorem for products of normal ordered products: The radially ordered product of two normal ordered monomials $: \phi_1 \cdots \phi_m :$ and $: \phi_{m+1} \cdots \phi_{m+n} :$ of free fields ϕ_i is the sum over all pairwise contractions, including the case of zero contractions, between $\phi_1, \ldots, \phi_m$ and $\phi_{m+1}, \ldots, \phi_{m+n}$ in the normal ordered product $: \phi_1 \cdots \phi_{m+n} :.$

As an example, we compute the OPE $T(z)T(w)$ where $T(z) = \frac{1}{2} : \phi_z \phi_z :$, using the OPE (4.28) for ϕ_z. There are three types of terms: those with two, one and zero contractions. Upon combining terms we find

$$T(z)T(w) = 2 \times \frac{1}{4} : \underbrace{\phi_z \phi_z \phi_w \phi_w} : + 4 \times \frac{1}{4} : \underbrace{\phi_z \phi_z \phi_w \phi_w} : + \frac{1}{4} : \phi_z \phi_z \phi_w \phi_w :$$

$$= \frac{\frac{1}{2}}{(z-w)^4} + \frac{1}{(z-w)^2} : \phi_z \phi_w : + : T(z)T(w) : . \tag{4.40}$$

This contains operators depending on both z and w on the right hand side. By Taylor expanding the z-dependent operators we can express our results in terms of local operators depending on w and coefficient functions which depend on $z - w$. This way we obtain the standard form of the OPE,

$$T(z)T(w) = \frac{\frac{1}{2}}{(z-w)^4} + \frac{2T(w)}{(z-w)^2} + \frac{\partial_w T}{z-w} + \cdots, \tag{4.41}$$

where the omitted terms are regular for $w \to z$.

Similarly, one can show that

$$T(z)\phi_w = \frac{\phi_w}{(z-w)^2} + \frac{\partial_w \phi_w}{z-w} + \cdots. \tag{4.42}$$

Observe that, in both OPEs, the second factor of the product appears together with a second order pole, followed by its derivative combined with a first order pole.

This indicates that these OPEs encode how fields behave under the transformations generated by the energy momentum tensor.

Using the same prescription as in (4.33), we can convert the OPE (4.41) and (4.42) into commutation relations between the operators in the mode expansions of $T(z)$ and ϕ_z. We define the mode expansion of $T(z)$ by

$$T(z) = \sum_n L_n z^{-n-2}. \tag{4.43}$$

Exercise 4.3.3 Show that the coefficients L_n in (4.43) are given by

$$L_n = \frac{1}{2} \sum_m : \alpha_{n-m}\alpha_m : . \tag{4.44}$$

Using the contour integration method[7] it is straightforward to show that (4.42) is equivalent to

$$[L_m, \alpha_n] = -n\alpha_{m+n}, \tag{4.45}$$

while (4.41) is equivalent to

$$[L_m, L_n] = (m - n)L_{m+n} + \frac{1}{12}m(m - 1)(m + 1)\delta_{m+n,0}. \tag{4.46}$$

Due to the presence of a second term on the right-hand side, (4.46) is an extension of the Witt algebra, called the *Virasoro algebra*. The additional term is a consequence of the *conformal anomaly*. A classical symmetry is called anomalous if it is modified at the quantum level. For local symmetries, which guarantee the decoupling of unphysical modes, anomalies make the quantum theory inconsistent. But here we are dealing with a rigid symmetry, where the presence of an anomaly just means that the action of the symmetry has been modified by quantum effects.[8] Note that the anomaly is absent for the finite-dimensional $\mathfrak{sl}(2, \mathbb{C})$ subalgebra spanned by L_{-1}, L_0, L_1, which generates the action of the Möbius group on our quantum field theory.

The commutation relations (4.45), equivalently the OPE (4.42), tell us how the quantum field ϕ_z behaves under conformal transformations. We will see later, when we look at conformal field theory from a top-down perspective, that the coefficient of the second order term tells us that ϕ_z is a 'conformal primary of weight one', while the energy momentum tensor $T(z)$ is a 'conformal quasi-primary of weight two'.

We had postponed to discuss the role of the undifferentiated scalar field $X(z, \bar{z})$. From (4.27) it is clear that the two-point function takes the form

$$\langle 0|X(z, \bar{z})X(w, \bar{w})|0\rangle = -\frac{1}{4}\log|z - w|^2 + \cdots . \tag{4.47}$$

---

[7] One could of course instead use the commutation relations (4.21), but this is more tedious. In particular one needs to be careful with the normal ordering of products.

[8] This remark applies to conformal field theories. In string theory the conformal symmetry is the 'left-over' of a local symmetry, and therefore the anomaly must cancel, in order to decouple unphysical degrees of freedom. To show that the quantum theory formulated in the conformal gauge is actually reparametrisation and Weyl invariant, one needs to include the so-called Faddeev–Popov ghost fields, which also contribute to the conformal anomaly. As a result, the CFT describing the physical degrees of freedom of a string has non-zero central charge ($c = 26$ for bosonic strings and $c = 15$ for superstrings) (see Section 12.2.3).

This correlation function does not behave properly under conformal transformations. Since $X(z, \bar{z})$ transforms as a function under reparametrisations, the correlator should be invariant under conformal transformations, including scale transformations $(z - w) \mapsto \lambda(z - w)$, $\lambda \in \mathbb{R}^{>0}$. However, the right-hand side of (4.47) is not scale invariant, but transforms with a shift. Therefore, the quantum field $X(z, \bar{z})$ is not a conformal field, that is, it does not transform as a tensor field under conformal transformations. One clue about the nature of the problem is that the two-point function does not only diverge a short distances $|z - w| \to 0$, but also at large distances $|z - w| \to \infty$, which sets two-dimensional massless scalar fields apart from massless scalar fields in higher dimensions. Since large distances corresponds to low energies, this is an *infrared divergence*. Therefore, $X(z, \bar{z})$ itself is not among the physical fields of the conformal field theory. Instead, the physical fields are $\phi_z \propto \partial_z X$ and higher derivatives, and as we will see below, exponentials of $X(z, \bar{z})$. These fields are well behaved and we will see that they generate all string states.

Since the Taylor series of the exponential function has an infinite radius of convergence, we can define the exponential of the field $X(z, \bar{z})$ through its Taylor series. As for other composite fields, we indicate normal ordering by colons:

$$E_k(z, \bar{z}) :=: \exp(ikX(z, \bar{z})) :, \quad k \in \mathbb{R}. \tag{4.48}$$

To compute OPEs involving exponentials it is useful to decompose $X(z, \bar{z})$ as[9]

$$X(z, \bar{z}) = \frac{1}{2}\left(X(z) + \bar{X}(\bar{z})\right), \tag{4.49}$$

and to introduce the multi-valued OPEs

$$X(z)X(w) = -\log(z - w) + \cdots, \quad \bar{X}(\bar{z})\bar{X}(\bar{w}) = -\log(\bar{z} - \bar{w}) + \cdots, \tag{4.50}$$

which imply the correlation function (4.47). While these OPEs contain logarithms, they are useful for deriving other, single-valued OPEs by taking derivatives and exponentials. This way one can obtain the following important OPEs involving exponentials:

$$\phi_z E_k(w, \bar{w}) = \frac{k/2}{z - w} E_k(w, \bar{w}) + \cdots, \tag{4.51}$$

$$E_k(z, \bar{z})E_l(w, \bar{w}) = |z - w|^{kl/2}\left(E_{k+l}(w, \bar{w}) + \cdots\right), \tag{4.52}$$

$$T(z)E_k(w, \bar{w}) = \left(\frac{k^2/8}{(z - w)^2} + \frac{\partial_w}{z - w}\right)E_k(w, \bar{w}). \tag{4.53}$$

The first two equations (4.51) and (4.52) show, together with (4.28), that the fields ϕ_z and $E_k(z, \bar{z})$ form a closed algebra with respect to OPEs. We will see later that these operators encode the whole excitation spectrum of the string. The third equation (4.53) looks similar to (4.42) and we will see in Chapter 5 that it tells us that the exponential transforms as a conformal field with weight $k^2/8$.

Exercise 4.3.4 Derive the OPE (4.51) from the OPE (4.50).
Exercise 4.3.5 Derive the OPE (4.53) from the OPE (4.50).

---

[9] The normalisation is chosen for later convenience.

Exercise 4.3.6   Given that the singular part of the OPE between two operators A and B only contains terms proportional to the unit operator, show that

$$: \exp(A) :: \exp(B) := \exp(\underbrace{AB}) : \exp(A + B) : . \qquad (4.54)$$

This is a version of the *Baker–Campbell–Hausdorff* ('BCH') formula.

Exercise 4.3.7   Use (4.54) to derive the OPE (4.52).

4.4 From Operators to States

We now turn to the relation between the conformal fields $\phi_z, E_k(z, \bar{z})$ and the physical states of a closed string with a one-dimensional target space. This will illustrate the *state-operator correspondence* of two-dimensional conformal field theories. The idea is that the insertion of a conformal field at a point $P \in \Sigma$ of the world-sheet corresponds to the creation or annihilation of a state. A closed string corresponds to a loop, and if we trace it either back to the infinite past or to the infinite future, we obtain a semi-infinite cylinder, which we can map conformally to a punctured disk (see Figure 4.2).

Instead of specifying initial conditions, we fill the puncture by a local operator which carries the asymptotic momentum and other properties of this state. We start with inserting a single operator on the world-sheet, and choose a local coordinate z such that the point P corresponds to $z = 0$. Then, we define a state $|k\rangle$ by

$$|k\rangle = \lim_{z \to 0} E_k(z, \bar{z})|0\rangle, \qquad (4.55)$$

where $|0\rangle$ is the ground state of our theory, defined by the property $\alpha_m|0\rangle = 0$ for $m \geq 0$. We claim that $|k\rangle$ is an eigenstate of the operator $p = 2\alpha_0 = 2\tilde{\alpha}_0$, that is, in the string theory interpretation, a momentum eigenstate. To prove this we must show that $p|k\rangle = k|k\rangle$. One way to do this is to use the formalism of OPEs, and compute

$$\begin{aligned}
p|k\rangle &= \lim_{w \to 0} 2 \oint_{z=w} \frac{dz}{2\pi i} \phi_z E_k(w, \bar{w})|0\rangle = \lim_{w \to 0} \oint_{z=w} \frac{dz}{2\pi i} \frac{k}{z - w} E_k(w, \bar{w})|0\rangle \\
&= k \lim_{w \to 0} E_k(w, \bar{w})|0\rangle = k|k\rangle .
\end{aligned}$$

In the first step, we replace p by a contour integral over ϕ_z. Note that the contour only encloses the pole at $z = w$, because $p|0\rangle = 0$. Then, we use the OPE (4.51), carry out the contour integral, and, finally, take the limit $w \to 0$. Alternatively, one can show that

$$\lim_{z \to 0} E_k(z, \bar{z})|0\rangle = e^{ikx}|0\rangle, \qquad (4.56)$$

and use the relation $[x, p] = i$ to show that $p e^{ikx}|0\rangle = k e^{ikx}|0\rangle$.

Fig. 4.2   An incoming or outgoing string is represented by a semi-infinite cylinder, which can be mapped conformally to a punctured disk. A vertex operator representing the string is inserted at the puncture.

Exercise 4.4.1 Write down an explicit formula for $E_k(z,\bar{z}) =: \exp(ikX(z,\bar{z})) :$ by putting all annihilation operators to the right of all creation operators. Use the resulting formula to derive (4.56).

Exercise 4.4.2 Show that $e^{ikx}|0\rangle$ is a momentum eigenstate.

Next, let us see what kind of state the operator $\phi_z = 2i\partial_z X$ creates. We claim that

$$\lim_{z\to 0} \phi_z|0\rangle = \alpha_{-1}|0\rangle, \tag{4.57}$$

and, more generally, that

$$\lim_{z\to 0} \partial_z^{m-1}\phi_z = (m-1)!\,\alpha_{-m}|0\rangle. \tag{4.58}$$

Therefore, by applying exponentials and derivatives of $X(z,\bar{z})$, that is the operators $E_k(z,\bar{z})$, ϕ_z, $\partial_z\phi_z$, ..., and $\bar{\phi}_{\bar{z}}, \partial_{\bar{z}}\bar{\phi}_{\bar{z}}, \ldots$, we obtain a basis

$$\alpha_{-1}^{m_1}\alpha_{-2}^{m_2}\cdots\tilde{\alpha}_{-1}^{\tilde{m}_1}\tilde{\alpha}_{-2}^{\tilde{m}_2}\cdots|k\rangle \tag{4.59}$$

for the Hilbert space of a closed string with a one-dimensional target space. The corresponding local operators, which are, somewhat misleadingly, called *vertex operators*, have the form[10]

$$V(z,\bar{z}) =: (\partial_z X)^{m_1}(\partial_z^2 X)^{m_2}\cdots(\partial_{\bar{z}}X)^{\tilde{m}_1}(\partial_{\bar{z}}^2 X)^{\tilde{m}_2}\cdots\exp(ikX(z,\bar{z})) :, \tag{4.60}$$

where $V(z,\bar{z})$ depends on data $(m_1,\ldots,\tilde{m}_1,\ldots,k)$, which characterise the asymptotic string state.

If we insert more than one vertex operator, then the situation is locally the same around each insertion. Consider the case where two vertex operators are inserted. The natural quantity associated with this situation is the vacuum correlator or two-point function of the two vertex operators. By moving their positions to $z = 0$ and $z = \infty$,

$$\lim_{z_1\to\infty}\lim_{z_2\to 0}\langle 0|V_1(z_1,\bar{z}_1)V_2(z_2,\bar{z}_2)|0\rangle = \langle V_1|V_2\rangle, \tag{4.61}$$

we can interpret this as a probability amplitude for finding state 1 in the asymptotic future if state 2 has been prepared in the asymptotic past.[11] Interactions, that is decay and scattering processes correspond to strings splitting and joining, and are thus encoded in the global structure of the world-sheet. There only is one basic interaction process, the splitting of one string into two, or, read in time-reversed order, the joining of two strings into one (see Figure 4.3).

One crucial difference to point particles is that the world-sheet locally looks the same everywhere, that is, there are no distinct points corresponding to interaction vertices. Compare Figures 1.3 and 4.3.

The simplest example of a scattering process is a two-to-two process with two initial and two final states (see Figure 4.4). The most natural way of representing this is by a 'pair of pants' type diagram, where the boundary loops represent asymptotic states (which we usually represent by semi-infinite cylinders which are however

---

[10] Note that 'vertex operators' correspond to asymptotic states, not to vertices as such. However, we will see below that the insertion of an additional vertex operator indeed creates a vertex.

[11] Whether states are ingoing or outgoing is actually determined by their momenta. While it is helpful to use world-sheet time to interpret space-time processes, space-time physics must ultimately be independent of how we parametrise the world-sheet.

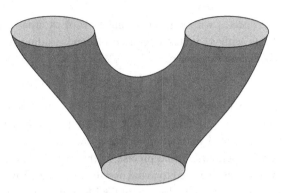

Fig. 4.3   World-sheet geometry of the basic closed string interaction process.

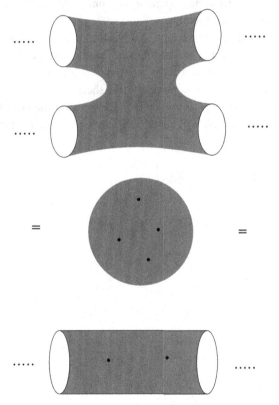

Fig. 4.4   A 2-to-2 tree level scattering process can be represented equivalently by a pair of pants type diagram, a sphere with four vertex operators, or an infinite cylinder with two vertex operators.

hard to draw). Using conformal symmetry, we can map this to a sphere with four vertex operator insertions. Alternatively, we can put two of the vertex operators at the positions $z = 0$ and $z = \infty$ and replace them by the corresponding bra- and ket-states. This parametrisation, where a single closed string propagates through and interacts with two other strings is useful for computations. Comparing to Figure 4.3, we see that a two-to-two scattering diagram contains two basic closed string vertices, and that adding one further vertex operator adds one closed string vertex to the world-sheet.

Processes with a general number M of external states are represented by world-sheets with M vertex operators inserted, leading to M-point world-sheet vacuum correlation functions

$$\langle 0|V_1(z_1,\bar{z}_1)\cdots V_M(z_M,\bar{z}_M)|0\rangle. \tag{4.62}$$

S-matrix elements for string interactions cannot depend on the world-sheet coordinates $z_1,\ldots,z_M$, and, therefore, the natural candidates for string S-matrix elements are integrated correlation functions, where we integrate over the positions of all vertex operators. To be precise, if the world-sheet has globally defined conformal symmetries, we need to avoid over-counting. We have seen that if Σ is a sphere, then there is a well defined action of the Moebius group $SL(2,\mathbb{C})$. This allows one to map three points to arbitrary positions, which conventionally are taken to be $0, 1, \infty$. The positions of vertex operators are only integrated over the sphere starting with the fourth vertex operator.

The diagram we have drawn does not contain handles ('loops'), and thus is a *tree-level diagram*. Adding one handle changes the topology of the world-sheet from a sphere to a torus, and adds two closed string vertices. When keeping the same vertex operators, this world-sheet represents a one-loop correction to the tree level scattering amplitude. By adding more handles we can also define multiloop amplitudes. We will come back to general aspects of interactions in Chapter 12.

Two-Dimensional Conformal Field Theories

5.1 Some Remarks on Conformal Field Theories in General Dimension

In this chapter, we look at conformal field theories from a complementary top-down perspective. We start by providing some background on conformal symmetry in general dimension. A transformation $x^\mu \mapsto x'^\mu$ is called *conformal* if the metric is invariant up to a positive function,

$$h'_{\mu\nu}(x') = e^{2\Lambda(x)} h_{\mu\nu}(x). \tag{5.1}$$

Such *conformal isometries* preserve angles, but, in general, not length. Conformal transformations can be classified by identifying their generating vector fields $\xi^\mu(x)$, where

$$x^\mu \mapsto x'^\mu_t = x^\mu + t\xi^\mu(x) + \cdots, \quad |t| \ll 1 \tag{5.2}$$

is the corresponding one-parameter family of infinitesimal conformal transformations. Since

$$h'_{\mu\nu}(x') = h_{\mu\nu}(x) - t\left(\partial_\mu \xi_\nu + \partial_\nu \xi_\mu\right) + \cdots \tag{5.3}$$

a vector field generates a conformal transformation if and only if it satisfies the *conformal Killing equation*

$$\partial_\mu \xi_\nu + \partial_\nu \xi_\mu = \lambda(x) h_{\mu\nu}, \tag{5.4}$$

where, as we can see by taking the trace, $\lambda(x) = \frac{2}{d} \partial_\mu \xi^\mu$, and where d is the number of dimensions. It can be shown[1] that, in dimensions $d > 2$, the conformal Killing equation has up to $\frac{1}{2}(d+2)(d+1)$ linearly independent solutions, and that this bound is saturated for flat metrics. In the following, we consider d-dimensional Minkowski metrics specifically. In this case, the general solution for the generating vector field of a conformal transformation can be parametrised as

$$\xi_\mu = a_\mu + b_{\mu\nu} x^\nu + c_{\mu\nu\rho} x^\nu x^\rho, \tag{5.5}$$

where $b_{\mu\nu}$ has an antisymmetric part (the Lorentz transformations) and a trace part (the dilatations): $b_{\mu\nu} = m_{[\mu\nu]} + \alpha\eta_{\mu\nu}$. The parameters $c_{\mu\nu\rho}$, which correspond to the so-called *special conformal transformations*, are determined by a vector b_μ. The most general conformal transformation depends on $d + \frac{1}{2}d(d-1) + 1 + d = \frac{1}{2}(d+2)(d+1)$ independent parameters, which can be taken to be $\alpha, a_\mu, b_\mu, m_{\mu\nu} = -m_{\nu\mu}$. The corresponding finite transformations are listed in Table 5.1.

---

[1] See, e.g., Di Francesco et al. (1997) which we closely follow in this section.

Table 5.1. Finite conformal transformations

Transformation	Parameter	Exponentiated action	Generator
Translation	a_μ	$x'^\mu = x^\mu + a^\mu$	P_μ
Lorentz	$m_{\mu\nu}$	$x'^\mu = \Lambda^\mu{}_\nu x^\nu$	$L_{\mu\nu}$
Dilatation	α	$x'^\mu = \alpha x^\mu$	D
Special conformal	b_μ	$x'^\mu = \frac{x^\mu - b^\mu x^\nu x_\nu}{1 - 2b^\mu x_\mu + b^\mu b_\mu x^\nu x_\nu}$	K_μ

The generators $P_\mu, L_{\mu\nu}, D$ of translations, Lorentz transformations and dilatations close among themselves and generate a subgroup of the conformal group which acts on Minkowski space by affine transformations. In contrast, the special conformal transformations generated by K_μ act non-affinely, and map points to infinity. To have a well defined action of the full conformal group, one needs to extend Minkowski space by adding a point at infinity. This is similar to extending the complex plane to the Riemann sphere. The resulting space is called the *conformal compactification* of Minkowski space. By taking suitable linear combinations of the generators, one can show that the full conformal group of Minkowski space is isomorphic to $SO(2, d)$. For a flat metric of general signature $(n, d - n)$ the conformal group is $SO(n + 1, d - n + 1)$.

Since Poincaré transformations and dilatations (scale transformations) form a subgroup of the conformal group, it is a non-trivial question under which conditions *scale invariance* implies conformal invariance. The following can be shown:

1. The Noether current for scale invariance is $j_D^\mu = T^{\mu\nu} x_\nu$.
2. If the energy momentum tensor is traceless, $T^\mu{}_\mu = 0$, then the theory is conformally invariant.

One subtlety is that the energy momentum tensor is not unique, but can be modified by *improvement terms*, $T^{\mu\nu} \mapsto T^{\mu\nu} + \partial_\rho B^{\rho\mu\nu}$, where $B^{\rho\mu\nu} = -B^{\mu\rho\nu}$. Thus, given a theory which is Poincaré and scale invariant, the question whether the theory is conformally invariant is equivalent to the question whether its energy momentum tensor can be made traceless by adding an improvement term.

One characteristic feature of scale and conformal symmetry is that, in general, they are broken by quantum effects. This breaking is measured by the so-called *trace anomaly*, the expectation value $\langle T^\mu{}_\mu \rangle$ of the trace of the energy-momentum tensor. One example is Yang–Mills theory with massless matter. This theory is classically conformally invariant, but renormalisation introduces a scale which manifests itself in *running couplings*, that is in a dependence of the coupling 'constant' on the energy scale. The scale dependence of couplings is measured by the so-called β-functions. Conformally invariant quantum field theories can be obtained by choosing the matter content of the theory such that its β-functions vanish. Such theories correspond to fixed points of the renormalisation group flow. Examples are provided by theories with extended supersymmetry, for example, $\mathcal{N} = 4$ super-Yang–Mills theory, and $\mathcal{N} = 2$ super-Yang–Mills theoris with specific matter content.

If we extrapolate the above discussion to two dimensions, we predict that the resulting conformal groups are $SO(2,2) \cong SO(2,1) \times SO(1,2) \cong SL(2,\mathbb{R}) \times SL(2,\mathbb{R})$ in Lorentz signature and $SO(1,3) \cong SL(2,\mathbb{C})$ in Euclidean signature.[2] These are, indeed, the conformal groups which have a well defined finite action on the conformal compactifications of Minkowski space and of Euclidean space, the latter being the two-sphere. However, the specific feature of two-dimensional conformal field theories is the existence on an infinite dimensional Lie algebra of infinitesimal conformal transformations. While almost all of these transformations do not exponentiate to globally defined finite transformations, invariance under infinitesimal transformations still imposes powerful constraints, called *conformal Ward identities*. These take the form of differential equations which relate the correlation functions of the theory.

5.2 Conformal Primaries

Let us now revisit the topics of Chapter 4 from a top-down perspective. We begin with general definitions and recover previous results as examples.

5.2.1 Definition of Conformal Primaries, Holomorphic, and Chiral Fields

A conformal primary $\phi(z,\bar{z})$ is a field which transforms as a co-tensor under conformal transformations $z \mapsto w = f(z)$, where $\partial_{\bar{z}}f = 0$. For $h \geq 0, \bar{h} \geq 0$ such tensors take the form $\Phi = \phi(z,\bar{z})dz^h d\bar{z}^{\bar{h}}$, where $(h, \bar{h})$ are called the *conformal weights*.[3]

Due to the restriction to conformal and, thus, holomorphic transformations, such tensors only have one component $\phi(z,\bar{z})$. If we interpret conformal transformations passively, then

$$\phi(z,\bar{z})dz^h d\bar{z}^{\bar{h}} = \phi'(w,\bar{w})dw^h d\bar{w}^{\bar{h}} \Rightarrow \phi'(w,\bar{w}) = \phi(z,\bar{z}) \left(\frac{dw}{dz}\right)^{-h} \left(\frac{d\bar{w}}{d\bar{z}}\right)^{-\bar{h}}. \quad (5.6)$$

To obtain the infinitesimal version of (5.6) we write

$$z \mapsto w = z + t\xi(z) + \cdots, \quad (5.7)$$

where the parameter t is assumed to be infinitesimal and real, and where $\xi = (\xi(z), \bar{\xi}(\bar{z}))$ is the generating vector field of the transformation.[4] Working at infinitesimal level amounts to neglecting terms of higher order in t, which is equivalent to neglecting terms of higher order in ξ. The variation of a conformal primary under an infinitesimal conformal transformation is (see Exercise 5.2.1):

---

[2] Here $\cong$ denotes local group isomorphisms, that is the groups have the same Lie algebra but may be related by a non-trivial covering map.

[3] The case of negative $h, \bar{h}$ will not be relevant to us, since we will see later that in a unitary conformal field theory $h \geq 0, \bar{h} \geq 0$. See, however, Exercise 5.2.3 for how such fields can be interpreted as tensor components.

[4] See, e.g., Frankel (2004) for background on vector fields, transformation groups and Lie derivatives.

$$(\delta_\xi \phi)(z, \bar{z}) := \lim_{t \to 0} \frac{\phi'(z, \bar{z}) - \phi(z, \bar{z})}{t}.$$

$$= -\left(\xi(z)\partial_z + \bar{\xi}(\bar{z})\partial_{\bar{z}} + h\partial_z\xi + \bar{h}\partial_{\bar{z}}\bar{\xi}\right)\phi(z, \bar{z}). \tag{5.8}$$

This is, up to sign, the *Lie derivative* $L_\xi \Phi$ of the tensor $\Phi = \phi(z, \bar{z})dz^h d\bar{z}^{\bar{h}}$ with respect to the vector field ξ (Exercise 5.2.4).

We now introduce some more terminology. Fields with zero antiholomorphic weight, $\bar{h} = 0$, will be called *holomorphic fields*, fields with $h = 0$ *antiholomorphic fields*. Note that these conditions still allow that the component field $\phi(z, \bar{z})$ is a general complex-valued function. If the component field is a holomorphic function, $\partial_{\bar{z}}\phi = 0$, then the field $\phi(z)$ is called a *chiral field*, and if it is antiholomorphic, $\partial_z\phi = 0$, then $\phi(\bar{z})$ is called an *antichiral* field. Chiral and antichiral fields are related by Wick rotation to left- and right-moving fields in Lorentz signature.

The conformal transformations include dilatations $z \mapsto \lambda z$, $\lambda \in \mathbb{R}^{>0}$, which act on conformal primaries by

$$\phi'(\lambda z, \lambda \bar{z}) = \lambda^{-(h+\bar{h})}\phi(z, \bar{z}). \tag{5.9}$$

The quantity $h + \bar{h}$ is called the *scaling dimension* or *scaling weight* of ϕ. Another class of conformal transformations are rotations $z \mapsto e^{i\theta}z$, which act on conformal primaries by

$$\phi'(e^{i\theta}z, e^{-i\theta}\bar{z}) = e^{-i\theta(h-\bar{h})}\phi(z, \bar{z}). \tag{5.10}$$

The quantity $h - \bar{h}$ is called the *conformal spin*. This is motivated by the fact that this quantity classifies representations of the two-dimensional rotation group $SO(2) \cong U(1)$. Note, however, that there is no concept of spin in a two-dimensional space-time, since space is one-dimensional.

We would like to be able to go from conformal fields and OPEs to operators and their commutation relations. This is done through the mode expansion, which is a formal power series in z and z^{-1}. For a chiral primary of weight h, the expansion takes the form

$$\phi(z) = \sum_{n=-\infty}^{\infty} \phi_n z^{-n-h}. \tag{5.11}$$

The coefficients ϕ_n can be projected out using the residue theorem:

$$\phi_n = \oint_{z=0} \frac{dz}{2\pi i} \phi(z) z^{n+h-1}, \tag{5.12}$$

where the integration contour encircles $z = 0$ counter-clockwise.

Exercise 5.2.1 Derive the infinitesimal variation (5.8) from the formula (5.6) for finite transformations.

Exercise 5.2.2 How does the formula (5.8) change if we interpret conformal transformations actively, as diffeomorphisms, rather than passively? *Instruction:* If we interpret transformations actively, then z and w are coordinates of different points, rather than expressions for the same point relative to two coordinate systems. When working infinitesimally, the transformation takes the form $z \mapsto z + t\xi(z)$, where $\xi = (\xi(z), \bar{\xi}(\bar{z}))$ is the infinitesimal vector field generating the transformation.

Exercise 5.2.3 Show that conformal fields $\phi(z, \bar{z})$ of negative weights $h < 0, \bar{h} < 0$ can be interpreted as components of the tensors of the form $\Phi = \phi(z, \bar{z})\partial_z^{-h}\partial_{\bar{z}}^{-\bar{h}}$.

Exercise 5.2.4 The infinitesimal change of a tensor field $\Phi = \phi(z, \bar{z})dz^h d\bar{z}^{\bar{h}}$ under the (active) transformation generated by a vector field ξ is given by the Lie derivative

$$L_\xi \Phi = (L_\xi \phi)dz^h d\bar{z}^{\bar{h}} := \lim_{t \to 0} \frac{1}{t}\left(\Phi(z + t\xi, \bar{z} + t\bar{\xi}) - \Phi(z, \bar{z})\right) \qquad (5.13)$$

of Φ with respect to ξ. Show that

$$(L_\xi \phi)(z, \bar{z}) = -\delta_\xi \phi, \qquad (5.14)$$

where $\delta_\xi \phi$ is given by (5.8). *Remark:* If you first show that

$$L_\xi(\phi(z, \bar{z})) = \xi \partial_z \phi + \bar{\xi} \partial_{\bar{z}} \phi, \quad L_\xi(dz) = \partial_z \xi dz, \quad L_\xi(d\bar{z}) = \partial_{\bar{z}} \bar{\xi} d\bar{z}, \quad (5.15)$$

then the result follows by applying the Leibniz rule (product rule).[5] This shows that by imposing linearity and the Leibniz rule the relations (5.15) allow you to compute the Lie derivatives of arbitrary co-tensors ($h \geq 0, \bar{h} \geq 0$). By working out the Lie derivatives of the vector fields $\partial_z, \partial_{\bar{z}}$ you can then compute the Lie derivatives of arbitrary tensors.

5.2.2 Hermitian Conjugation

Hermitian conjugation in the Euclidean plane is defined in such a way that it is consistent with Hermitian conjugation in Minkowski signature. In terms of modes this means that Hermitian conjugation acts such that

$$(\phi_n)^\dagger = (\phi^\dagger)_{-n}, \qquad (5.16)$$

where $(\phi_n)^\dagger$ is the Hermitian conjugate of the n-th mode of the field $\phi(z)$, while $(\phi^\dagger)_{-n}$ is the $(-n)$-th mode of the Hermitian conjugate field $\phi^\dagger(z)$. The relation between $\phi(z)$ and $\phi^\dagger(z)$ which correctly induces (5.16) is (see Exercise 5.2.5)

$$(\phi(z))^\dagger = \bar{z}^{-2h}\phi^\dagger\left(\frac{1}{\bar{z}}\right). \qquad (5.17)$$

Note that the Hermitian conjugate of a Euclidean chiral primary is again a chiral primary, and not an antichiral primary. We can, therefore, as in Minkowski signature, impose a reality condition which respects chirality. A chiral primary is called a *Hermitian chiral primary*, if

$$(\phi_n)^\dagger = \phi_{-n}. \qquad (5.18)$$

Hermitian fields defined in this way arise from Wick-rotating real fields. The left-moving string coordinate $X(z)$ is a chiral Hermitian field. In the following, we take chiral primaries to be Hermitian, unless stated otherwise.

---

[5] Note that $(L_\xi \phi)(z, \bar{z})$ is the single component of the tensor $L_\xi \Phi$, while $L_\xi(\phi(z, \bar{z}))$ is the Lie derivative of the function $\phi(z, \bar{z})$. The reason why this distinction is relevant will become clear through the exercise.

Exercise 5.2.5   The mode expansion of a chiral conformal primary of weight h on the Minkowski cylinder is

$$\phi(\sigma) = \sum_{n=-\infty}^{\infty} \phi_n e^{-2in(\sigma^0 + \sigma^1)}. \tag{5.19}$$

The Hermitian conjugate field $\phi^\dagger(\sigma)$ on the Minkowski cylinder is defined by $\phi^\dagger(\sigma) = (\phi(\sigma))^\dagger$. Show that this implies

$$(\phi_n)^\dagger = (\phi^\dagger)_{-n}. \tag{5.20}$$

Apply a Wick rotation $\sigma^0 = -i\sigma^2$ and perform a conformal transformation from the Euclidean cylinder with complex coordinate $u = 2(\sigma^2 + i\sigma^1)$ to the complex plane with complex coordinate $z = e^u$. Thus, show that the mode expansion on the complex plane is

$$\phi(z) = \sum_{n=-\infty}^{\infty} \phi_n z^{-n-h}. \tag{5.21}$$

Remark: Here, we use $u = 2w$ instead of $w = \sigma^2 + i\sigma^1$ to suppress irrelevant numerical factors 2^h.

Compare the Hermitian conjugate $(\phi(z))^\dagger$ of (5.21) to the field $\phi^\dagger(z)$, which is defined by its mode expansion

$$\phi^\dagger(z) = \sum_{n=-\infty}^{\infty} (\phi^\dagger)_n z^{-n-h}, \tag{5.22}$$

where $(\phi^\dagger)_n = (\phi_{-n})^\dagger$ are the same modes as in the expansion of $\phi(z)$. Thus, show that $\phi(z)$ and $\phi^\dagger(z)$ are related by (5.17).

5.2.3 Operator Products and Commutators

We now state the definitions of operator products and commutators in Euclidean signature. These definitions are motivated by the examples provided in Chapter 4. The radially ordered operator product of two Euclidean field operators is defined by[6]

$$R(\phi_1(z)\phi_2(w)) = \begin{cases} \phi_1(z)\phi_2(w), & \text{if } |z| > |w|, \\ \phi_2(w)\phi_1(z), & \text{if } |z| < |w|. \end{cases} \tag{5.23}$$

Since only radially ordered products are defined, the radial ordering symbol 'R' will be omitted in the following. Commutators[7] are defined by

$$[\phi_1(z), \phi_2(w)]_{|z|=|w|} = \lim_{\epsilon \to 0+} [(\phi_1(z)\phi_2(w))_{|z|=|w|+\epsilon} - (\phi_2(w)\phi_1(z))_{|z|=|w|-\epsilon}]. \tag{5.24}$$

---

[6] We restrict ourselves to bosonic operators. For fermionic operators, this definition includes a relative minus sign.

[7] More precisely, these are 'equal radius commutators', which are the analogues of equal time commutators.

5.3 Energy-Momentum Tensor and Virasoro Algebra

In a general CFT, we will not know how the energy-momentum tensor can be expressed in terms of the conformal primaries. We make the following natural assumptions:

1. The energy momentum tensor is traceless:

$$T_{z\bar{z}} = 0. \tag{5.25}$$

2. The energy momentum tensor is conserved:

$$\partial_{\bar{z}} T_{zz} = 0, \quad \partial_z T_{\bar{z}\bar{z}} = 0. \tag{5.26}$$

We will now show that under these assumptions the energy-momentum tensor generates the infinitesimal conformal transformations of conformal fields. In general, a conserved current

$$\partial_z j_{\bar{z}} + \partial_{\bar{z}} j_z = 0 \tag{5.27}$$

defines a conserved charge Q by integration over space, which in our case is the integration over a circle around the origin in the complex plane:

$$Q = \oint \frac{dz}{2\pi i} j_z + \oint \frac{d\bar{z}}{2\pi i} j_{\bar{z}}. \tag{5.28}$$

The conserved charge Q generates the infinitesimal symmetry transformations of fields,

$$\delta\phi(z, \bar{z}) = -[Q, \phi(z, \bar{z})]. \tag{5.29}$$

In a setting where holomorphic and antiholomorphic sectors decouple, conservation laws involve conserved chiral and antichiral currents, $\partial_{\bar{z}} j_z = 0$ and $\partial_z j_{\bar{z}} = 0$, and the associated chiral and antichiral charges. We focus on chiral fields in the following. Chiral currents need to have weight $(1, 0)$ in order that the associated conserved charge is a scalar. Since $T(z)$ has weight $(2, 0)$, it does not give rise to a single conserved current, but to infinitely many, which are obtained by multiplying $T(z)$ with a chiral holomorphic vector field $\xi(z) = \xi^z$.[8] The chiral conserved charge T_ξ associated with the conformal transformation $z \mapsto z + t\xi(z)$ is

$$T_\xi = \oint_{z=0} \frac{dz}{2\pi i} \xi(z) T(z), \tag{5.30}$$

and acts on chiral conformal primaries $\phi(w)$ by

$$\delta_\xi \phi(w) = -[T_\xi, \phi(w)] = -\oint_{z=w} \frac{dz}{2\pi i} \xi(z) T(z) \phi(w). \tag{5.31}$$

Given (5.8), this implies that the OPE between the energy momentum tensor and a chiral conformal primary of weight h takes the form (Exercise 5.3.1)

---

[8] A chiral holomorphic vector field $\xi = \xi^z \partial_z = \xi(z)\partial_z$ has weight $(-1, 0)$.

$$T(z)\phi(w) = \frac{h\phi(w)}{(z-w)^2} + \frac{\partial_w\phi}{z-w} + \cdots . \tag{5.32}$$

Looking back to (4.42) and (4.53) we note that $\phi_z \propto \partial_z X$ is a conformal primary of weight $(1,0)$, while the exponential $E_k(z,\bar{z}) =: \exp(ikX(z,\bar{z})) :$ is a conformal primary of weight $(\frac{1}{8}k^2, \frac{1}{8}k^2)$.

The OPE of the energy-momentum tensor with itself is determined by imposing that the conformal transformations operate on fields as a Lie algebra, which implies that the Jacobi identity

$$[[T_{\xi_1}, T_{\xi_2}], \phi] = [T_{\xi_1}, [T_{\xi_2}, \phi]] - [T_{\xi_2}, [T_{\xi_1}, \phi]] \tag{5.33}$$

must hold. Since the r.h.s. only depends on the action of T_ξ on ϕ, this provides information about the commutator $[T_{\xi_1}, T_{\xi_2}]$. The commutator of two conformal transformations of a chiral primary is

$$[\delta_{\xi_1}, \delta_{\xi_2}]\phi(z) = \delta_{\xi_2\partial_z\xi_1 - \xi_1\partial_z\xi_2}\phi(z). \tag{5.34}$$

For the action of $[T_{\xi_1}, T_{\xi_2}]$ on $\phi(z)$ to have the same result, the OPE $T(z)T(w)$ must have the form

$$T(z)T(w) = \frac{c/2}{(z-w)^4} + \frac{T(w)}{(z-w)^2} + \frac{\partial_w T}{z-w} + \cdots , \tag{5.35}$$

where c is a constant. The value of c is not determined by (5.33) since the fourth order pole does not contribute to the commutator on the left hand side and thus is not visible in the transformation properties of primaries (Exercise 5.33). We have already seen in an explicit example that such terms exist, and, therefore, we treat c as a model dependent constant. In Section 5.2.2, we will show that unitarity requires $c > 0$, in particular there is no non-trivial unitary CFT with $c = 0$. Without the fourth order term $T(z)$ would be a conformal primary of weight $(2,0)$. Note that the three singular terms in the OPE (5.35) are precisely all the terms which are consistent with $T(z)$ having scaling weight 2. This indicates that $T(z)$ still transforms covariantly under part of the conformal transformations. And, indeed, the infinitesimal conformal transformation of $T(z)$ found from (5.35) is (see Exercise 5.3.6)

$$\delta_\xi T(z) = -(2\partial_z\xi + \xi\partial_z)T(z) - \frac{c}{12}\partial^3\xi(z). \tag{5.36}$$

The last term vanishes precisely for infinitesimal conformal transformations in the subalgebra $\mathfrak{sl}(2,\mathbb{C})$. It can be shown that the corresponding finite conformal transformations are (see Exercise 5.3.7)

$$T'(z) = (\partial_z w)^2 T(w(z)) + \frac{c}{12}S(w,z), \tag{5.37}$$

where $S(w,z)$ is the so-called Schwarzian derivative

$$S(w,z) := \frac{\partial_z w \partial_z^3 w - \frac{3}{2}(\partial_z^2 w)^2}{(\partial_z w)^2}. \tag{5.38}$$

The Schwarzian derivative is related to the theory of automorphic functions. Note that it vanishes for Möbius transformations:

$$S(w,z) = 0 \Leftrightarrow w = \frac{az+b}{cz+d}, \quad \begin{pmatrix} a & b \\ c & d \end{pmatrix} \in \mathrm{SL}(2,\mathbb{C}). \tag{5.39}$$

Therefore, $T(z)$ still transforms as a tensor under the group of global conformal transformations of the two-sphere, the Möbius group $\mathrm{SL}(2,\mathbb{C})$.

This gives rise to the following definitions:

- A field is called a *primary* field if it transforms as a tensor under conformal transformations.
- A field is called a *secondary* field if it is not a primary field.
- Secondary fields which transform as tensors under Möbius transformations are called *quasi-primary* fields.

We will see later that secondary fields arise from primary fields as so-called descendants. For example, higher derivatives, $\partial_z^m X$ with $m > 1$ of scalar fields 'descend' from $\partial_z X$, and $T(z)$ descends from the unit operator (see Exercise 5.4.5).

Using the formal Laurent expansion

$$T(z) = \sum_n L_n z^{-n-2} \Leftrightarrow L_n = \oint_{z=0} \frac{dz}{2\pi i} T(z) z^{n+1} \tag{5.40}$$

one obtains the commutation relations (see Exercise 5.3.3)

$$[L_m, L_n] = (m-n)L_{m+n} + \frac{c}{12} m(m-1)(m+1)\delta_{m+n,0}. \tag{5.41}$$

The Lie algebra (5.41) differs from the Witt algebra by the extra term proportional to c. This term is proportional to the unit operator, and, therefore, corresponds to a *central extension* of the Witt algebra, called the *Virasoro algebra*. The parameter c is called the *central charge* and takes a fixed value for a given conformal field theory. For $c \neq 0$, the conformal symmetry is anomalous, that is the symmetry algebra is modified by quantum effects, which introduces a sensitivity on the scale at which the theory is probed. Depending on context, this scale can be the renormalisation scale, the effect of boundaries, or the effect of curvature.

We can also convert the OPE (5.32) between $T(z)$ and a chiral primary of weight h into commutators (see Exercise 5.3.4):

$$[L_n, \phi(z)] = z^n(z\partial_z + (n+1)h)\phi(z), \tag{5.42}$$

$$[L_n, \phi_m] = (n(h-1) - m)\,\phi_{n+m}. \tag{5.43}$$

Exercise 5.3.1 Show that the transformation defined by (5.31) is an infinitesimal conformal transformation (5.8) with $\bar{h} = 0$ if and only if the OPE between the energy momentum tensor and conformal primary of weight $(h, 0)$ takes the form (5.32).

Exercise 5.3.2   Verify (5.34).

Exercise 5.3.3   Show that the OPE (5.35) implies the Virasoro algebra (5.41).

Exercise 5.3.4   Show that the commutator relations (5.42) and (5.43) are equivalent to the transformation formula (5.8) (with $\bar{h} = 0$) for chiral primaries. In particular, show that L_n generates a conformal transformation with $\xi(z) = -z^{n+1}$.

Exercise 5.3.5   Show that (5.33) implies that the singular part of the OPE of T with itself must take the form (5.35). Check in particular that the term proportional to c does not contribute to the transformation of primary fields and is therefore an allowed modification of the Witt algebra.

Exercise 5.3.6   Derive the infinitesimal conformal transformation (5.36) of $T(z)$ from the OPE (5.35).

Exercise 5.3.7   Show that the finite conformal transformation (5.37) implies the infinitesimal transformation (5.36).[9]

5.4 The State – Operator Correspondence

5.4.1 The $\mathrm{SL}(2, \mathbb{C})$ Vacuum – Primary States

We now formulate the state – operator correspondence in generality. The first assumption we make is that there is a ground state $|0\rangle$, from which the Hilbert space $\mathcal{H}$ is generated by applying field operators evaluated at $z = 0$. This state is interpreted as the *in-vacuum* in string theory. The corresponding *out-vacuum* $\langle 0|$ is acted upon by fields evaluated at $z = \infty$.

What properties should a reasonable ground state have? One natural requirement is that the insertion of the energy momentum tensor defines a regular, finite norm state,

$$\lim_{z \to 0} T(z)|0\rangle = \text{regular.} \tag{5.44}$$

This implies (see Exercise 5.4.2)

$$L_n|0\rangle = 0, \quad n > -2. \tag{5.45}$$

Note that these are precisely the conformal transformations which are regular at $z = 0$. Similarly, a finite action of the energy momentum tensor on the 'out-vacuum' $\langle 0|$ implies

$$\lim_{z \to \infty} \langle 0|T(z) = \text{regular} \Rightarrow \langle 0|L_n = 0 , \quad n < 2. \tag{5.46}$$

This condition is the Hermitian conjugate of $L_n|0\rangle = 0$ for $n < 2$, since $L_n^\dagger = L_{-n}$. The conformal transformations generated by L_n, $n < 2$ are precisely those which are regular on $z = \infty$. Combining (5.44) and (5.46), we see that the generators L_{-1}, L_0, L_1

---

[9] See Schottenloher (1997) for the exponentiation of infinitesimal transformations.

of the Möbius group $SL(2, \mathbb{C})$ annihilate both the in- and the out-vacuum. Therefore, $|0\rangle$ is called the $SL(2, \mathbb{C})$ *vacuum*.

The second condition we impose is that the insertion of a chiral primary $\phi(z)$ at $z = 0, \infty$ creates a state of finite norm. This tells us which modes of a Hermitian field $\phi(z)$ are annihilation operators:

$$\phi_n|0\rangle = 0 , \qquad n > -h,$$
$$\langle 0|\phi_n = 0 , \qquad n < h. \tag{5.47}$$

Note that for fields with negative weights, $h < 0$, there are modes which neither annihilate the in- nor the out-vacuum. We will see later that in a unitary conformal theory $h \geq 0$, so that this situation does not arise. Therefore, we will not discuss this case. However, it is relevant for non-unitary conformal field theories, including the so-called $b - c$ system which describes the Fadeev–Popov ghost fields of string theory.

We can now define in-states by inserting conformal primaries at the origin,

$$|\phi\rangle = \lim_{z \to 0} \phi(z)|0\rangle = \phi_{-h}|0\rangle. \tag{5.48}$$

States generated by primary fields are called *primary states*. The insertion of a field operator at $z \neq 0$ should create the same state at the corresponding position. In other words, $|\phi\rangle_z = \phi(z)|0\rangle$ should be related to $|\phi\rangle = \phi(0)|0\rangle$ by the action of the translation operator. On the complex plane, infinitesimal translations are generated L_{-1} and $\bar{L}_{-1}$. Therefore (see Exercise 5.4.1),

$$\phi(z) = e^{-zL_{-1}} \phi(0) e^{zL_{-1}} \Rightarrow |\phi\rangle_z = e^{-zL_{-1}}|\phi\rangle. \tag{5.49}$$

Out-states are defined by Hermitian conjugation $\langle\phi| = |\phi\rangle^\dagger$:

$$\langle\phi| = \langle 0|(\phi^\dagger)_h. \tag{5.50}$$

Exercise 5.4.1   How are the operators T_x and T_y, which generate translations in the x- and y-direction on the complex plane with complex coordinate $z = x+iy$ related to L_{-1} and $\bar{L}_{-1}$? Use your result to derive (5.49).

Exercise 5.4.2   Show that (5.44) implies (5.45). Also show (5.46) directly, without using Hermiticity.

Exercise 5.4.3   Show that the requirement that the insertion of a conformal primary at $z = 0, \infty$ creates a state of finite norm implies (5.47).

5.4.2 Highest Weight States

Primary states $|\phi\rangle$ satisfy

$$L_0|\phi\rangle = h|\phi\rangle, \quad L_n|\phi\rangle = 0 \quad \text{for} \quad n > 0. \tag{5.51}$$

Such states are called *highest weight states*, because by application of the operators $L_{-n}, n > 0$ one obtains the basis

$$|\phi^{(k_1,\ldots,k_m)}\rangle = L_{-k_1} \cdots L_{-k_m}|\phi\rangle, \quad k_1 \geq k_2 \geq \ldots k_m > 0 \tag{5.52}$$

of a representation $V(c, h)$ of the Virasoro algebra. Representations $V(c, h)$ are called *Verma modules*. The terminology *highest weight representation* and highest weight state is standard in mathematical representation theory, while from the physical point of view $|\phi\rangle$ is the ground state or lowest energy state of the representation because the application of the raising operators $L_{-n}, n > 0$ raises the eigenvalue of the world-sheet Hamiltonian L_0:

$$L_0(L_{-n}|\phi|\rangle) = (n+h)(L_{-n}|\phi\rangle) \tag{5.53}$$

$$\Rightarrow L_0\left|\phi^{(k_1,\dots,k_m)}\right\rangle = \left(h + \sum_{i=1}^{m} k_i\right)\left|\phi^{(k_1,\dots,k_m)}\right\rangle. \tag{5.54}$$

The difference $\sum_{i=1}^{m} k_i$ of the L_0-eigenvalues of $|\phi\rangle$ and $|\phi^{(k_1,\dots,k_m)}\rangle$ is called the *level* of the state $|\phi^{(k_1,\dots,k_m)}\rangle$.

A conformal field theory can contain finitely or infinitely many primary fields $|\phi_i\rangle$. If the primary states have different conformal weights, $h_i \neq h_j$, they are orthogonal to each other, for any scalar product for which L_0 is Hermitian (see Exercise 5.4.4):

$$\langle\phi_i|\phi_j\rangle = 0, \quad \text{if } i \neq j. \tag{5.55}$$

If conformal primaries have the same conformal weights, we can use the Gram-Schmidt algorithm to find new linear combinations which are mutually orthogonal. It follows that states in different Verma modules are orthogonal. By the same argument, within each Verma module states at different level are automatically orthogonal, while for states which are the same level we can choose orthogonal linear combinations.

Exercise 5.4.4 Show that states with different eigenvalues for L_0 are orthogonal for any scalar product for which L_0 is Hermitian.

5.4.3 Descendant Fields

States which are obtained by the application of $L_{-n}, n > 0$ from a primary state are called *descendant states*, and the corresponding conformal fields are called *descendant fields* and denoted

$$\phi^{(k_1,\dots,k_m)}(z) = \hat{L}_{-k_1}\cdots\hat{L}_{-k_m}\phi(z). \tag{5.56}$$

In general, descendants are secondary fields which don't transform as conformal tensors. The descendants which are obtained by the application of a single operator L_{-n} appear in the OPE between the energy momentum tensor and the primary:

$$T(z)\phi(w) = \sum_{k=0}^{\infty}(z-w)^{-2+k}\phi^{(k)}(z) \Leftrightarrow \phi^{(k)}(w) = \oint_{z=w}\frac{dz}{2\pi i}(z-w)^{1-k}T(z)\phi(w). \tag{5.57}$$

Note:

$$\hat{L}_0\phi(z) = h\phi(z), \quad \hat{L}_{-1}\phi(z) = \partial_z\phi(z),\dots \tag{5.58}$$

and thus all derivatives $\partial_z^k\phi(z)$ of $\phi(z)$ appear as descendants. Descendants of the form $\phi^{(k_1,k_2)}(z)$ are obtained from the OPE of $T(z)$ with descendants $\phi^{(k)}(z)$, and so on. A primary field ϕ and its descendants form a *conformal family* denoted $[\phi]$.

Exercise 5.4.5 Show that the energy momentum tensor $T(z)$ is a descendant and belongs to the conformal family $[I]$ of the identity I.
Hint: work out $\hat{L}_{-n}I(z)$ for $n = 1, 2, \ldots$

5.5 General Aspects of Two-Dimensional Conformal Field Theories

5.5.1 General Discussion of OPEs

To describe a conformal field theory, it is sufficient to know a complete set $\{O_i\}$ of local operators with conformal weights $h_i, \bar{h}_i$. OPEs of such operators take the form

$$O_i(z, \bar{z})O_j(w, \bar{w}) = \sum_k (z - w)^{h_k - h_i - h_j}(\bar{z} - \bar{w})^{\bar{h}_k - \bar{h}_i - \bar{h}_j} C_{ij}^k O_k(w, \bar{w}). \tag{5.59}$$

The OPE between primary fields takes the form

$$\phi_i(z, \bar{z})\phi_j(w, \bar{w}) = \sum_k \sum_{\{l, \bar{l}\}} C_{ij}^k \beta_{ij}^{\{l\}k} \beta_{ij}^{\{\bar{l}\}k}(z - w)^{h_k + \sum_n l_n - h_i - h_j}$$

$$(\bar{z} - \bar{w})^{\bar{h}_k + \sum_n \bar{l}_n - \bar{h}_i - \bar{h}_j} \phi_k^{\{l, \bar{l}\}}(w, \bar{w}), \tag{5.60}$$

where $\{l, \bar{l}\}$ is short for the indices $\{l_1, l_2, \ldots, \bar{l}_1, \bar{l}_2, \ldots\}$ labelling descendants. In (5.60) the coefficients C_{ij}^k are the 'Clebsch–Gordon coefficients' between the involved conformal families, $[\phi_i] \times [\phi_j] = C_{ij}^k[\phi_k]$. The coefficients $\beta_{ij}^{\{l\}k}, \beta_{ij}^{\{\bar{l}\}k}$ encode how the descendants of the primaries ϕ_k enter into the OPE.

It can be shown that conformal invariance implies that:

- The coefficients $\beta_{ij}^{\{l\}k}, \beta_{ij}^{\{\bar{l}\}k}$ are determined by $h_i, h_j, h_k, \bar{h}_i, \bar{h}_j, \bar{h}_k$ and the central charge c.
- All correlation functions are determined by the three-point functions between primary fields, which are encoded in the representation theoretic data C_{ij}^k.

Therefore, a two-dimensional conformal field theory is completely determined by its central charge c, the weights $h_i, \bar{h}_i$ of its chiral primaries, and the coefficients C_{ij}^k.

5.5.2 Unitary Conformal Field Theories

We now explore the consequences of the additional condition that the scalar product between physical states in a unitary conformal field theory must be positive definite. For a highest weight state $|\phi_j\rangle$ of weight h_j this implies the inequality

$$\langle \phi_j | L_n L_{-n} | \phi_j \rangle = \left(2nh_j + \frac{c}{12}(n^3 - n)\right)\langle \phi_j | \phi_j \rangle \geq 0, \tag{5.61}$$

where equality is only possible if $L_{-n}\phi = 0$. This, in turn, implies

- In a unitary conformal field theory $c > 0$.
- In a unitary conformal field theory the weights h_j of primary states are non-negative, $h_j \geq 0$, and any state with $h_j = \bar{h}_j = 0$ corresponds to the SL$(2, \mathbb{C})$-vacuum.

Exercise 5.5.1 Derive (5.61) and the resulting conditions on c and h_j which are satisfied in unitary conformal field theories.

5.5.3 Null Vectors

As discussed in Section 5.4.3, descendants are normally secondary states. A special situation arises when a Verma module $V(c, h)$ contains descendants $|\chi\rangle \neq 0$ which are annihilated by all L_n, $n > 0$. These states are primaries and generate a Verma submodule within $V(c, h)$, which therefore is reducible.[10] Descendants which are primaries are orthogonal to all states in $V(c, h)$ including themselves. They are called *null vectors* or *singular vectors*. If null vectors exist, the Verma module $V(c, h)$ cannot be positive definite but only positive semi-definite. A positive definite representation is obtained by identifying any two states which differ by a null vector $|\chi\rangle$:

$$|\phi\rangle \simeq |\phi'\rangle \Leftrightarrow |\phi\rangle - |\phi'\rangle = |\chi\rangle. \tag{5.62}$$

Observe that adding a null vector is akin to making a gauge transformation. In fact we will see that in string theory null vectors correspond to space-time gauge transformations. Mathematically, (5.62) corresponds to taking the quotient of $V(c, h)$ by the Verma module generated by a null vector.

Finding all possible null vectors for given $c, h_i, \bar{h}_i$ plays an important role in the classification of unitary conformal field theories. The problem of identify all null vectors at a given level N can be phrased as the problem of finding all zeros of the level N *Kac determinant*

$$M_N(h, c) = \det\left(\left\langle h\left|\prod_i L_{k_i} \prod_j L_{-m_j}\right|h\right\rangle\right), \quad \sum_i k_i = \sum_j m_j = N. \tag{5.63}$$

Exercise 5.5.2 Show that a null state (that is, a descendant which also is a primary) is orthogonal to all states in its Verma module.

Exercise 5.5.3 Show that $L_{-1}|h\rangle$ is a null state if and only if $h = 0$. Find the zeros of the Kac determinant at level $N = 1, 2$.

5.5.4 Results on Classification

Unitary conformal field theories have been classified for $c < 1$. It turns out that only certain discrete values of c and h are allowed, and that theories with $c < 1$ contain a finite number of conformal primaries. These theories form an infinite series, called the *unitary minimal series*, labelled by the conformal charge c, which takes values $\frac{1}{2} \leq c < 1$. The case $c = \frac{1}{2}$ corresponds to a conformal field theory which describes the two-dimensional Ising model at its critical point. Since all correlation functions can be obtained from a finite number of three-point functions between conformal primaries, these theories are exactly solvable. The allowed values of c, h are rational, which makes the minimal models examples of *rational conformal field theories*.

---

[10] A representation is called *reducible* if it has an invariant subspace. In general, this does not imply that the representation is *decomposable* into a direct sum of representations.

For $c \geq 1$, the allowed values of c and h are continuous, and theories have an infinite number of primary fields. However, it can happen that a conformal field theory has a so-called *extended chiral algebra*, which extends the Virasoro algebra into a larger algebra with a finite number of primaries. In this case solving the theory reduces again to knowing a finite number of three-point functions which encode how the finitely many representations of the extended chiral algebra couple to one another. We will discuss the classification of $c = 1$ conformal field theories in Chapter 13.

5.6 Literature

A comprehensive treatment of two-dimensional conformal field theory is given in Di Francesco et al. (1997). It also reviews higher-dimensional conformal field theories and includes a careful discussion of the relation between scale and conformal invariance. An excellent introduction to conformal field theory applied to string theory are the lecture notes Ginsparg (1988), while a more comprehensive and up to date treatment including boundary conformal field theory (open strings) is Blumenhagen and Plauschinn (2009). For mathematically inclined readers, Schottenloher (1997) provides a treatment which carefully distinguishes between local and global aspects, explains the details of conformal compactifications, and covers the representation theory of the Virasoro algebra. See also Frenkel and Ben-Zvi (2001) where vertex operator algebras are discussed from a mathematical point of view.

In this chapter, we discuss *partition functions* for particles and closed strings from a world-line/world-sheet perspective. Partition functions are very important: they encode the states of a theory[1] and act as generating functionals for correlation functions. We also introduce the concept of *modular invariance*, which sets (closed) strings apart from particles, and is responsible for the absence of UV divergencies.

6.1 Partition Functions for Particles and Strings

The canonical partition function Z in quantum statistical mechanics is defined as

$$Z = \mathrm{tr}_{\mathcal{H}} e^{-\beta H} = \sum_s e^{-\beta E_s} = \sum_n d(n) e^{-\beta E_n},$$

where $\beta = (k_B T)^{-1}$ is the inverse temperature, k_B is the Boltzmann constant, H is the Hamiltonian, and the trace is taken over the Hilbert space $\mathcal{H}$ of the theory. In the second step, we have evaluated the trace using a basis of energy eigenstates, labelled by s, with energy E_s. This is the form the canonical partition function takes in classical statistical mechanics. In the last step, we write Z as a sum over the distinct eigenvalues E_n of H. The coefficients $d(n)$ specify the number of states with a given energy E_n. For simplicity, we have assumed that the spectrum of H is discrete.

The partition function of a quantum theory can be obtained by Wick rotation of the trace of the time evolution operator,

$$Z = \mathrm{tr}_{\mathcal{H}} U(t_2, t_1) = \mathrm{tr}_{\mathcal{H}} e^{-i(t_2 - t_1)H} = \sum_{|\phi\rangle} \left\langle \phi | e^{-i(t_2 - t_1)H} | \phi \right\rangle \to \mathrm{tr}_{\mathcal{H}} e^{-\beta H},$$

where the sum is over an orthonormal basis $|\phi\rangle$ of $\mathcal{H}$. The trace corresponds to a finite time evolution where the initial and final state are the same. The finite propagation time $t_2 - t_1$ is identified with the inverse temperature.

Diagrammatically, the time evolution operator $U(t_2, t_1)$ is represented by a line, and taking the trace corresponds to identifying the initial and final state, as well as summing over states, thus obtaining a loop (see Figure 6.1). This partition function is represented by a one-loop diagram and therefore also called the one-loop partition function. While Feynman diagrams are usually considered in momentum space, here we think in terms of position space. When forming the loop we have to choose its length, which corresponds to the proper time t the particle needs to circle around. As we will discuss in Chapter 11, this is the so-called *Stückelberg proper time*

---

[1] The commonly used symbol Z stands for the German 'Zustandssumme', which literally means 'sum over states'.

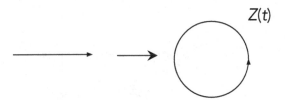

Fig. 6.1 The world-line partition $Z(t)$ function is obtained from the propagator by identifying the initial and final state, and summing over all states. It depends on the proper time t needed to complete the loop.

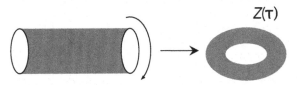

Fig. 6.2 The world-sheet partition $Z(\tau)$ function of a closed string is obtained from the propagator by identifying the initial and final state, and summing over all states. It depends on the length of the cylinder (propagator) and the shift applied to the final state, which combine into a complex parameter τ.

parametrisation of the partition function. While from the space-time perspective we should integrate over all values of proper time t, we can also consider an unintegrated partition function $Z(t)$ which is a function of proper time. We will call this the *world-line partition function.*

If we extend this concept to a closed string, then at the diagrammatical level the propagator is a cylinder instead of a line, and by identifying the initial with the final state we obtain a torus. Compared to the point particle case, a closed string has the additional option of twisting the string before gluing the cylinder to a torus. In other words, the initial and final state may differ by a translation along the string (see Figure 6.2).

The corresponding partition function is

$$Z = \text{tr}_{\mathcal{H}} e^{-tH + isP}, \tag{6.1}$$

where $H = L_0^{\text{cyl}} + \bar{L}_0^{\text{cyl}}$ is the world-sheet Hamiltionian and $P = L_0^{\text{cyl}} - \bar{L}_0^{\text{cyl}}$ the world-sheet momentum operator. Note that $L_0^{\text{cyl}}, \bar{L}_0^{\text{cyl}}$ are the Laurent modes of the energy-momentum tensor on the cylinder. We need to distinguish them from their counterparts $L_0, \bar{L}_0$ on the complex plane, because the energy momentum tensor is only a quasi-primary field, and the conformal transformation between cylinder and plane does not belong to the Möbius group. The transformation takes the form

$$T_{\text{(cyl)}} = 4\left(z^2 T(z) - \frac{c}{24}\right), \tag{6.2}$$

and therefore

$$L_0^{\text{(cyl)}} = L_0 - \frac{c}{24}. \tag{6.3}$$

See Exercise 6.1.1. Observe that the L_0-eigenvalues shift proportional to the central charge.

Setting $s = 2\pi\tau_1, t = 2\pi\tau_2, \tau = \tau_1 + i\tau_2, q = e^{2\pi i\tau}, \bar{q} = e^{-2\pi i\bar{\tau}}$ we obtain

$$Z(\tau) = \text{tr}_{\mathcal{H}}\left(q^{L_0^{\text{cyl}}} \bar{q}^{\bar{L}_0^{\text{cyl}}}\right) = \text{tr}_{\mathcal{H}}\left(q^{L_0 - \frac{c}{24}} \bar{q}^{\bar{L}_0 - \frac{\bar{c}}{24}}\right).$$

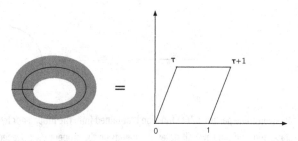

A torus can be represented by a parallelogram with opposing sides identified.

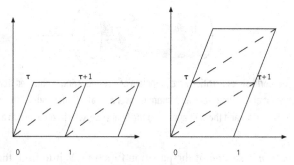

The dashed parallelograms generated from the standard parallelogram by $T : \tau \mapsto \tau + 1$ and by $U : \tau \mapsto \frac{\tau}{\tau+1}$ can be mapped back to the standard parallelogram by 'cut and paste'. Therefore they are different representations of the same torus.

Note that we have allowed that the holomorphic and antiholomorphic sectors have different central charges. This occurs in important applications, including the heterotic string, which we will briefly introduce in Chapter 14.

The *modular parameter* τ is related to the classification of tori up to conformal transformations. A torus can be realised as a parallelogram in the complex plane, with opposite sides identified, $z \cong z + 1 \cong z + \tau$ (see Figure 6.3).

Since conformal transformations include scale transformations, the size of the parallelogram is not relevant, only its shape. To represent all conformally inequivalent tori, it is therefore sufficient to consider parallelograms with edges $0, 1, \tau, \tau + 1$, and we may choose $\text{Im}(\tau) > 0$, since tori with $\text{Im}(\tau) < 0$ are obtained by reflection. If τ varies over the upper half plane $\mathcal{H} = \{\tau \in \mathbb{C} | \text{Im}(\tau) > 0\}$, we will obtain all conformally inequivalent tori, though more than once. On a torus one can perform *large reparametrisations*, which are not continuously connected to the identity. These are generated by the *Dehn twists* where the torus is cut along a non-contractible curve, twisted by an angle of 2π, and glued back. The resulting discrete group, called the *modular group* is generated by two transformations T, U, which act by $T : \tau \mapsto \tau + 1$ and $U : \tau \mapsto \frac{\tau}{\tau+1}$ (see Figure 6.4).

The modular group $\text{PSL}(2, \mathbb{Z}) = \text{SL}(2, \mathbb{Z})/\mathbb{Z}_2$ acts by

$$\tau \to \frac{a\tau + b}{c\tau + d}, \quad \text{where} \quad \begin{pmatrix} a & b \\ c & d \end{pmatrix} \in SL(2, \mathbb{Z}). \tag{6.4}$$

When taking into account the action of the modular group, conformally inequivalent tori are parametrised by a fundamental domain $\mathcal{F} \subset \mathcal{H}$. To identifty a fundamental domain, it is convenient to replace the generator U by

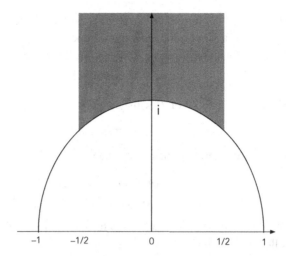

Fig. 6.5 The standard fundamental domain of the modular group of the torus.

$$S = T^{-1}UT^{-1} \; : \; \tau \mapsto -\frac{1}{\tau}, \tag{6.5}$$

which acts on τ by inversion at the unit circle followed by reflection at the real axis (to move the point obtained by inversion back to the upper half plane). The standard fundamental domain

$$\mathcal{F} = \left\{ \tau \in \mathcal{H} \, \middle| \, -\frac{1}{2} < \mathrm{Re}(\tau) < \frac{1}{2}, \; |\tau| > 1 \right\} \cup \left\{ \tau \in \mathcal{H} \, \middle| \; \mathrm{Re}(\tau) = -\frac{1}{2}, \; |\tau| \geq 1 \right\}$$

$$\cup \left\{ \tau \in \mathcal{H} \, \middle| \; |\tau| = 1, \; -\frac{1}{2} < \mathrm{Re}(\tau) \leq 0 \right\} \tag{6.6}$$

is obtained by restricting to a strip of width one parallel to the imaginary axis to account for the identifications by T, and cutting out the unit circle to account for the identifications by S, see Figure 6.5. Note that one has to avoid double-counting points on the boundary, which makes the description slightly complicated.

Exercise 6.1.1 Derive (6.2) and (6.3). *Hint:* In our conventions the holomorphic energy-momentum tensor on the Euclidean cylinder has the form

$$T(w) = 4 \sum_n L_n^{\mathrm{cyl}} e^{-2nw}. \tag{6.7}$$

This equation follows from (2.56) by applying the analytical continuation described in Section 4.1. We have set $T = \frac{1}{\pi}$ and re-labelled $\tilde{L}_n \mapsto L_n$. Note that with our conventions there is a factor 4 in (6.7), which will ensure that there is no relative factor between L_n on the cylinder and on the plane.

6.2 The Chiral Partition Function of a Free Boson

In this section, we consider a specific example, the chiral partition function

$$Z(\tau) = \mathrm{tr}_{\mathcal{H}} \left(q^{L_0 - \frac{c}{24}} \right)$$

of the chiral conformal primary $\phi(z) \propto i\partial_z X$. This is related to the chiral partition function of a string propagating on a one-dimensional space. $\mathcal{H}$ is the Fock space associated with the modes α_m, $m \in \mathbb{Z}$, with relations

$$[\alpha_m, \alpha_n] = m\delta_{m+n,0}.$$

The ground state $|0\rangle$ of $\mathcal{H}$ satisfies

$$\alpha_m|0\rangle = 0, \quad m \geq 0, \quad \langle 0|0\rangle = 1.$$

We choose an orthonormal basis of $\mathcal{H}$:

$$|n_1, n_2, \ldots\rangle = N_{\{n_k\}}\alpha_{-1}^{n_1}\alpha_{-2}^{n_2} \cdots |0\rangle, \quad n_k \geq 0, \tag{6.8}$$

$$\langle n_1, n_2, \ldots | n_1', n_2', \ldots\rangle = \delta_{n_1,n_1'}\delta_{n_2,n_2'} \cdots . \tag{6.9}$$

We will not need the normalisation factors $N_{\{n_k\}}$ explicitly. For reference, we note the following formulas

$$L_0 = \frac{1}{2}\alpha_0^2 + \sum_{n=1}^{\infty}\alpha_{-n}\alpha_n, \quad [L_0, \alpha_m] = -m\alpha_m, \quad L_0|n_1, n_2, \ldots\rangle = \sum_{k=1}^{\infty} kn_k|n_1, n_2, \ldots\rangle. \tag{6.10}$$

Now, we compute

$$Z(\tau) = \sum_{n_1=0}^{\infty}\sum_{n_2=0}^{\infty} \cdots \left\langle n_1, n_2, \ldots | q^{L_0 - \frac{1}{24}} | n_1, n_2, \ldots\right\rangle$$

$$= q^{-\frac{1}{24}}\sum_{n_1=0}^{\infty}\sum_{n_2=0}^{\infty} \cdots q^{\sum_{k=1}^{\infty} kn_k} = q^{-\frac{1}{24}}\sum_{n_1=0}^{\infty}\sum_{n_2=0}^{\infty} \cdots \prod_{k=1}^{\infty} q^{kn_k}$$

$$= q^{-\frac{1}{24}}\prod_{k=1}^{\infty}\sum_{n_k=0}^{\infty} q^{kn_k} = q^{-\frac{1}{24}}\prod_{k=1}^{\infty}\frac{1}{1-q^k}. \tag{6.11}$$

We used that $c = 1$, that L_0 and q^{L_0} have the same eigenstates, and we performed a geometric sum (which converges for $|q| < 1$).

The result is the inverse of the Dedekind η-function:[2]

$$\eta(\tau) := q^{\frac{1}{24}}\prod_{k=1}^{\infty}(1-q^k).$$

Adding the corresponding antiholomorphic sector, we obtain

$$Z'(\tau, \bar{\tau}) = \mathrm{tr}_{\mathcal{H}}\left(q^{L_0 - \frac{1}{24}}\bar{q}^{\bar{L}_0 - \frac{1}{24}}\right) = |\eta(\tau)|^{-2}.$$

Is this partition function invariant under modular transformations? To answer this it is sufficient to find the behaviour under the two generating transformations T and S. The result is

$$\eta(\tau + 1) = e^{\frac{i\pi}{12}}\eta(\tau), \quad \eta\left(-\frac{1}{\tau}\right) = \sqrt{-i\tau}\,\eta(\tau).$$

The T-transformation follows straight from the definition, while the S-transformation requires to use the *Poisson resummation formula*.[3]

---

[2] See Appendix D.1.
[3] See Section 13.2.8, where we perform a similar computation for T-duality, which is an $SL(2, \mathbb{Z})$ transformation acting on space-time.

Thus,

$$Z'(\tau + 1, \bar{\tau} + 1) = Z'(\tau, \bar{\tau}), \quad Z'\left(-\frac{1}{\tau}, -\frac{1}{\bar{\tau}}\right) = \frac{1}{|\tau|}Z'(\tau, \bar{\tau}), \qquad (6.12)$$

and the partition function is not modular invariant. To obtain a modular invariant partition function we need to add further degrees of freedom. In string theory a modular invariant partition function is non-optional, since modular transformations are part of the reparametrizations, under which the theory must be invariant. In the case at hand it is obvious that we forgot to include the degrees of freedom corresponding to the centre of mass motion of the string, so the lack of modular invariance reminds us to include them. Note that the creation and annihiliation operators α_m, $m \neq 0$ associated with the conformal primaries $i\partial_z X$ and $i\partial_{\bar{z}} X$ only account for the oscillations of the string but not for its centre of mass motion. We have included the momentum operator $p = 2\alpha_0$ but, since the conjugate operator x of p is missing, we only have states with zero momentum in our Hilbert space.

To include the momentum eigenstates we compute

$$|k\rangle := e^{ikx}|0\rangle \Rightarrow p|k\rangle = k|k\rangle \Rightarrow q^{L_0}\bar{q}^{\bar{L}_0}|k\rangle = (q\bar{q})^{\frac{1}{8}p^2}|k\rangle = e^{-\frac{\pi}{2}\tau_2 k^2}|k\rangle. \qquad (6.13)$$

Since the momentum operator has continuous eigenvalues, the 'trace' over its eigenstates is an integral

$$\text{tr}\left(e^{-\frac{\pi}{2}\tau_2 k^2}\right) \simeq \int_{-\infty}^{\infty} dk\, e^{-\frac{\pi}{2}\tau_2 k^2} \simeq \tau_2^{-1/2}.$$

We have ignored constant factors, which are irrelevant for the modular transformation properties. Adding the integral over momentum eigenstates to our partition functions, we obtain

$$Z(\tau, \bar{\tau}) = \tau_2^{-1/2}|\eta(\tau)|^{-2}, \qquad (6.14)$$

which is modular invariant.

Exercise 6.2.1 Verify the transformations (6.12) and show that (6.14) is modular invariant.

PART III

THE SPACE-TIME PERSPECTIVE

7 Covariant Quantisation I

7.1 Outline of Covariant Quantisation

We now return from the world-sheet perspective to the space-time perspective and formulate the quantum theory of open and closed strings. We will follow a Lorentz covariant approach, where no gauge fixing conditions are imposed apart from the conformal gauge. The price that we have to pay for manifest Lorentz covariance is that states are subject to residual gauge transformations. Covariant quantisation proceeds in three steps:

1. We construct a representation space $\mathcal{F}$ for the canonical commutation relations of the string coordinates $X^\mu(\sigma)$ and their conjugate momenta. The space $\mathcal{F}$, which we will call the Fock space, carries a scalar product between states which is Lorentz invariant but indefinite.
2. We impose the constraints $L_m = 0$, $\tilde{L}_m = 0$, which defines the subspace $\mathcal{F}_{\text{phys}} \subset \mathcal{F}$ of physical states. The operators $L_0, \tilde{L}_0$ have an ordering ambiguity, which we parametrise by a constant a, corresponding to a shift of the ground state energy.
3. We identify any two states which differ by a null state with one another, since adding null states corresponds to the residual gauge transformations we still have in the conformal gauge. The resulting quotient space $\mathcal{H} = \mathcal{F}_{\text{phys}}/\sim$ is the Hilbert space of states.

To obtain a well defined quantum theory, the indefinite scalar product on $\mathcal{F}$ must become positive semi-definite by restriction to $\mathcal{F}_{\text{phys}}$, so that by dividing out null states we obtain a positive definite scalar product on $\mathcal{H}$. The *no ghost theorem* asserts that this happens if and only if the dimension of space-time is $D = 26$, and also fixes the shift of the ground state energy, $a = 1$.

7.2 The Fock Space

Let us recapitulate where we were when we left the space-time perspective at the end of Section 3.3. The canonical commutation relations between the modes of a closed string are

$$[x^\mu, p^\nu] = i\eta^{\mu\nu}, \quad [\alpha_m^\mu, \alpha_n^\nu] = m\delta_{m+n,0}\eta^{\mu\nu}, \quad [\tilde{\alpha}_m^\mu, \tilde{\alpha}_n^\nu] = m\delta_{m+n,0}\eta^{\mu\nu}, \tag{7.1}$$

with Hermiticity properties $(x^\mu)^\dagger = x^\mu$, $(p^\mu)^\dagger = p^\mu$, $(\alpha_m^\mu)^\dagger = \alpha_{-m}^\mu$, $(\tilde{\alpha}_m^\mu)^\dagger = \tilde{\alpha}_{-m}^\mu$. The ground state $|0\rangle$ for the Fock space $\mathcal{F}$ is defined by[1]

$$p^\mu|0\rangle = 0, \quad \alpha_m^\mu|0\rangle = 0, \quad \tilde{\alpha}_m^\mu|0\rangle = 0 \quad \text{for} \quad m > 0. \tag{7.2}$$

For momentum eigenstates

$$p^\mu|k\rangle = k^\mu|k\rangle.$$

where $k = (k^\mu) \in \mathbb{R}^D$, the scalar product is

$$\langle k|k'\rangle = \delta^D(k - k').$$

Momentum eigenstates are not normalisable. Normalisable states are obtained by forming square-integrable superpositions. String oscillations are described by harmonic oscillators, one for each space-time dimension and frequency. The standard basis which we choose for the oscillation states is

$$\alpha_{-m_1}^{\mu_1}\alpha_{-m_2}^{\mu_2}\cdots|0\rangle, \quad \mu_j = 0, \ldots, D-1, \quad 0 < m_1 \leq m_2 \leq \cdots.$$

Since the creation and annihilation operators transform as Lorentz vectors, the scalar product between states is indefinite (see (3.46)). This indefiniteness is a necessary consequence of manifest Lorentz invariance. However, this is no immediate problem since only the Hilbert space $\mathcal{H} = \mathcal{F}_{\text{phys}}/\sim$ needs to be positive definite.

Combining momentum and oscillation states, we obtain the following basis for $\mathcal{F}$:[2]

$$\mathfrak{B} = \left\{ \alpha_{-m_1}^{\mu_1}\alpha_{-m_2}^{\mu_2}\cdots|k\rangle \in \mathcal{F} \mid k \in \mathbb{R}^D, \mu_i = 0, \ldots, D-1, m_j = 1, 2, \ldots \right\}$$

7.3 Implementation of the Constraints

The constraints of the classical theory are $L_m = 0$ for open strings, and $L_m = \tilde{L}_m = 0$ for closed strings. In the following we will consider closed strings for definiteness, but for simplicity we will suppress the right-moving oscillators $\tilde{\alpha}_n^\nu$ unless they are relevant. Note that one chiral sector of a closed string looks similar to the open string, but as we have seen the contributions of oscillators to the mass differ by a numerical factor.

Classically, the constraints are conditions that we impose on solutions of the dynamical equations as initial conditions. In the quantum theory the constraints select a subspace $\mathcal{F}_{\text{phys}} \subset \mathcal{F}$, called the space of *physical states*. For $m \neq 0$ we impose

$$|\phi\rangle \in \mathcal{F}_{\text{phys}} \subset \mathcal{F} \Rightarrow L_m|\phi\rangle = 0, \ m > 0. \tag{7.3}$$

Since $L_m^\dagger = L_{-m}$, this is implies that on the subspace $\mathcal{F}_{\text{phys}}$ the operators L_m are *weakly zero* in the sense that their matrix elements vanish between physical states:

$$\langle\phi_1|L_m|\phi_2\rangle = 0, \quad \text{for} \quad |\phi_1\rangle, |\phi_2\rangle \in \mathcal{F}_{\text{phys}}. \tag{7.4}$$

---

[1] Compared to Part II we see that $|0\rangle$ is the $\mathfrak{sl}(2, \mathbb{C})$ vacuum of the world-sheet conformal field theory.
[2] For closed strings we have to add the second set $\tilde{\alpha}_n^\nu$ of oscillators.

The operator L_0 is special, not only because it is Hermitian, $L_0^\dagger = L_0$, but also because it has an *ordering ambiguity*. The normal ordering of a monomial consisting of creation and annihilation operators amounts to putting all annihilation operators to the right. For products of α_m^μ this means

$$: \alpha_m^\mu \alpha_n^\nu := \begin{cases} \alpha_m^\mu \alpha_n^\nu, & \text{if } m < 0, n > 0, \\ \alpha_n^\nu \alpha_m^\mu, & \text{if } m > 0, n < 0, \\ \text{any ordering}, & \text{else}. \end{cases}$$

Since α_m^μ and α_n^ν commute unless $m + n = 0$, classical ordering (CO) and normal ordering (NO) agree for L_m with $m \neq 0$:

$$L_m^{\text{CO}} = \frac{1}{2} \sum_{n=-\infty}^{\infty} \alpha_{m-n} \cdot \alpha_n = \frac{1}{2} \sum_{n=-\infty}^{\infty} : \alpha_{m-n} \cdot \alpha_n = L_m^{\text{NO}} = L_m.$$

For L_0 we need to be careful. The normal ordered version of L_0 is

$$L_0^{\text{NO}} = \frac{1}{2} \sum_{n=-\infty}^{\infty} : \alpha_{-n} \cdot \alpha_n := \frac{1}{8} p^2 + N, \tag{7.5}$$

where we used that $\frac{1}{2}\alpha_0 \cdot \alpha_0 = \frac{1}{8}p^2$ for closed strings, and defined the *number operator*

$$N := \sum_{n=1}^{\infty} \alpha_{-n} \cdot \alpha_n.$$

Comparing this to the classical ordering

$$L_0^{\text{CO}} = \frac{1}{2} \sum_{n=-\infty}^{\infty} \alpha_{-n} \cdot \alpha_n = \frac{1}{8} p^2 + \frac{1}{2} \sum_{n=1}^{\infty} (\alpha_n \cdot \alpha_{-n} + \alpha_{-n} \cdot \alpha_n), \tag{7.6}$$

we find that the difference is a divergent series:

$$L_0^{\text{CO}} - L_0^{\text{NO}} = \frac{D}{2} \sum_{n=1}^{\infty} n. \tag{7.7}$$

Since $L_0 + \tilde{L}_0$ is the closed string world-sheet Hamiltonian, this is an ambiguity of the ground state energy. The standard choice made in quantum field theory in a Minkowski space-time is to impose normal ordering. Then the ground state energy is zero, which is the only value consistent with Poincaré invariance of the ground state. If one considers quantum field theory in a finite volume, or quantum field theory on a curved space-time, one needs to be more careful. The world-sheet cylinder has finite spatial extension, and in this case the ground state energy of a quantum field theory is, in general, volume dependent. This is the famous *Casimir effect*.

We will compute the ground state energy later using light-cone quantisation. Here we parametrise a possible shift by introducing two constants $a, \tilde{a}$:

$$|\phi\rangle \in \mathcal{F}_{\text{phys}} \Rightarrow (L_0 - a)|\phi\rangle = 0 = (\tilde{L}_0 - \tilde{a})|\phi\rangle. \tag{7.8}$$

These constraints can be re-organised into

$$(L_0 + \tilde{L}_0 - a - \tilde{a})|\phi\rangle = 0, \quad (L_0 - \tilde{L}_0 - a + \tilde{a})|\phi\rangle = 0.$$

In the first relation the shift of $L_0 + \tilde{L}_0$ corresponds to a shift in the spectrum of the world-sheet Hamiltonian, which we expect since the world-sheet has finite spatial extension. Since $L_0 - \tilde{L}_0$ is the world-sheet momentum operator, a shift of its spectrum by $a - \tilde{a} \neq 0$ would correspond to a string ground state which is not translation invariant on the world-sheet, that is, dependent on the choice of the origin of the spatial coordinate σ^1. This is unphysical, and we conclude that in a physically meaningful string theory $a = \tilde{a}$.

Taking into account all constraints, we arrive at the following definition of physical states for the closed string:

$$|\phi\rangle \in \mathcal{F}_{\text{phys}} \Leftrightarrow \begin{cases} (L_0 - a)|\phi\rangle = 0, & (\tilde{L}_0 - a)|\phi\rangle = 0, \\ L_m|\phi\rangle = 0, m > 0, & \tilde{L}_m|\phi\rangle = 0, m > 0. \end{cases} \tag{7.9}$$

Exercise 7.3.1 Verify the expressions (7.5), (7.6), and (7.7) for the normally and classically ordered forms of L_0 and their difference.

7.4 Mass Eigenstates

We will now analyse the physical meaning of the constraints related to $L_0, \tilde{L}_0$. We use that $L_0 = \frac{1}{8}p^2 + N$, where the number operator N satisfies

$$[N, \alpha^{\mu}_{-m}] = m\alpha^{\mu}_{-m}, \quad m > 0.$$

Our standard basis $\mathcal{B}$ consists of eigenstates of N:

$$N\alpha^{\mu_1}_{-m_1}\alpha^{\mu_2}_{-m_2}\cdots|k\rangle = (m_1 + m_2 + \cdots)\alpha^{\mu_1}_{-m_1}\alpha^{\mu_2}_{-m_2}\cdots|k\rangle.$$

We will use the same symbol N for the number operator and its eigenvalues, called *total excitation numbers*. Now we impose the L_0-constraint on a physical state $|\phi\rangle \in \mathcal{F}_{\text{phys}}$ which we take, without loss of generality, to be an eigenstate of p and N:

$$(L_0 - a)|\phi\rangle = 0 \Rightarrow \left(\frac{1}{8}p^2 + N\right)|\phi\rangle = |\phi\rangle \Rightarrow \frac{1}{8}k^2 + N = a. \tag{7.10}$$

Using $k^2 = -M^2$, and also imposing $(\tilde{L}_0 - a)|\phi\rangle = 0$, we obtain

$$\frac{1}{8}M^2 = N - a = \tilde{N} - a.$$

This can be reorganised into:

- The *mass shell condition*:
$$\frac{1}{4}M^2 = N + \tilde{N} - 2a.$$

- The *level matching condition*:
$$N = \tilde{N}.$$

The level matching condition implies that for a closed string left- and right-moving excitations contribute equally to the mass. This is a consequence of translation invariance in σ^1.

Let us restore the units of mass and length in the mass formula. Conventionally the mass formula is not presented using the string tension but the so-called *Regge slope parameter*

$$\alpha' = \frac{1}{2\pi T}. \tag{7.11}$$

Then the mass shell condition for closed strings is:

$$\alpha' M^2 = 2(N + \tilde{N} - 2a). \tag{7.12}$$

Note that α' has dimension length-squared. *String units* are defined by $\pi T = 1 \Leftrightarrow \alpha' = \frac{1}{2}$.

We will now turn to a more detailed investigation of the properties of physical states. For simplicity this will be done for the open string. We will return to the closed string spectrum in Chapter 9.

7.5 Physical States of the Open String

The above analysis can be repeated for the open string. We omit the details. Using that for open strings $\alpha_0^\mu = p^\mu$ and $L_0 = \frac{1}{2}p^2 + N$, the constraint $(L_0 - a)|\phi\rangle = 0$ gives rise to the open string mass shell condition:

$$\alpha' M^2 = N - a. \tag{7.13}$$

The mass eigenstates with $N = 0, 1, 2$ are:

N	$\alpha' M^2$	State	Representation of Lorentz group
0	$-a$	$|k\rangle$	Scalar
1	$-a + 1$	$\alpha_{-1}^\mu |k\rangle$	Vector
2	$-a + 2$	$\alpha_{-2}^\mu |k\rangle$	Vector
		$\alpha_{-1}^\mu \alpha_{-1}^\nu |k\rangle$	Symmetric Tensor

The state $|k\rangle$ with $N = 0$ contains no oscillations and is the ground state of the string. Note that this state is different from the $\mathfrak{sl}(2, \mathbb{C})$ vacuum of the conformal field theory, because it carries momentum. The higher we go in N, the more options we have to build states using creation operators. The number of states is related to the problem of *partitions* of the integer N. For example, $N = 2 = 1 + 1$ has two partitions while $N = 4 = 3 + 1 = 2 + 2 = 2 + 1 + 1 = 1 + 1 + 1+$ has five partitions. We will explore this further in Section 11.2. Representations of the Lorentz group have been listed in the last column.

Since we still have to impose the constraints $L_m|\phi\rangle = 0$, $m > 0$ we do not yet know which of the mass eigenstates are physical. Also, without being able to fix a

we do not quite know the mass spectrum. By inspection, there are two distinguished choices for a:

1. For $a = 0$ the spectrum of M^2 is non-negative. This is what we want in a physical theory. However we expect (and will verify later) that there is a Casimir effect which results in $a \neq 0$.
2. For $a = 1$ the state with $N = 1$, which is a Lorentz vector, is massless. We will show below that this state is photon-like, that is, it is a massless vector with the corresponding gauge symmetry. A later computation will show that the Casimir effect indeed implies that $a = 1$, provided that the number of space-time dimensions is $D = 26$. As a consequence the state with $N = 0$ is a *tachyon*, a state with $M^2 < 0$.

From now on, we will work with the value $a = 1$. We will first show that the $N = 1$ state has the kinematical properties of a photon. Afterwards, we will come back to the issues raised by the ground state being a tachyon.

7.6 The Photon

For the tachyon $|k\rangle$ the conditions $L_m|\phi\rangle = 0, m > 0$ are satisfied automatically. This is different for $N = 1$. Let us consider a general linear combination of level-1 basis states, $|\phi\rangle = \zeta_\mu \alpha^\mu_{-1}|k\rangle$, where the vector ζ encodes the polarisation of the state. Given that $a = 1$, the L_0-constraint gives $k^2 = 0$. The other constraints $L_m|\phi\rangle = 0$ imply (see Exercise 7.6.1) that the polarisaton vector ζ must be Lorentz-orthogonal to the momentum vector, so that the complete set of conditions for level-1 states is

$$k^2 = 0 , \quad k^\mu \zeta_\mu = 0. \tag{7.14}$$

One special type of solution for the polarisation vector is $\zeta_\mu = \lambda k_\mu$. The corresponding states

$$|\psi\rangle = \lambda k_\mu \alpha^\mu_{-1}|k\rangle, \quad \lambda = \text{const.}$$

are null states, as we will show now.

- Firstly, their norm is zero. Formally:

$$\langle\psi|\psi\rangle = \lambda^2 k_\mu k_\nu \langle k|[\alpha^\mu_1, \alpha^\nu_{-1}]|k\rangle = \lambda^2 k^2 \langle k|k\rangle = 0, \tag{7.15}$$

since $k^2 = 0$. Since $\langle k|k'\rangle = \delta^D(k - k')$ on actually needs to verify this using normalisable 'wave packages' (see Exercise 7.6.2).

- More generally, states of the form $|\psi\rangle$ are orthogonal to all physical states. Since states at different level are automatically orthogonal, we only need to check this for physical level-one states:

$$\langle k'|\alpha^\mu_1 \zeta_\mu k_\nu \alpha^\nu_{-1}|k\rangle = \zeta \cdot k \langle k'|k\rangle = 0, \tag{7.16}$$

since, (very) formally, 'either $k \neq k'$ or $k = k'$ and therefore $\zeta \cdot k = 0$'. Like in the previous case, a proper derivation is obtained using normalisable states (see Exercise 7.6.2).

Any two physical states which differ by a null state are gauge equivalent:

$$\zeta_\mu \alpha_{-1}^\mu |k\rangle \sim \zeta_\mu \alpha_{-1}^\mu |k\rangle + \lambda k_\mu \alpha_{-1}^\mu |k\rangle.$$

In summary, level-1 states satisfy the following kinematical relations:

$$k^2 = 0, \quad \zeta_\mu k^\mu = 0, \quad \zeta_\mu \sim \zeta_\mu + \lambda k_\mu. \qquad (7.17)$$

We will now show that these are precisely the relations which characterise a 'photon', that is a massless vector particle with a local gauge symmetry. The Maxwell equations in an arbitrary number of space-time dimensions are:

$$\partial^\mu F_{\mu\nu} = j_\nu, \qquad (7.18)$$

$$\epsilon^{\mu_1\mu_2\mu_3\cdots} \partial_{\mu_1} F_{\mu_2\mu_3} = 0. \qquad (7.19)$$

The homogeneous Maxwell equations (7.19) are *Bianchi identities* (integrability conditions), which guarantee the local existence of a vector potential A_μ for the field strength tensor $F_{\mu\nu}$:

$$F_{\mu\nu} = \partial_\mu A_\nu - \partial_\nu A_\mu.$$

The inhomogeneous Maxwell equations (7.18) are dynamical equations which determine the field strength in terms of the electromagnetic current j_μ. They are the Euler–Lagrange equations of the Maxwell Lagrangian

$$S[A] = \int d^D x \left(-\frac{1}{4} F_{\mu\nu} F^{\mu\nu} - j_\mu A^\mu \right). \qquad (7.20)$$

The field strength $F_{\mu\nu}$ is invariant under local gauge transformations of the vector potential,

$$A_\mu \to A_\mu + \partial_\mu \chi \Rightarrow F_{\mu\nu} \to F_{\mu\nu}. \qquad (7.21)$$

Using the existence of a vector potential, the free Maxwell equations ($j_\nu = 0$) become

$$\partial^\mu F_{\mu\nu} = \Box A_\mu - \partial_\mu \partial^\nu A_\nu = 0.$$

This can be simplified by imposing the Lorenz gauge $\partial^\nu A_\nu = 0$, which is a Lorentz covariant gauge:[3]

$$\Box A_\mu = 0, \quad \text{if} \quad \partial^\nu A_\nu = 0.$$

The Lorenz gauge does not fix the gauge freedom for A_μ completely. There is a residual gauge symmetry:

$$\partial^\nu A_\nu \to \partial^\nu (A_\nu + \partial_\nu \chi) = \partial^\nu A_\nu + \Box \chi = \partial^\nu A_\nu, \quad \text{if} \quad \Box \chi = 0.$$

This looks increasingly similar to the properties of the level-1 state of the open string. To make the relation explicit, we switch from position space to momentum space. The vector potential can be expanded in plane waves (=momentum eigenstates):

$$A_\mu(x) = \int d^D k \, \zeta_\mu(k) e^{ik\cdot x} + \text{complex conjugate}.$$

---

[3] No misspelling here. Lorentz transformations have been named after Hendrik Lorentz, the Lorenz gauge after Ludvig Lorenz.

Now we can translate the relations imposed on $A_\mu(x)$ to momentum space:

$$\Box A_\mu = 0 \Leftrightarrow k^\mu k_\mu = 0,$$
$$\partial^\mu A_\mu = 0 \Leftrightarrow \zeta^\mu k_\mu = 0,$$
$$A_\mu \simeq A_\mu + \partial_\mu \chi \Leftrightarrow \zeta_\mu \simeq \zeta_\mu + \lambda k_\mu. \tag{7.22}$$

Thus the kinematical relations characterising a free photon in a covariant gauge are precisely the same as for the level-1 state of the open string.

An explicit decomposition of A_μ or ζ_μ into unphysical, physical, and null (pure gauge) components is possible, but requires us to make an explicit choice for the momentum vector. We choose a momentum vector corresponding to a plane wave/photon momentum eigenstate moving in the positive $(D-1)$-direction:

$$k = (k^0, 0, \ldots, 0, k^0).$$

This choice breaks manifest Lorentz symmetry but is still invariant under the subgroup $SO(D-2) \subset SO(1, D-1)$ corresponding to rotations around $\vec{k} = (0, \ldots, 0, k^0)$.[4] We can now work out the explicit form of physical polarisation vectors:

$$k \cdot \zeta = 0 \Rightarrow -k^0 \zeta^0 + k^0 \zeta^{D-1} = 0 \Rightarrow \zeta^0 = \zeta^{D-1} \tag{7.23}$$

$$\Rightarrow \zeta = (\zeta^0, \zeta^1, \ldots, \zeta^{D-2}, \zeta^0).$$

Since the residual gauge symmetry is $\zeta \to \zeta + \lambda k$, we can decompose ζ into a transversal part and a pure gauge part:

$$\zeta = \zeta_{\text{transv}} + \zeta_{\text{gauge}} = (0, \zeta^1, \ldots, \zeta^{D-2}, 0) + (\alpha, 0, \cdots, 0, \alpha),$$

where α is arbitrary. To fix the gauge completely, we need to choose a value for α, the natural choice beging $\alpha = 0$.

This example illustrates that a complete gauge fixing in a Lorentz covariant gauge theory breaks manifest Lorentz covariance. In the free Maxwell theory the choice $\alpha = 0$ corresponds to imposing the Coulomb gauge on top of the Lorenz gauge. Using the notation $V = (V^\mu) = (V^0, \vec{V}) = (V^0, V^i)$ for Lorentz vectors, we have

$$\partial^\mu A_\mu = 0 \quad \Leftrightarrow \quad k^\mu \zeta_\mu = 0,$$
$$\nabla \cdot \vec{A} = \partial^i A_i = 0 \quad \Leftrightarrow \quad \vec{k} \cdot \vec{\zeta} = k^i \zeta_i = 0.$$

For a plane wave along the $(D-1)$–direction this implies

$$\vec{k} = (0, 0, \cdots, k^0) \Rightarrow \vec{\zeta} = (\zeta^1, \zeta^2, \cdots, \zeta^{D-2}, 0),$$

which indeed corresponds to $\alpha = 0$. In this gauge $\vec{k}$ and $\vec{\zeta}$ are orthogonal, and we see explicitly that photons have transversal polarisation.

Let us consider the case $D = 4$, which has some special features. Here the subgroup of transversal rotations is $SO(2) \subset SO(1, 3)$, which is abelian. Under this group transversal polarisation vectors $\zeta_{\text{transv}} = (\zeta^1, \zeta^2)$ transform as

---

[4] Actually, the invariance group can be shown to be a subgroup $E(D-2) \subset SO(1, D-1)$, which is isomorphic to the $(D-2)$-dimensional Euclidean group $SO(2) \ltimes \mathbb{R}^2$, see Weinberg (1995) or Sexl and Urbantke (2001).

$$\begin{pmatrix} \zeta^1 \\ \zeta^2 \end{pmatrix} \to \begin{pmatrix} \cos\varphi & -\sin\varphi \\ \sin\varphi & \cos\varphi \end{pmatrix} \begin{pmatrix} \zeta^1 \\ \zeta^2 \end{pmatrix}.$$

By taking complex linear combinations, corresponding to a circular polarisation basis, we obtain a pair of complex conjugated numbers

$$\zeta^\pm = \zeta^1 \pm i\zeta^2,$$

which transform as

$$\begin{pmatrix} \zeta^+ \\ \zeta^- \end{pmatrix} \to \begin{pmatrix} e^{i\varphi} & 0 \\ 0 & e^{-i\varphi} \end{pmatrix} \begin{pmatrix} \zeta^+ \\ \zeta^- \end{pmatrix}.$$

Mathematically, we have used the group isomorphism $SO(2) \simeq U(1)$, which allows us to diagonalise rotation matrices over the complex numbers. States which transform with eigenvalue $e^{ih\varphi}$ are said to have *helicity h*. Thus, the circular polarisation states of a photon have helicity $h = \pm1$.

Exercise 7.6.1   Verify that the physical state conditions for a general level-1 open string state $\zeta_\mu \alpha^\mu_{-1} |k\rangle$ imply (7.14), and that this is sufficient to satisfy all physical state conditions.

Exercise 7.6.2   Justify the relations (7.15) and (7.16) using normalisable states ('wave packets') instead of momentum eigenstates. *Hint:* Adapt the treatment of the relativistic particle and scalar field from Chapter 3. The result is, essentially, the covariant ('Gupta–Bleuler') quantisation of the photon field.

7.7 The Tachyon

While it is nice that for $a = 1$ the open string spectrum contains a photon, we should comment on the presence of a state of negative M^2, called the tachyon. While naively negative M^2 corresponds to a particle propagating faster than the speed of light, the proper interpretation is given by quantum field theory. Consider the action

$$S[\phi] = \int d^D x \left(-\frac{1}{2} \partial_\mu \phi \partial^\mu \phi - V(\phi) \right) \tag{7.24}$$

for a real scalar field ϕ, where V is a polynomial in ϕ. To identify the particle spectrum of the theory, we first need to identify a ground state, and then apply perturbative quantisation, that is, we expand around the ground state and treat higher order terms as perturbations. For the above example, ground states correspond to setting $\phi = $ const. and then choosing the constant value to minimise V:

$$\phi(x) = \phi_* = \text{constant}, \quad \text{where} \quad V'(\phi_*) := \frac{dV}{d\phi}(\phi_*) = 0, \quad V''(\phi_*) \geq 0.$$

For general V one needs to distinguish local and global minima. Local minima are unstable against thermal or quantum tunnelling, though they may be long-lived. For simplicity, we assume that there is a single, non-degenerate minimum. We expand around the minimum $\phi = \phi_*$ and obtain

$$V = V(\phi_*) + \frac{1}{2} V''(\phi_*)(\phi - \phi_*)^2 + \cdots$$

where the omitted terms are of higher than second order in $(\phi - \phi_*)$. ϕ_* is called the *vacuum expectation value* or *condensate* of ϕ, also denoted $\langle \phi \rangle$. We can define a shifted field $\varphi = \phi - \phi_*$ with vanishing expectation value. Replacing ϕ by φ the action (7.24) takes the standard form

$$S[\varphi] = \int d^D x \left(-\frac{1}{2} \partial_\mu \varphi \partial^\mu \varphi - \frac{1}{2} m^2 \varphi^2 - V_I(\varphi) \right) \qquad (7.25)$$

of the action of a real scalar field with mass m, where $m^2 = V''(\phi_*)$, and where the potential V_I contains terms of order higher than two in φ.[5] Thus the mass-squared of the particle associated with the field ϕ is given by the curvature of its potential at the minimum.

Consider now the following example for $V(\phi)$:

$$S[\phi] = \int d^D x \left(-\frac{1}{2} \partial_\mu \phi \partial^\mu \phi + \frac{1}{2} \mu^2 \phi^2 - \frac{\lambda}{2} \phi^4 \right). \qquad (7.26)$$

This does not describe a particle with $m^2 = -\mu^2 < 0$. The function V has three extrema: a maximum at $\phi = 0$ and two degenerate minima at values $\phi = \pm \phi_*$, where $\phi_* > 0$. Since we find two degenerate minima, this theory shows *spontaneous symmetry breaking*: the Lagrangian is invariant under $\phi \to -\phi$, while the ground state is not. For perturbative quantisation we have to choose one ground state to expand around, say, ϕ_*.[6] Defining $\varphi = \phi - \phi_*$ we find

$$S[\varphi] = \int d^D x \left(-\frac{1}{2} \partial_\mu \varphi \partial^\mu \varphi - \mu^2 \varphi^2 - V_I(\varphi) \right). \qquad (7.27)$$

Thus the theory describes a particle of mass m, where $m^2 = 2\mu^2 \geq 0$.

Exercise 7.7.1   Find the ground states $\pm \phi_*$ of the action (7.26), show that by expanding around the minimum at ϕ_* one obtains the action (7.27) and determine the interaction terms.

Something similar happens for the open string tachyon. While most results in string theory have been obtained using the 'first quantised' world-sheet formalism, the proof of *tachyon condensation* is an application of *string field theory*. A string field Φ is, roughly speaking, a collection of infinitely many conventional quantum fields, one for each particle-like excitation of the string, starting with the tachyon field T. Computation of the tachyon potential $V(T)$ shows that $T = 0$ is a local maximum, and that there exists a minimum at some value $T = T_* \neq 0$. Moreover, it has been shown that the ground state of the open string is the closed string, in the following sense: firstly, quantum theories of open strings always contain closed strings. This follows from the observation that open string loop diagrams contain intermediate states which are closed strings. This can be shown explicitly by computing open string scattering amplitudes. Here, we will, instead, give a plausibility argument based on an open string world-sheet, which allows us to visualise the intermediate closed

---

[5] We have assumed that $V(\phi_*) = 0$. Constant terms contribute to the cosmological constant when the theory is coupled to gravity.

[6] Note that there might be interesting non-perturbative effects which are not captured by the perturbative expansion around one of the classical ground states. This field theory is similar to the anharmonic oscillator, where non-perturbative effects modify the ground state energy and lift its degeneracy.

string states: an annulus, which is an open string one-loop diagram, can also be interpreted as a cylinder, which is a closed string propagator. Secondly, open strings with Neumann boundary conditions correspond to the presence of a space-filling D-25-brane. Thus instability of the open string vacuum is equivalent to the instability of the D-25 brane, and once the D-25 brane has decayed, open strings have disappeared and only closed strings are left. The proof of tachyon condensation shows that the tension (energy per world-volume) of the D-25-brane exactly corresponds to the difference of the tachyon potential between the local maximum and the minimum.

Given that bosonic open string theory is unstable, why did we bother to show that the level-1 open string state is a photon? The open and closed string theories that we have defined so far only have states in tensor representations of the Lorentz group, that is, bosonic states. Hence they are called *bosonic string theories*. To describe matter we need to generalise them to strings with fermionic excitations, which ultimately leads us to *superstrings*. Like bosonic strings, superstrings contain massless excitations which can be interpreted as photons and gravitons, but for them these are ground states, and tachyons are absent. We will, therefore, continue to study bosonic string theories, and focus on those aspects which they share with superstrings, which will be introduced and discussed in Chapter 14.

What Have We Learned So Far?

At this point, we have seen that the L_0 and $\tilde{L}_0$-constraints allow us to write down the mass spectrum for the open and closed string. Given that $a = \tilde{a} = 1$, the lowest mass eigenstates are listed in Tables 7.1 and 7.2. Since we are using a

Table 7.1. Lowest mass eigenstates of the open string			
N	$\alpha'M^2$	State	Representation of the Lorentz group
0	-1	$\|k\rangle$	Scalar
1	0	$\alpha^\mu_{-1}\|k\rangle$	Vector
2	1	$\alpha^\mu_{-2}\|k\rangle$	Vector
		$\alpha^\mu_{-1}\alpha^\nu_{-1}\|k\rangle$	Symmetric Tensor

Table 7.2. Lowest mass eigenstates of the closed string			
$N = \tilde{N}$	$\alpha'M^2$	State	Representation of Lorentz group
0	-4	$\|k\rangle$	Scalar
1	0	$\alpha^\mu_{-1}\tilde{\alpha}^\nu_{-n}\|k\rangle$	2nd rank tensor
2	4	$\alpha^\mu_{-1}\alpha^\nu_{-1}\tilde{\alpha}^\rho_{-1}\tilde{\alpha}^\sigma_{-1}\|k\rangle$	4th rank tensor
		$\alpha^\mu_{-1}\alpha^\nu_{-1}\tilde{\alpha}^\rho_{-2}\|k\rangle$	3rd rank tensor
		$\alpha^\mu_{-2}\tilde{\alpha}^\rho_{-1}\tilde{\alpha}^\sigma_{-1}\|k\rangle$	3rd rank tensor
		$\alpha^\mu_{-2}\tilde{\alpha}^\rho_{-2}\|k\rangle$	2nd rank tensor

covariant quantisation scheme, the states organise themselves into representations of the Lorentz group. To identify the particle content we have to impose the remaining constraints. As we have seen for the massless open string state, these then determine the spin or helicity of the state. To make this analysis systematic, we will review the relation between one-particle states and unitary irreducible representations of the Poincaré group in Chapter 8, before looking at massless closed string states and massive open and closed string states in Chapter 9.

7.8 Literature

In this chapter, we have touched upon *string field theory* when discussing tachyon condensation. String field theory is less developed and more complicated than the world-sheet formulation, and literature is relatively sparse. There is the monograph Siegel (1988) and two very recent books, one introductory, Erbin (2021), and one exploring the algebraic structure in detail, Doubek et al. (2020). Moreover, there are the excellent lecture notes Erler (2019), Erler (2020). Some string theory textbooks, including Kaku (1988), Kaku (1991), Polchinski (1998a), Andu Becker et al. (2007) and West (2012) contain sections on string field theory. For tachyon condensation, see Sen (1998), Berkovits et al. (2000), and Erler (2013).

8 Intermezzo – Representations of the Poincaré Group

8.1 Review of Representations of the Poincaré Group

Intuitively, an elementary particle is a localised object which is completely characterised by a small number of properties, including its mass, spin, momentum, energy and charges. Among these quantities, some depend on its state of motion and thus are *extrinsic*, while its mass, spin and charges are inherent or *intrinsic* properties which characterise the 'species' the particle belongs to. Mass and spin characterise how a particle behaves under Poincaré transformations, while charges are related to internal symmetries, which we will discard in this chapter. In quantum field theory 'particles', that is *one-particle states*, are identified by decomposing the Hilbert space into unitary, irreducible representations of the Poincaré group.

The Poincaré group is the semi-direct product of the Lorentz group with the translation group: while Lorentz transformations close among themselves, the generators of translations transform as vectors under the Lorentz group. The group of translations is not only abelian, but an *invariant abelian subgroup* of the Poincaré group. Therefore, the Poincaré group is not semi-simple, which makes its representation theory more complicated than the one of simple and semi-simple groups.[1]

By a classical result of Wigner, irreducible representations of the Poincaré group are determined by representations of certain subgroups, namely of the translation group and of the so-called *little group*. Translations are generated by the momentum operator P_μ. Its square $P^\mu P_\mu$ is a *Casimir operator* of the Poincaré group, that is it commutes with all generators. By *Schur's lemma*, it is, therefore, proportional to the unit operator on irreducible representations. Since

$$P^\mu P_\mu = -M^2,$$

we recognise that the invariant which labels representations of the translation subgroup is the square of the mass. Representations with $M^2 < 0$ are mathematically well defined, but would correspond to one-particle states which are tachyons. As discussed in Section 7.7, excitations of quantum fields with $M^2 < 0$ do not corre­spond to stable particles, and we therefore discard such tachyonic representations. This leaves us with *massive representations*, $M^2 > 0$, and *massless representations*, $M^2 = 0$.

The irreducible representations of an abelian group are one-dimensional. In our Fock space, irreducible representations of the translation group are spanned by momentum eigenstates $|k^\mu\rangle$ which in the position space representation correspond

[1] See Appendix G for the definition of (semi-)simplicity and other group-theoretical concepts.

to plane waves e^{ikx}. Since the spectrum of the momentum operator P_μ on Minkowski space is continuous, these are generalised eigenvectors, which are not normalisable. To obtain normalisable states, we have to form superpositions. Such 'direct integrals' of representations are given concretely by the familiar momentum space wave packets. The Fourier decomposition of wave functions or fields is a decomposition into irreducible representations of the translation group.

The second step is to deal with the Lorentz subgroup of the Poincaré group. Lorentz transformations act non-trivially on the momenta k^μ. Under the action of the Lorentz group, momenta organise themselves into *orbits*, which are partially classified by the invariant M^2. Since time-reversal and parity are not symmetries of all laws of nature, we only consider the connected part of the Lorentz group, which for non-spacelike vectors preserves the distinction between future-pointing and past-pointing vectors. This leads to the following classification of orbits:

1. Time-like future pointing orbits, with standard representative

$$(k^0, \vec{0}), \quad k^0 > 0.$$

   This corresponds to the choice of a *rest frame* where the momentum does not have spatial components.
2. Time-like past pointing orbits, with standard representative

$$(k^0, \vec{0}), \quad k^0 < 0.$$

   Following Section 3.1, these orbits are used to represent antiparticles (unless particle and antiparticle are the same).
3. Null future-pointing orbits

$$(k^0, \underline{0}, k^0), \quad k^0 > 0,$$

   where $\underline{0}$ is the zero vector in the $(D-2)$ directions transverse to the light-cone spanned by the first and last component. This is the standard form for the momentum of a massless particle propagating in the positive $(D-1)$-direction.
4. Null past-pointing orbits

$$(k^0, \underline{0}, k^0), \quad k^0 < 0.$$

   These orbits are used to represent massless antiparticles.
5. Space-like orbits

$$(0, \underline{0}, k^{D-1}).$$

   These orbits correspond to tachyonic representations which we discard.
6. The orbit $\{(0, \underline{0}, 0)\}$ of the zero vector, which represents the vacuum (zero particle state).

The choice of a standard representative for the momentum does not complete fix our freedom to perform Lorentz transformations. The subgroup $G_{\text{little}} \subset \mathrm{SO}(1, n-1)$ which preserves the representative is called the *isotropy group*, and in the context of representation theory, the *little group*. The little group of massive representations is isomorphic to the rotation group $\mathrm{SO}(D-1)$. For the standard representatives $(k^0, \vec{0})$ this is the subgroup $\mathrm{SO}(D-1) \subset \mathrm{SO}(1, D-1)$ of rotations acting on vectors $(k^0, \vec{k})$ with a fixed value of k^0. The choice of an $\mathrm{SO}(D-1)$ representation determines, together with the mass, a representation of the Poincaré group. In four dimensions representations of the massive little group SO(3) are labelled by one invariant, the spin. Thus, in four dimensions a massive one-particle state is characterised by its mass and spin. In higher dimensions the rotation subgroup has rank higher than one, and its representations are labelled by more than one invariant. We will use Young tableaux, which are briefly explained in Appendix E, to handle these representations.

For massless particles the little group is the $(D - 2)$-dimensional Euclidean group $SO(D-2) \ltimes \mathbb{R}^{D-2}$. This is a bit larger than the obvious invariance group of the standard representative $(k^0, \underline{0}, k^0)$. Since the little group is not semi-simple, one has to iterate the method. Only those representations where the Casimir operator of the translation subgroup $\mathbb{R}^{D-2}$ acts trivially turn out to be relevant for describing particles. In this case, the little group of the Euclidean group is $SO(D - 2)$, and the choice of an $SO(D-2)$ representation determines a massless representation of the Poincaré group. For terminological simplicity, we will refer to this $SO(D - 2)$ subgroup as the *little group of massless particles* or *helicity group* or *transverse rotation group*. For the standard representative $(k^0, \underline{0}, k^0)$, this is the subgroup $SO(D - 2) \subset SO(1, D - 1)$ of rotations acting on the transversal components $k^1, \ldots, k^{D-2}$. In four dimensions the little group of massless particles is the abelian group $SO(2) \cong U(1)$. Its irreducible representations are one-dimensional and labelled by a single integer, the helicity λ.[2] In dimensions $D > 4$ the group $SO(D - 2)$ is non-abelian and in dimensions $D > 5$ it has rank higher than one. We will again use Young tableaux to handle these representations.

We remark that we have restricted ourselves to *tensor representations*, which in four dimensions are the representations with integer spin or helicity. These describe bosons. There are also fermions, which in four dimensions belong to representations with half-integer spin or helicity. These *spinor representations* of the Poincaré group are, strictly speaking, not representations but *projective representations*, that is 'representations up to sign'. Equivalently, they are proper representations of double coverings of the Poincaré group and of the respective little groups. We will discuss spinor representations, for general dimension $D \geq 2$, in Chapter 14 where we will explain how to generalise string theory such that it has excitations which are fermions.

8.2 Group Theoretical Interpretation of the Photon State

Let us apply the theory reviewed above to the level-1 state or photon of the open string,

$$\zeta_\mu \alpha^\mu_{-1} |k\rangle, \quad k^2 = 0, \quad k\zeta = 0, \quad \zeta \simeq \zeta + \alpha k.$$

The L_0-constraint has fixed the mass to be zero. To determine the representation of the helicity group we need to decompose the vector representation of the Lorentz group $SO(1, D - 1)$ into representations of $SO(D - 2)$. This is an example of a

---

[2] Due to CPT invariance, in physical theories such representations must come in pairs which combine any λ with its negative.

so-called *branching rule*. For tensor representations of pseudo-orthogonal groups
$SO(p, q)$ *Young tableaux* are a convenient way to manipulate representations. A brief
summary of the notation and of the rules we will use in this book can be found
in Appendix E. By applying the branching rules for the chain $SO(1, D - 1) \supset$
$SO(D - 1) \supset SO(D - 2)$ of subgroups we find:

$$\square \rightarrow \square \oplus \bullet$$

$$\rightarrow \square \oplus \bullet \oplus \bullet$$

$$D = \underbrace{(D - 2)}_{\text{physical}} + \underbrace{1}_{\text{null}} + \underbrace{1}_{\text{unphysical}}$$

For general D a photon transforms as a vector of the helicity group. The physical state
condition and the residual gauge symmetry remove the two helicity scalars which are
also contained in the original Lorentz vector. In four dimensions the helicity group
$SO(2) \subset SO(1, 3)$ is abelian and its irreducible representations are one-dimensional:

$$\square \rightarrow \bullet \oplus \bullet \oplus \bullet \oplus \bullet$$

$$4 = \underbrace{1}_{h=1} + \underbrace{1}_{h=-1} + \underbrace{1}_{h=0} + \underbrace{1}_{h=0}$$

While the two physical helicity states $h = \pm 1$ are irreducible under $SO(2)$ they
combine into an irreducible two-dimensional representation of $O(2)$ once we include
reflections.

8.3 Virasoro Constraints, Poincaré Representations, and Effective Field Theory

Our discussion of the photon state illustrates a general point. The Virasoro constraints
which select the physical states assign to them their mass and their spin or helicity,
hence a representation of the Poincaré group. This fixes the linear part of the field
equations and hence the bilinear part of the space-time effective action for this
particle. For the photon we have seen explicitly how the conditions on momentum
and polarisation are related to the free Maxwell equations by Fourier transformation.
More generally, Lorentz covariant linear field equations, such as the Klein–Gordon,
Dirac, Maxwell, and Proca equations characterise representations of the Poincaré
group. That is, the free part of the field equations specifies the mass and spin or
helicity of the corresponding particle. In this way, we can translate information from
the world-sheet perspective (Virasoro constraints) into information form the space-
time perspective (mass and spin/helicity of string excitations, free part of the effective
action). To make contact with particle physics it is useful to express the information
about the most relevant, that is, lightest, string modes in terms of a space-time
effective field theory.

8.4 Literature

The representation theory of the Poincaré group and its application to particle physics and quantum field theory is explained in detail in Sexl and Urbantke (2001) and Woit (2017). See also Weinberg (1995) for a detailed treatment of massless particles. Hamermesh (1962) explains how Young tableaux work for orthogonal groups.

Covariant Quantisation II

We now turn to the closed string spectrum. The lowest mass eigenstates are listed in Table 7.2. Like for the open string, the closed string ground state is a tachyon, indicating that we have not identified the right ground state. For the closed bosonic string the true ground state is not known. We will ignore this issue and focus on the features that the closed bosonic string shares with tachyon-free closed superstring theories, in particular, the existence of a massless level-1 state which can be identified with the graviton.

9.1 The Graviton

The general form of a massless closed string state is

$$|\zeta, k\rangle = \zeta_{\mu\nu} \alpha_{-1}^{\mu} \tilde{\alpha}_{-1}^{\nu} |k\rangle. \tag{9.1}$$

Having imposed the mass shell condition and level matching, the only constraints which lead to additional conditions are

$$L_1 |\zeta, k\rangle = \tilde{L}_1 |\zeta, k\rangle = 0, \tag{9.2}$$

which imply

$$k^{\mu} \zeta_{\mu\nu} = 0 = \zeta_{\mu\nu} k^{\nu}. \tag{9.3}$$

We decompose $\zeta_{\mu\nu}$ into its symmetric and its antisymmetric part, $\zeta_{\mu\nu} = s_{\mu\nu} + b_{\mu\nu}$, where

$$s_{\mu\nu} = \zeta_{(\mu\nu)} = \frac{1}{2} \left(\zeta_{\mu\nu} + \zeta_{\nu\mu} \right), \quad b_{\mu\nu} = \zeta_{[\mu\nu]} = \frac{1}{2} \left(\zeta_{\mu\nu} - \zeta_{\nu\mu} \right), \tag{9.4}$$

satisfying

$$k^{\mu} s_{\mu\nu} = 0, \quad k^{\mu} b_{\mu\nu} = 0. \tag{9.5}$$

We postpone looking at the antisymmetric part. States of the form

$$s_{\mu\nu} \alpha_{-1}^{\mu} \tilde{\alpha}_{-1}^{\nu} |k\rangle, \quad s_{\mu\nu} = s_{\nu\mu} \tag{9.6}$$

are physical if

$$k^2 = 0, \quad k^{\mu} s_{\mu\nu} = 0. \tag{9.7}$$

Similar to the photon, one can verify that states with a polarisation of the form

$$\sigma_{\mu\nu} = k_{\mu} \zeta_{\nu} + k_{\nu} \zeta_{\mu}, \quad \text{where} \quad k^{\mu} \zeta_{\mu} = 0, \tag{9.8}$$

are null. The corresponding residual gauge symmetry is

$$s_{\mu\nu} \rightarrow s_{\mu\nu} + k_\mu \zeta_\nu + k_\nu \zeta_\mu, \quad \text{where} \quad k^\mu \zeta_\mu = 0. \tag{9.9}$$

Since the trace s^μ_μ of $s_{\mu\nu}$ is a Lorentz scalar, we decompose the symmetric tensor $s_{\mu\nu}$ into a traceless symmetric tensor and a scalar. To do this explicitly we choose an auxiliary null vector $\bar{k}$ which is linearly independent from the null vector k, by imposing $k^\mu \bar{k}_\mu = 1$. We make the following decomposition:

1. The *traceless part* of $s_{\mu\nu}$ is

$$\psi_{\mu\nu} = s_{\mu\nu} - \frac{1}{D-2} s^\rho_\rho \left(\eta_{\mu\nu} - k_\mu \bar{k}_\nu - k_\nu \bar{k}_\mu \right). \tag{9.10}$$

2. The *trace part* of $s_{\mu\nu}$ is

$$\phi_{\mu\nu} = \frac{1}{D-2} s^\rho_\rho \left(\eta_{\mu\nu} - k_\mu \bar{k}_\nu - k_\nu \bar{k}_\mu \right). \tag{9.11}$$

To check that this decomposition is correct, we observe that

$$s_{\mu\nu} = \psi_{\mu\nu} + \phi_{\mu\nu}, \quad \eta^{\mu\nu} \psi_{\mu\nu} = 0, \quad \eta^{\mu\nu} \phi_{\mu\nu} = s^\rho_\rho. \tag{9.12}$$

It is important to note that the trace part of $s_{\mu\nu}$ is physical,

$$k^\mu \phi_{\mu\nu} = 0, \tag{9.13}$$

and not null. Therefore, it corresponds to physical scalar field, called the *dilaton*. We will see later that the vacuum expectation value of the dilaton is a free parameter ('modulus') at tree level, and that it determines the *dimensionless string coupling constant* g_S. We will come back to the dilaton later.

The physical state conditions imply that $\psi_{\mu\nu}$ must be transversal, $k^\mu \psi_{\mu\nu} = 0$. For the standard representative $k = (k^0, 0, \cdots, k^0)$ this implies:

$$k^0 \psi_{00} + k^0 \psi_{0,D-1} = 0 \Rightarrow \psi_{0,D-1} = -\psi_{00}, \tag{9.14}$$

etc. Combining this with $\psi_{\mu\nu} = \psi_{\nu\mu}$ we obtain

$$(\psi_{\mu\nu}) = \begin{pmatrix} \psi_{00} & \psi_{01} & \psi_{02} & \cdots & \psi_{0,D-2} & -\psi_{00} \\ \psi_{01} & \psi_{11} & \psi_{12} & \cdots & \psi_{1,D-2} & -\psi_{01} \\ \vdots & & & & & \vdots \\ \psi_{0,D-2} & \psi_{1,D-2} & \psi_{2,D-2} & \cdots & \psi_{D-2,D-2} & -\psi_{0,D-2} \\ -\psi_{00} & -\psi_{01} & -\psi_{02} & \cdots & -\psi_{0,D-2} & \psi_{00} \end{pmatrix}. \tag{9.15}$$

The null part is $k_\mu \zeta_\nu + \zeta_\mu k_\nu$ where $k^\mu \zeta_\mu = 0$. Therefore, we can decompose $\psi_{\mu\nu}$ into a transverse part, which represents the physical states, and a null part,

$$\psi_{\mu\nu} = \psi^{\text{transv.}}_{\mu\nu} + \psi^{\text{null}}_{\mu\nu}. \tag{9.16}$$

Explicit:

$$(\psi^{\text{transv.}}_{\mu\nu}) = \begin{pmatrix} 0 & 0 & 0 & \cdots & 0 & 0 \\ 0 & \psi_{11} & \psi_{12} & \cdots & \psi_{1,D-2} & 0 \\ \vdots & & & & & \vdots \\ 0 & \psi_{1,D-2} & \psi_{2,D-2} & \cdots & \psi_{D-2,D-2} & 0 \\ 0 & 0 & 0 & \cdots & 0 & 0 \end{pmatrix}, \tag{9.17}$$

with $\psi_{11} + \psi_{22} + \cdots + \psi_{D-2,D-2} = 0$, and

$$(\psi_{\mu\nu}^{\text{spur.}}) = \begin{pmatrix} \lambda_0 & \lambda_1 & \lambda_2 & \cdots & \lambda_{D-2} & -\lambda_0 \\ \lambda_1 & 0 & 0 & \cdots & 0 & -\lambda_1 \\ \vdots & & & & & \vdots \\ \lambda_{D-2} & 0 & 0 & \cdots & 0 & -\lambda_{D-2} \\ -\lambda_0 & -\lambda_1 & -\lambda_2 & \cdots & -\lambda_{D-2} & \lambda_0 \end{pmatrix}. \tag{9.18}$$

The total number of states distributes as follows:

$$\underbrace{\frac{1}{2}D(D+1) - 1}_{\text{symmetric/traceless}} = \underbrace{\left(\frac{1}{2}(D-2)(D-1) - 1\right)}_{\text{physical}} + \underbrace{(D-1)}_{\text{null}} + \underbrace{D}_{\text{unphysical}}. \tag{9.19}$$

We can interpret this as follows:

- The constraints $k^\mu s_{\mu\nu} = 0$ remove D unphysical components.
- The residual gauge symmetries $s_{\mu\nu} \to s_{\mu\nu} + k_\mu \zeta_\nu + k_\nu \zeta_\mu$, depend on a vector ζ_μ, which is restricted by $k^\mu \zeta_\mu = 0$, leaving $D-1$ independent gauge degrees of freedom. These are the null states.
- This leaves us with $\frac{1}{2}D(D+1) - 1$ independent physical states, which are represented by the symmetric traceless matrix ($\psi_{\mu\nu}^{\text{transv.}}$).

In $D = 4$ the helicity group is SO(2), under which the two independent physical degrees of freedom transform as a rank 2 symmetric traceless tensor:[1]

$$\begin{pmatrix} \phi_{11} & \phi_{12} \\ \phi_{12} & -\phi_{11} \end{pmatrix} \to \begin{pmatrix} \cos\varphi & -\sin\varphi \\ \sin\varphi & \cos\varphi \end{pmatrix} \begin{pmatrix} \phi_{11} & \phi_{12} \\ \phi_{12} & -\phi_{11} \end{pmatrix} \begin{pmatrix} \cos\varphi & \sin\varphi \\ -\sin\varphi & \cos\varphi \end{pmatrix}. \tag{9.20}$$

As for the photon, we can define circular polarisation states

$$\phi_{\pm\pm} = \phi_{11} \pm i\phi_{12}, \tag{9.21}$$

which transform as

$$\phi_{++} \to e^{2i\varphi}\phi_{++}, \quad \phi_{--} \to e^{-2i\varphi}\phi_{--}. \tag{9.22}$$

Thus, physical states carry helicity $h = \pm 2$.

We can formulate this in terms of group theory using Young tableaux. The decomposition of a symmetric traceless tensor $\square\square$ of the Lorentz group SO$(1, D-1)$ into representations of the little group SO$(D-2)$ is:

$$\square\square \to \square\square \oplus \square \oplus \bullet$$

$$\to \square\square \oplus (\square \oplus \bullet) \oplus (\square \oplus \bullet \oplus \bullet)$$

$$\frac{1}{2}D(D+1) - 1 = \underbrace{\frac{1}{2}(D-2)(D-1) - 1}_{\text{physical}} + \underbrace{(D-1)}_{\text{null}} + \underbrace{D}_{\text{unphysical}}.$$

In $D = 4$ the decomposition of the tensor $\square\square$ of SO$(1,3)$ into irreducible representations of SO$(2) \simeq$ U(1) is:

---

[1] We are using matrix notation: $\phi'_{ij} = R_i{}^k R_j{}^l \phi_{kl} \Leftrightarrow \phi' = R\phi R^T$.

$$\square\square \rightarrow 9 \times \bullet$$

$$[9] = [h = 2] + [h = -2] + 2[h = 1] + 2[h = -1] + 3[h = 0].$$

Thus $\psi_{\mu\nu}$ describes a massless symmetric tensor which in $D = 4$ corresponds to a 'massless spin-2 particle' with helicity eigenstates $h = \pm2$. Such a state has the kinematic properties of a graviton, which is defined as the linearised fluctuation of the space-time metric $g_{\mu\nu}$ relative to a background metric $g_{\mu\nu}^{(0)}$, which satisfies the classical field equations ('on-shell background'). We take the background to be Minkowksi space and decompose

$$g_{\mu\nu} = \eta_{\mu\nu} + \tilde{\psi}_{\mu\nu}. \tag{9.23}$$

In order to show that such a field has the same kinematical properties as the above level-1 closed string state, one needs to perform a linearisation of the vacuum Einstein equations

$$R_{\mu\nu} - \frac{1}{2}Rg_{\mu\nu} = 0 \tag{9.24}$$

around Minkowski space. This is more complicated than for the photon and we will omit details of the computation. The linearised version of the diffeomorphism invariance of general relativity is the gauge invariance

$$\tilde{\psi}_{\mu\nu} \rightarrow \tilde{\psi}_{\mu\nu} + \partial_\mu\Lambda_\nu + \partial_\nu\Lambda_\mu, \tag{9.25}$$

where Λ_μ is a vector field. Indices are raised and lowered with the background metric $\eta_{\mu\nu}$. Expanding (9.24) to linear order in $\tilde{\psi}_{\mu\nu}$ gives

$$\square\tilde{\psi}_{\mu\nu} + \partial_\mu\partial_\nu\tilde{\psi}_\rho^\rho - \partial_\mu\partial_\rho\tilde{\psi}_\nu^\rho - \partial_\nu\partial_\rho\tilde{\psi}_\mu^\rho + \eta_{\mu\nu}\partial_\rho\partial_\sigma\tilde{\psi}^{\rho\sigma} - \eta_{\mu\nu}\square\tilde{\psi}_\rho^\rho = 0. \tag{9.26}$$

Similar to the Maxwell field, we can reduce this to the massless wave equation by imposing gauge conditions. The analogue of the Lorenz gauge is the *harmonic gauge*

$$\partial^\mu\tilde{\psi}_{\mu\nu} = \frac{1}{2}\partial_\nu\tilde{\psi}_\rho^\rho. \tag{9.27}$$

Imposing the harmonic gauge reduces the field equation to

$$\square\tilde{\psi}_{\mu\nu} - \frac{1}{2}\eta_{\mu\nu}\square\tilde{\psi}_\rho^\rho = 0. \tag{9.28}$$

By taking the trace of this equation,

$$\square\tilde{\psi}_\mu^\mu - \frac{D}{2}\square\psi_\rho^\rho = \left(1 - \frac{D}{2}\right)\square\tilde{\psi}_\rho^\rho = 0, \tag{9.29}$$

we see that for $D > 2$ the field equation reduces to the wave equation, $\square\tilde{\psi}_{\mu\nu} = 0$. The resulting set of equations, and their Fourier transforms are

$$\square\tilde{\psi}_{\mu\nu} = 0 \Leftrightarrow k^2 = 0,$$

$$\partial^\mu\tilde{\psi}_{\mu\nu} = \frac{1}{2}\partial_\mu\psi_\rho^\rho \Leftrightarrow k^\mu\tilde{s}_{\mu\nu} = \frac{1}{2}k_\nu\tilde{s}_\rho^\rho,$$

$$\tilde{\psi}_{\mu\nu} \cong \tilde{\psi}_{\mu\nu} + \partial_\mu\Lambda_\nu + \partial_\nu\Lambda_\mu, \quad \square\Lambda_\mu = 0 \Leftrightarrow \tilde{s}_{\mu\nu} \cong \tilde{s}_{\mu\nu} + k_\mu\tilde{\zeta}_\nu + k_\nu\tilde{\zeta}_\mu.$$

The leaves us with the same number $\frac{1}{2}D(D + 1) - 2D = \frac{1}{2}(D - 2)(D - 1) - 1$ of independent physical degrees of freedom. The above conditions are similar to but not exactly the same as those we found for the graviton string state. The difference is that

$\psi_{\mu\nu}$ is traceless, while $\tilde{\psi}_{\mu\nu}$ is not. However, we can impose $\psi^{\mu}_{\mu} = 0$ as an additional Lorentz covariant gauge condition. This does not change the number of degrees of freedom, because to preserve the harmonic gauge we must impose the additional condition $\partial_{\mu}\Lambda^{\mu} = 0$ on the parameter of the residual gauge transformation. This removes one gauge transformation, leaving the number of degrees of freedom the same. The resulting set of conditions is now the same as for the graviton string state.

9.2 The Kalb–Ramond Field (*B*-Field)

9.2.1 Physical States

We now turn to the antisymmetric part of the level-1 state. States of the form

$$b_{\mu\nu}\alpha^{\mu}_{-1}\tilde{\alpha}^{\nu}_{-1}|k\rangle \tag{9.30}$$

are physical if

$$k^2 = 0, \quad k^{\mu}b_{\mu\nu} = 0. \tag{9.31}$$

States of the special form

$$\beta_{\mu\nu} = k_{\mu}\zeta_{\nu} - k_{\nu}\zeta_{\mu}, \quad \text{where} \quad k^{\mu}\zeta_{\mu} = 0 \tag{9.32}$$

are null. The corresponding residual gauge symmetry is

$$b_{\mu\nu} \to b_{\mu\nu} + k_{\mu}\zeta_{\nu} - k_{\nu}\zeta_{\mu}, \quad \text{where} \quad k^{\mu}\zeta_{\mu} = 0. \tag{9.33}$$

The counting of unphysical, null and independent physical states is (see Exercise 9.2.1):

$$\underbrace{\frac{1}{2}D(D-1)}_{\text{antisymmetric}} = \underbrace{\frac{1}{2}(D-2)(D-3)}_{\text{physical}} + \underbrace{(D-2)}_{\text{null}} + \underbrace{(D-1)}_{\text{unphysical}}. \tag{9.34}$$

The corresponding antisymmetric tensor state is known as 'the *B*-field' or 'Kalb–Ramond field'. The *B*-field is the simplest example of a *higher rank gauge field*. A new feature compared to Maxwell theory is that higher rank gauge theories have 'gauge symmetries for gauge symmetries'. The field strength

$$H_{\mu\nu\rho} = 3\partial_{[\mu}B_{\nu\rho]} = \partial_{\mu}B_{\nu\rho} + \partial_{\nu}B_{\rho\mu} + \partial_{\rho}B_{\mu\nu} \tag{9.35}$$

is invariant under gauge transformations

$$B_{\mu\nu} \to B_{\mu\nu} + \partial_{\mu}\Lambda_{\nu} - \partial_{\nu}\Lambda_{\mu}, \tag{9.36}$$

where the parameter Λ_{μ} is a vector field. If we change Λ_{μ} by a gradient, $\Lambda_{\mu} \to \Lambda_{\mu} + \partial_{\mu}\phi$, then $B_{\mu\nu}$ is invariant. Thus, the gauge parameter Λ_{μ} is subject to a Maxwell-like gauge symmetry. This needs to be taken into account when counting independent degrees of freedom. Gauge fields of rank $r > 2$ appear in superstring theories, and have an even higher tower of nested gauge transformations.

9.2.2 Dualisation of Antisymmetric Tensor Fields

Why are antisymmetric tensor fields rarely mentioned in standard quantum field theory courses? In four dimensions, a massless antisymmetric rank 2 tensor field is equivalent to an 'axion-like' scalar, that is a scalar which only appears through its derivative, and thus has a symmetry under constant shifts. We illustrate this using the free theory, which has a Maxwell-like Lagrangian

$$L[B] = -\frac{1}{6}H_{\mu\nu\rho}H^{\mu\nu\rho}. \tag{9.37}$$

The field equations are

$$\partial^{\mu}H_{\mu\nu\rho} = 0, \quad \epsilon^{\mu\nu\rho\sigma}\partial_{\mu}H_{\nu\rho\sigma} = 0. \tag{9.38}$$

The first equation is the Euler–Lagrange equation obtained by variation of the action with respect to the gauge potential $B_{\mu\nu}$. The second equation is the local integrability condition or *Bianchi identity* which the field strength must satisfy in order to admit a gauge potential.

We define the *dual field strength F_{μ}* as the Hodge dual of $H_{\mu\nu\rho}$:

$$F_{\mu} = \frac{1}{6}\epsilon_{\mu\nu\rho\sigma}H^{\nu\rho\sigma} \Leftrightarrow H_{\mu\nu\rho} = \epsilon_{\mu\nu\rho\sigma}F^{\sigma}. \tag{9.39}$$

In terms of the dual field strength the field equations become

$$\epsilon^{\mu\nu\rho\sigma}\partial_{\rho}F_{\sigma} = 0, \quad \partial^{\mu}F_{\mu} = 0. \tag{9.40}$$

The Euler–Lagrange equation for $H_{\mu\nu\rho}$ can be interpreted as a Bianchi identity for the rank 1 field strength F_{μ}. If we introduce a scalar potential ϕ by

$$F_{\mu} = \partial_{\mu}\phi, \tag{9.41}$$

then we can interprete the Bianchi identity for $H_{\mu\nu\rho}$ as an Euler–Langrange equation for ϕ:

$$\partial^{\mu}\partial_{\mu}\phi = 0, \tag{9.42}$$

with dual Lagrangian

$$L[\phi] = -\frac{1}{2}\partial_{\mu}\phi\partial^{\mu}\phi. \tag{9.43}$$

In the dual formulation the gauge symmetry of $B_{\mu\nu}$ has become a shift symmetry $\phi \mapsto \phi + \text{const.}$ for the dual potential ϕ. Note that while the field strengths $H_{\mu\nu\rho}$ and F_{μ} are related algebraically, the relation between the potentials involves integration and thus is non-local. We will see in Chapter 13 how four-dimensional string theories can be obtained by *compactification*. One then usually dualises the *B*-field into an axion-like scalar, called the *universal string axion*.

Duality transformations can be implemented at the level of the Lagrangian by promoting the Bianchi identities to field equations using a Lagrange multiplier. Upon eliminating the original field strength by its equation of motion, which now is algebraic, one obtains the dual Lagrangian. The Lagrange multiplier becomes the dual gauge potential (see Exercise 9.2.2).

If one applies this procedure to the four-dimensional electromagnetic field strength $F_{\mu\nu}$, and exchanges $F_{\mu\nu}$ with its Hodge-dual $\tilde{F}_{\mu\nu} = \frac{1}{2}\epsilon_{\mu\nu\rho\sigma}F^{\rho\sigma}$, this amounts to exchanging the electric and magnetic components $\vec{E}, \vec{B}$ of the electromagnetic field. Hence, this is called *electric-magnetic duality*. The dual, 'magnetic' vector potential $\tilde{A}_\mu$, where $\tilde{F}_{\mu\nu} = \partial_\mu \tilde{A}_\nu - \partial_\nu \tilde{A}_\mu$, can be used to introduce a gauge coupling of the electromagnetic field to magnetic charges. In interacting theories, including non-abelian gauge theories, this transformation can be very useful because it maps the strong coupling regime of the original 'electric' theory to a dual, 'magnetic' theory, which is weakly coupled. The relation between electric and magnetic couplings is illustrated by the relation between the coupling constants $g, \tilde{g}$ in Exercise 9.2.2.

The dualisation procedure can be extended to general dimension D and works in the presence of gravity. Interaction terms for higher-rank gauge fields do not obstruct dualisation, as long as there are no minimal couplings between the gauge potential and a field which is charged under the higher rank gauge symmetry. In D dimensions a rank p field strength is Hodge dual to a rank $D-p$ field strength. The corresponding gauge potentials have ranks $p - 1$ and $D - p - 1$, respectively. We leave it to the reader to work out the details (see Exercise 9.2.2). The higher rank gauge fields of superstring theories other than the B-field can be dualised, after compactification to four dimensions into further *non-universal axions*. The world-sheet formulation of T-duality, to be discussed in Section 9.2.2, applies a similar dualisation procedure to the two-dimensional world-sheet field theory.

Exercise 9.2.1 Show that the gauge symmetry (9.33) parametrised by ζ_μ has its own gauge symmetry. Explain the decomposition (9.34) into inequivalent physical, gauge, and unphysical modes by carefully counting independent gauge symmetries. Compare to the decomposition of the tensor $b_{\mu\nu}$ using branching rules.

Exercise 9.2.2 Consider the generalised Maxwell action

$$S[A] = \int d^D x \left(-\frac{1}{2p!\, g^2} F_{\mu_1 \cdots \mu_p} F^{\mu_1 \cdots \mu_p} \right) \tag{9.44}$$

in D-dimensional Minkowski space, where the rank p field strength $F_{\mu_1 \cdots \mu_p}$ is the generalised curl (exterior derivative) of a rank $(p-1)$ antisymmetric tensor potential $A_{\mu_1 \cdots \mu_{p-1}}$,

$$F_{\mu_1 \cdots \mu_p} = p\, \partial_{[\mu_1} A_{\mu_2 \cdots \mu_p]}, \tag{9.45}$$

and where g is a constant which can be interpreted as a coupling constant once interaction terms are added. Generalise the discussion of Section 9.2.2 and construct a dual action which involves a dual rank $(D - p)$ field strength $H_{\mu_1 \cdots \mu_{D-p}}$ and a rank $(D - p - 1)$ gauge potential. *Instruction:* The Bianchi identity

$$\partial_{[\mu_1} F_{\mu_2 \cdots \mu_{p+1}]} = 0 \tag{9.46}$$

can be promoted to an Euler–Lagrange equation using a rank $(D - p - 1)$ antisymmetric Langrange multiplier field $B_{\mu_1 \cdots \mu_{D-p-1}}$. The resulting action is (with normalisation chosen for later convenience):

$$S[F, B] = \int d^D x \left(-\frac{1}{2p!\, g^2} F_{\mu_1 \cdots \mu_p} F^{\mu_1 \cdots \mu_p} \right.$$

$$\left. -(-1)^p \frac{1}{p!\,(D-p-1)!} \epsilon^{\mu_1 \cdots \mu_D} \partial_{\mu_1} F_{\mu_2 \cdots \mu_{p+1}} B_{\mu_{p+2} \cdots \mu_D} \right). \tag{9.47}$$

Table 9.1. Vertex operators for closed string tachyonic and massless states	
State $|\phi\rangle$	Vertex Operator $V_\phi(z, \bar{z})$
$|k\rangle$	$: \exp(ik \cdot X) :$
$\zeta_{\mu\nu} \alpha_{-1}^\mu \tilde{\alpha}_{-1}^\nu |k\rangle$	$\zeta_{\mu\nu} : \partial_z X^\mu \partial_{\bar{z}} X^\nu \exp(ik \cdot X) :$

Obtain the dual action by eliminating $F_{\mu_1 \cdots \mu_p}$ from $S[F, B]$ using its equation of motion. How are the coupling constants $g, \tilde{g}$ of the dual actions related? Show that if you replace Minkowski space by a space of arbitrary signature (t, s) the overall sign of the dual action depends on the number of time-like dimensions, but not on the space-time dimension or on the rank of the gauge field. Does the dualisation still work if you consider the action on a curved space-time? While the computation can be done in components, readers familiar with differential forms will find it helpful to use them, in particular when including a Riemannian metric. For some applications boundary terms obtained when integrating by parts are relevant, so you might want to keep them.

9.3 Vertex Operators

Using results from Section 4.4, we can immediately write down the local operators which create any string state from our list at any point of the world sheet.[2] The vertex operators for closed string tachyonic and massless states are listed in Table 9.1.

One important question is how the Virasoro constraints, which select physical states, translate into conditions which select physical vertex operators. Using the state-operator correspondence we see that the condition is simply that physical vertex operators must have conformal weight $(1, 1)$. In terms of OPEs, this means that physical vertex operators have the following OPE with the energy-momentum tensor:

$$T(z)V(w, \bar{w}) = \frac{V(w, \bar{w})}{(z - w)^2} + \frac{\partial_w V}{(z - w)} + \cdots,$$
$$\bar{T}(\bar{z})V(w, \bar{w}) = \frac{V(w, \bar{w})}{(\bar{z} - \bar{w})^2} + \frac{\partial_{\bar{w}} V}{\bar{z} - \bar{w}} + \cdots.$$

It is instructive to check that for specific states one obtains the same physical state conditions on momenta k^μ and polarisations $\zeta_{\mu_1 \cdots}$ as from imposing the Virasoro constraints (see Exercise 9.3.1). Note that apart from getting the correct coefficients of the second and first order pole one needs to impose the vanishing of any higher order poles.

The condition that physical vertex operators must have weight $(1, 1)$ ensures that by integrating them over the world-sheet one obtains a conformally invariant quantity. Since the positions of vertex operators do not have an invariant meaning

---

[2] We will only consider closed string vertex operators, which are inserted in the interior of the world-sheet. For open strings vertex operators are inserted at the boundaries, see Section 9.6 for literature.

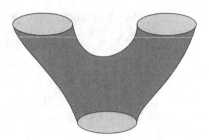

Fig. 9.1   The closed string vertex.

from the space-time point of view, string scattering amplitudes cannot depend on them. As we will discuss in more detail in Chapter 12, string scattering amplitudes are defined by integrating correlation functions of vertex operators over the world-sheet:

$$\mathcal{A}(1,\ldots) \propto \int_\Sigma dz_1 d\bar{z}_1 \cdots \langle V_1(z_1, \bar{z}_1) \cdots \rangle. \tag{9.48}$$

Exercise 9.3.1   The vertex operator for the Kalb–Ramond field is

$$V[X|b_{\mu\nu}, k_\rho] = b_{\mu\nu} : \partial_w X^\mu \partial_{\bar{w}} X^\nu, \exp(ik_\rho X^\rho(\sigma)) : \tag{9.49}$$

where $b_{\mu\nu} = -b_{\nu\mu}$ is the polarisation tensor and k_ρ the momentum. Show that by imposing that $V[X|b_{\mu\nu}, k_\rho]$ is a conformal primary of weight $(1, 1)$, you precisely recover the conditions (9.31) that $b_{\mu\nu}$ and k_μ must satisfy in order that $b_{\mu\nu}\alpha^\mu_{-1}\tilde{\alpha}^\nu_{-1}|k\rangle$ is a physical state.

9.4 The Dilaton

We now return to the dilaton and its relation to scales and couplings in string theory. In the world-sheet approach interactions corresponding to the splitting of one string into two, or the joining on two strings into one. If we restrict ourselves to theories of closed strings, this leaves us with one single type of vertex involving three closed strings (see Figure 9.1).

One important difference compared to point particles is that this vertex does not have a distinguished interaction point. In contrast, if we think about point particle interactions in terms of world-lines, we realise that graphs are not one-dimensional manifolds, but decompose into lines and distinguished points, the vertices. This leaves us a lot of freedom in defining interactions, since we can assign different coupling constants depending on how many lines meet and what species of particles the lines represent (see, e.g., Figure 1.3). For strings there is no such choice, and therefore there is only a single closed string coupling constant.[3] Moreover, since the graviton is among the closed string states, the closed string coupling constant must

---

[3] In theories with closed and open strings there are different types of vertices, but the corresponding coupling constants are related, due to the factorisation properties of scattering amplitudes. I.p. the coupling constant of three closed strings is proportional to the square of the coupling constant of three open strings. As a plausibility argument, note that we can cut the vertex in Figure 9.1 into two open string vertices. Any consistent string theory has only one independent coupling constant.

be proportional to the gravitational coupling constant κ. In dimensions $D > 2$ the gravitational coupling is dimensionful:

$$[\kappa] = \text{Mass}^{-(D-2)/2} = \text{Length}^{(D-2)/2}. \tag{9.50}$$

Instead of the gravitational coupling κ we can use other dimensionful constants, for example, Newton's constant G_N, where $\kappa = \sqrt{8\pi G_N}$, the Planck mass M_P, or the Planck length L_P. The world-sheet action already contains a fundamental dimensionful constant, the Regge slope α', with dimension

$$[\alpha'] = \text{Mass}^{-2} = \text{Length}^2. \tag{9.51}$$

Equivalently, we can use the string tension $T = \frac{1}{2\pi\alpha'}$, the string mass $M_S = \alpha'^{-1/2}$, or the string length $l_S = \alpha'^{1/2}$.

This raises the question whether and how these two fundamental scales – the *Planck scale* and the *string scale* – are related. If they were independent we would not have a single string theory but a one-parameter family of inequivalent string theories, parametrised by the dimensionless ratio M_S/M_P. However, it turns out that this ratio is determined by the vacuum expectation value of the dilaton. Therefore we get a single string theory, and the ratio M_S/M_P parametrises a one-parameter family of ground states.

To explain this further, we use a result which can be derived by studying string interactions explicitly.[4] To leading order in an expansion in derivatives, the interactions of stringy gravitons and dilatons are captured by the following space-time effective action:

$$S[g_{\mu\nu}, \phi] = \frac{1}{2\kappa^2} \int d^D x \sqrt{g}\, e^{-2\phi} \left(R + 4\partial_\mu \phi \partial^\mu \phi\right). \tag{9.52}$$

The corresponding equations of motions can be solved by setting $g_{\mu\nu} = \eta_{\mu\nu}$ and $\phi = \phi_0 = \text{const.}$, where the constant ϕ_0 is arbitrary. This shows that while Minkowksi space is a solution, this solution is not unique because we have to choose a vacuum expectation value ϕ_0 for the dilaton. This vacuum expectation value labels different vacua or groundstates of the action. Fields with arbitrary vacuum expectation values are called *moduli*, and the space of inequivalent ground states is referred to as the *moduli space of vacua*. We further observe that shifting the dilaton by a constant, $\phi \to \phi + \phi_0$, is equivalent to rescaling the gravitational coupling, $\kappa \to \kappa e^{-\phi_0}$, which changes the Planck scale M_P. To talk about a change of the Planck scale in a meaningful way, we must compare it to another scale. In our case the natural 'yard stick' is the string scale. Let us pick the solution of the field equations where the dilaton vacuum expectation value is zero, and for this particular solution we identify the string scale and Planck scale by setting

$$\kappa^2 = (\alpha')^{(D-2)/2}, \quad \text{for} \quad \phi_0 = 0. \tag{9.53}$$

The dilaton vacuum expectation value now parametrises the ratio between the two scales:

$$e^{2\phi_0} = \frac{\kappa^2}{(\alpha')^{(D-2)/2}}. \tag{9.54}$$

---

[4] String interactions will be discussed in Chapter 12.

We can regard κ as a function of ϕ_0 and eliminate it in favour α', or vice versa, keep κ and ϕ_0 while eliminating α'. In string perturbation theory one keeps α' and uses the dilaton vacuum expecation value to define a dimensionless string coupling constant:

$$g_S = e^{\phi_0}, \tag{9.55}$$

which replaces the dimensionful κ as the coupling constant associated to the basic vertex between three closed strings.

Up to numerical constants,[5] the relation of scales can be written as

$$\frac{M_S}{M_P} = e^{\phi_0} = \frac{L_P}{l_S}. \tag{9.56}$$

The existence of a degenerate ground state is an unphysical feature, which one expects to be lifted once interactions are taken into account, unless symmetries are imposed. In bosonic string theory higher order corrections create a potential for the dilaton. The same happens in superstring theories once supersymmetry is broken. While ideally this would lead to the selection of a unique stable ground state, new problems arise. One is that generically the corrected potential shows runaway behaviour which drives the string coupling to zero or infinite values. This is part of the problem of *moduli stabilisation*. Due to the existence of extra dimensions one needs to specify how the extra dimensions are shaped (and ensure that they are not visible at too low energies to avoid conflict with experimental and observational bounds). This leads to a large number of further moduli, which all need to be 'stablised'. While there are various proposals of how to stabilise most or all moduli, these create a new problem, since one is now left with a huge 'landscape' of discrete vacua which are stable or at least stable enough to define a 'possible universe'. It is an open question whether our universe is part of this landscape, and how a particular vacuum is chosen.

9.5 The No-Ghost Theorem

We return to the program of covariant quantisation. Unitarity is not manifest and must be verified. This amounts to showing the following:

1. The space of $\mathcal{F}_{phys}/\sim$ of gauge-inequivalent physical states must carry a positive definite scalar product.
2. The S-matrix, that is the asymptotic time evolution operator relating incoming and outgoing states in scattering processes, must be unitary.

We have seen that the value $a = 1$ is distinguished by the presence of photons and gravitons in the spectrum. Furthermore, one can show that for $D = 26, a = 1$ there is an enhanced number of physical null states, indicating that this case has enhanced symmetry (see Exercises 9.5.1 and 9.5.2). One can also show that for $D > 26$

---

[5] Our expressions for the relations between scales are given in a schematic way, neglecting conventional numerical factors, which become important when computing observables.

there always are negative norm states (see Exercise 9.5.3). These results can further be sharpened, though we will not show this within the covariant formulation: *the spectrum is free of negative norm states if $D = 26, a = 1$ or $D \leq 25, a \leq 1$.* The case $D = 26, a = 1$ is called *critical string theory*. At the level of the free theory it is not possible to exclude the case $D \leq 25$, since the truncation to a lower number of dimensions preserves the absence of negative norm states. However, the unitarity of the S-matrix requires $D = 26$, since otherwise negative norm states appear as intermediate states in loop diagrams, which leaves only the critical string, if we restrict ourselves to Minkowski backgrounds.[6] Light-cone quantisation, which we discuss Chapter 10, will allow us to give a complete derivation of the conditions $D = 26, a = 1$.

Exercise 9.5.1   A state $|\psi\rangle$ is called spurious, if it satisfies the mass shell condition and is orthogonal to all physical states.[7] Explain why a spurious state can be written as

$$|\psi\rangle = \sum_{n>0} L_{-n}|\chi_n\rangle, \quad \text{where} \quad (L_0 - a + n)|\chi_n\rangle = 0.$$

Explain why, without loss of generality, we can assume that

$$|\psi\rangle = L_{-1}|\chi_1\rangle + L_{-2}|\chi_2\rangle.$$

Show that states which are both spurious and physical have zero norm and thus are null states.

Exercise 9.5.2   Show that states of the form $|\psi\rangle = L_{-1}|\chi\rangle$, where $(L_0 - a + 1)|\chi\rangle = 0$ and $L_m|\chi\rangle = 0$, are both spurious and physical if and only if $a = 1$. This shows the existence of extra null states for $a = 1$. Check that the gauge degree of freedom of the photon belongs to this class.

Now assume that $a = 1$ and consider spurious states of the form

$$|\psi\rangle = \left(L_{-2} + \gamma L_{-1}^2\right)|\chi\rangle,$$

where $|\chi\rangle$ satisfies

$$L_m|\chi\rangle = 0, \ m > 0, \quad (L_0 + 1)|\chi\rangle = 0.$$

Show that these states are null states if $D = 26$, for a suitable choice of γ.

Exercise 9.5.3   Show that states of the form

$$|\phi\rangle = \left(c_1 \alpha_{-1} \cdot \alpha_{-1} + c_2 p \cdot \alpha_{-2} + c_3 (p \cdot \alpha_{-1})^2\right)|k\rangle$$

with $k^2 = -2$ are physical for suitable choices of c_1, c_2, c_3 and have negative norm for $D > 26$.

---

[6] *Non-critical string theories* which are consistent in $D \neq 26$ dimensions can be defined, but require non-trivial space-time backgrounds. They are less well understood than critical string theories. See Section 12.2.3 for some further remarks.

[7] This and the following two exercises have been adapted from Green et al. (1987). In the literature the term spurious is sometimes used as a synonym for a null state or state which is 'pure gauge'. Here we use a definition which does not require that a spurious state is a physical state, while we use the term null state for states which are both spurious and physical, and therefore represent the residual gauge freedom we have in describing physical states.

9.6 Further Remarks and Literature

A detailed account of the relation between the standard formulation of general relativity and the field theory of a massless spin-2 field in Minkowski space-time can be found in Ortin (2004). We have restricted ourselves to vertex operators for closed strings. Vertex operators for open strings are inserted at boundaries of the world-sheet, and to discuss them in the language of CFT one uses the boundary CFT formalism, see, for example, Blumenhagen and Plauschinn (2009), Blumenhagen et al. (2013). The approach to covariant quantisation which we have presented here is often referred to as *old covariant quantisation*. The so-called *new covariant quantisation* uses the BRST formalism, where the Fock space is extended by the Faddeev–Popov ghost fields. These ghost fields appear automatically in the path integral approach to quantisation. Both the BRST formalism and path integral quantisation of string theory are explained in the more comprehensive string theory textbooks that we mentioned in the Introduction. See also Section 12.2.3 for a short explanation of the role of Faddeev–Popov ghost fields.

10 Light-Cone Quantisation

10.1 Light-Cone Gauge and Light-Cone Quantisation for Particles

The Lorentz-covariant approach to quantisation does not permit a complete gauge fixing. Physical states are subject to residual gauge transformations, and verifying unitarity is difficult. In this chapter, we discuss a complementary approach, called *light-cone quantisation*, which gives up manifest Lorentz covariance, but provides unique representatives for physical states and makes unitarity manifest. We illustrate this using the relativistic particle before moving on to strings.

As a first step we introduce *light-cone coordinates* for space-time vectors by[1]

$$V^{\pm} = \frac{1}{\sqrt{2}} \left(V^0 \pm V^{D-1} \right).$$

The Lorentz scalar product in these coordinates is

$$V \cdot W = -V^- W^+ - V^+ W^- + \underline{V} \cdot \underline{W} = V_+ W^+ + V_- W^- + \underline{V} \cdot \underline{W},$$

where

$$\underline{V} \cdot \underline{W} = \sum_{i=1}^{d} = \delta_{ab} V^a W^b = V_a W^a, \quad d := D - 2,$$

is the part of the scalar product which is transverse to the light-cone directions $V^{\pm}$. Note that $V_{\pm} = -V^{\mp}$.

The null coordinate $x^+ = \frac{1}{\sqrt{2}}(x^0 + x^{D-1})$ can be used instead of the time-like coordinate x^0 to describe time evolution. Therefore, x^+ is referred to as *light-cone time*, and the corresponding component $p_+ = -p^-$ of the momentum is, up to a sign, the *light-cone energy* $E_{LC} = p^- = -p_+$. The minus sign has been included in the definition of E_{LC} to make it non-negative, as will become clear below. The *light-cone gauge* imposes that the world-line curve parameter τ is proportional to x^+:

$$x^+ = \frac{1}{m^2} p^+ \tau. \tag{10.1}$$

The light-cone energy p^- is determined by the mass shell condition as a function of $p^+, \underline{p}$ and of the mass m of the particle:

$$p^2 + m^2 = -2p^+ p^- + \underline{p} \cdot \underline{p} = 0 \Rightarrow p^- = \frac{1}{2p^+} \left(\underline{p} \cdot \underline{p} + m^2 \right).$$

---

[1] Note the conventional factor $1/\sqrt{2}$ which is different from the definition for world-sheet light-cone coordinates $\sigma^{\pm}$.

Note that we need to assume $p^+ \neq 0$, which is equivalent to $p^0 \neq p^1$ and thus amounts to a condition on the coordinate system we use.[2]

The x^+-component of the equation of motion $\ddot{x}^\mu = 0$ is satisfied automatically. The remaining components can be integrated to give

$$x^-(\tau) = x_0^- + \frac{p^-}{m^2}\tau, \quad \underline{x}(\tau) = \underline{x}_0 + \frac{\underline{p}}{m^2}\tau.$$

Since p^- is a dependent quantity, solving the x^--equation only requires to specify the initial condition x_0^-. The independent dynamical variables variables in the light-cone gauge are:

$$x^a, x_0^-, p^a, p^+, \quad a = 1, \ldots, D-2.$$

The gauge-fixed theory can now be quantised by imposing canonical commutation relations for the independent quantities:

$$[x_0^-, p^+] = i\eta^{-+} = -i, \quad [x^a, p^b] = i\eta^{ab} = i\delta^{ab}.$$

The Hilbert space of states is spanned by momentum eigenstates,

$$|k^+, \underline{k}\rangle,$$

that is by simultaneous eigenstates of p^+ and $\underline{p}$. One possible Hamiltonian for this system is the *light-cone Hamiltonian p^-* which generates translations in the light-cone time x^+. However, while we could use this operator to describe the time evolution of the system, the concepts of energy and mass for a relativistic particle are still tied to the standard Hamiltonian H, which generates translations in the world-line time τ. The definition (10.1) of the light-cone gauge implies

$$\frac{\partial}{\partial\tau} = \frac{\partial x^+}{\partial\tau}\frac{\partial}{\partial x^+} = \frac{p^+}{m^2}\frac{\partial}{\partial x^+},$$

and, therefore, the Hamiltonian takes the form

$$H = \frac{p^+ p^-}{m^2} = \frac{1}{2m^2}\left(\underline{p}\cdot\underline{p} + m^2\right)$$

in the light-cone gauge. The momentum eigenstates which form a basis of the light-cone Hilbert space $\mathcal{H}_{LC}$ are eigenstates of this operator:

$$H|k^+, \underline{k}\rangle = \frac{1}{2m^2}\left(\underline{k}\cdot\underline{k} + m^2\right)|k^+, \underline{k}\rangle.$$

10.2 The Light-Cone Gauge for Open Strings

We now extend the light-cone gauge to strings. For definiteness, we will consider open strings with Neumann boundary conditions. Space-time light-cone coordinates are denoted X^+, X^-, X^i, where

---

[2] For completeness, we point out that there can be subtleties related to zero momentum states which are not covered by the light-cone approach. In string theory there exist states which are physical at zero momentum, but not at non-zero momentum. Some of these states are important for proving the background independence of string theory under a change of the string coupling (dilaton background). See, e.g., Belopolsky and Zwiebach (1996); Astashkevich and Belopolsky (1997).

$$X^{\pm} = \frac{1}{\sqrt{2}} \left(X^0 \pm X^{D-1} \right) \qquad (10.2)$$

are light-like coordinates and X^i, $i = 1, \ldots, D - 2 = d$ are space-like coordinates which are transverse to the light-cone spanned by $X^{\pm}$. For strings the light-cone gauge is imposed on top of the conformal gauge. We have seen that in the conformal gauge we can still make conformal reparametrisations, $\sigma^+ \to \tilde{\sigma}^+(\sigma^+)$, $\sigma^- \to \tilde{\sigma}^-(\sigma^-)$. If we use this freedom to choose a new world-sheet time $\tilde{\sigma}^0 = \frac{1}{2}(\tilde{\sigma}^+(\sigma^0+\sigma^1)+\tilde{\sigma}^1(\sigma^0-\sigma^1))$, then the new world-sheet space coordinate $\tilde{\sigma}^1$ is already determined up to a constant, which for open string is fixed by the convention $\tilde{\sigma}^1 \in [0, \pi]$.[3] In the conformal gauge the string coordinates X^{μ} satisfy the two-dimensional wave equation, and, therefore, we can use conformal coordinate transformations to impose that σ^0 (where we have dropped the tilde) is proportional to X^+, up to an additive constant, which we choose to be zero. This condition defines the light-cone gauge:

$$X^+ = p^+ \sigma^0 \Leftrightarrow x^+ = 0 \, , \alpha_n^+ = 0 \, , n \neq 0. \qquad (10.3)$$

In terms of space-time light-cone coordinates, the constraints $L_m = \tilde{L}_m = 0$ are

$$(\dot{X} \pm X')^{\mu}(\dot{X} \pm X')_{\mu} = 0 \Rightarrow \dot{X}^- \pm X'^- = \frac{1}{2p^+}(\dot{X}^i \pm X'^i)^2. \qquad (10.4)$$

These equations can be solved for the oscillators $\alpha_n^-, n \neq 0$ and the momentum component $p^- = \alpha_0^-$:

$$\alpha_n^- = \frac{1}{2p^+} \sum_{i=1}^{d} \sum_{m=-\infty}^{\infty} \alpha_{n-m}^i \alpha_m^i =: \frac{1}{p^+} L_n^{\perp} \, , \quad n \in \mathbb{Z}. \qquad (10.5)$$

We have introduced the notation $L_n^{\perp}$ because these expressions take the same form as the Fourier components L_n of the energy-momentum tensor, except that the sum is restricted to the transverse directions.

For open strings in the light-cone gauge the remaining independent variables are $x^-, p^+, x^i, p^i, \alpha_n^i$, where $i = 1, \ldots, d$ and $n = \pm 1, \pm 2, \cdots$. Evaluating (10.5) for $n = 0$ we obtain the classical mass shell condition (in units where $\alpha' = \frac{1}{2}$):

$$M^2 = (2p^+ p^- - \underline{p} \cdot \underline{p}) = 2N^{\perp}, \qquad (10.6)$$

where

$$N^{\perp} = \sum_{i=1}^{d} \sum_{m=1}^{\infty} \alpha_{-m}^i \alpha_m^i$$

takes the same form as the total oscillation number N, but only involves transverse oscillations.

10.3 Light-Cone Quantisation for Open Strings

The quantum theory of open strings is defined by imposing the following commutation relations:

---

[3] For closed strings, the constant is not fixed and corresponds to the freedom of shifting the origin of the coordinate $\tilde{\sigma}^1$ along the string.

$$[x^-, p^+] = -i, \quad [x^i, p^j] = i\delta_{ij}, \quad [\alpha_m^i \alpha_n^j] = m\delta^{ij}\delta_{m+n,0}, \qquad (10.7)$$

where $i, j = 1, \ldots, d$ and $m, n = \pm 1, \pm 2, \ldots$. The first two relations are the light-cone commutation relations of a relativistic particle, and correspond to the centre of mass motion of the string. The last set of relations corresponds to an infinite set of harmonic oscillators labelled by the transverse directions $i, j = 1, \ldots, d$ and the modes $m, n \in \mathbb{Z}\backslash\{0\}$. Since this only involves the transverse coordinates, we obtain standard harmonic oscillator relations (up to normalisation), and it is clear that the Fock space carries a positive definite scalar product. The representation space of the relations (10.7) is called the light-cone Hilbert space $\mathcal{H}^{LC}$. It factorises as

$$\mathcal{H}^{LC} = \mathcal{H}^{LC}_{\text{CM}} \otimes \mathcal{H}^{LC}_{\text{osc}}$$

into the Hilbert space $\mathcal{H}^{LC}_{\text{CM}}$ of the centre of mass motion, which is isomorphic to the light-cone Hilbert space of a relativistic particle, and the Hilbert space $\mathcal{H}^{LC}_{\text{osc}}$ representing the infinitely many vibration modes.

For the oscillator part we choose a basis of the form

$$\alpha_{-m_1}^{i_1} \ldots |0\rangle_{\text{osc}}, \quad i = 1, \ldots, d, m_k > 0,$$

where the ground state $|0\rangle_{\text{osc}}$ is defined by

$$\alpha_m^i |0\rangle_{\text{osc}}, \quad m > 0.$$

Combining this with the momentum eigenstates $|k^+, \underline{k}\rangle$ for the centre of mass degrees of freedom, we obtain a basis

$$\alpha_{-m_1}^{i_1} \cdots |k^+, \underline{k}\rangle$$

for $\mathcal{H}^{LC}$.

The expression (10.6) for M^2 involves the operator $N^\perp$, which has an ordering ambiguity. In Section 10.6, we will see that this ambiguity is fixed by imposing Lorentz covariance. Alternatively, we can use the so-called ζ-function regularisation method, which gives the same answer and which we will explain next.

10.4 Ground State Energy via ζ-Function Method

Let us compare the classically ordered expression for M^2 to its normal ordered version. This computation is analogous to the one we did for L_0 previously, and again we find that both expressions differ by a divergent series:

$$M^2_{\text{class}} = \sum_{i=1}^{d} \sum_{m \neq 0} \alpha_{-m}^i \alpha_m^i = 2 \sum_{i=1}^{d} \sum_{m=1}^{\infty} \alpha_{-m}^i \alpha_m^i + \sum_{i=1}^{d} \sum_{m=1}^{\infty} m = 2N^\perp + \sum_{i=1}^{d} \sum_{m=1}^{\infty} m,$$

where $N^\perp$ is understood to be normal ordered. Introducing the operators

$$a_m^i = \frac{1}{\sqrt{m}} \alpha_m^i, \quad m > 0, \quad (a_m^i)^\dagger = \frac{1}{\sqrt{-m}} \alpha_m^i, \quad m < 0, \qquad (10.8)$$

which satisfy standard harmonic oscillator relations

$$[a_m^i, (a_n^j)^\dagger] = \delta^{ij}\delta_{mn},$$

we can rewrite this in the form

$$M^2_{\text{class}} = 2 \sum_{i=1}^{d} \sum_{m=1}^{\infty} m \left((a^i_m)^\dagger a^i_m + \frac{1}{2} \right).$$

Thus, up to an overall factor 2 (which we could read as α'^{-1} and move to the other side) the mass squared operator of the open string is a sum of infinitely many harmonic oscillator Hamiltonians with frequencies $\omega_m = m$. This expression is divergent because we try to add up the non-zero ground state energies of infinitely many harmonic oscillators. The same type of divergence arises when computing the ground state energy of a quantum field in a finite volume.

Mathematically the problem is to assign a finite value to the divergent series $S = \sum_{m=1}^{\infty} m$, which belongs to the family of series

$$S_s = \sum_{m=1}^{\infty} m^{-s}, \quad s \in \mathbb{C}. \tag{10.9}$$

S_s converges for $\text{Re}(s) > 1$, where it defines the so called Riemann ζ-function (see Appendix D.2). By analytic continuation, $\zeta(s)$ becomes a meromorphic function on the complex plane, which is holomorphic for $s \neq 1$ and has a simple pole at $s = 1$. This allows one to assign finite values to divergent sums of the form (10.9) with $s < 1$. In particular, the ζ-regularised sum relevant for our problem is

$$\left(\sum_{m=1}^{\infty} m \right)_\zeta = \zeta(-1) = -\frac{1}{12}.$$

Using this, the renormalised mass squared operator is

$$M^2 = 2 \left(N^\perp - \frac{d}{24} \right), \quad \text{where} \quad N^\perp = \sum_{i=1}^{d} \sum_{m>0} \alpha^i_{-m} \alpha^i_m = \sum_{i=1}^{d} \sum_{m>0} m (a^i_m)^\dagger a^i_m \tag{10.10}$$

is the sum of the normal ordered Hamiltonians for harmonic oscillators with frequencies $\omega_m = m$ for the d transverse directions. Anticipating that Lorentz covariance requires $D = 26$ and hence $d = 24$, we obtain the mass formula

$$\alpha' M^2 = N^\perp - 1, \tag{10.11}$$

where we have re-introduced the dimensionful parameter α'.

10.5 Open String Spectrum in the Light-Cone Gauge

One advantage of the light-cone gauge is that there are no residual gauge symmetries, so that we have a unique description of physical states. Using the mass formula (10.11), we list the lowest mass eigenstates of the open string in Table 10.1.

For massive states we need to combine representations of $SO(D - 2)$ into representations of the massive little group $SO(D - 1)$. As an example, consider the massive states at level $N^\perp = 2$. The states $\alpha^i_{-2} |k\rangle$ form a vector $\square$ of $SO(D-2)$, while $\alpha^i_{-1} \alpha^j_{-1} |k\rangle$ belong to a symmetric tensor of $SO(D - 2)$. Since the state is massive, we need to combine these $SO(D - 2)$ representations into representations of the little

Table 10.1. Lowest mass eigenstates of the open string (light-cone gauge)

$N^\perp$	$\alpha'M^2$	State	Rep. of $SO(D-2)$	Rep. of $SO(D-1)$
0	-1	$\lvert k\rangle$	$\bullet$	N/A
1	0	$\alpha^i_{-1}\lvert k\rangle$	$\square$	N/A
2	1	$\alpha^i_{-2}\lvert k\rangle$	$\square$	$\square\square$
		$\alpha^i_{-1}\alpha^j_{-1}\lvert k\rangle$	$\square\square + \bullet$	

group $SO(D-1)$. This can be done using the branching rule for the decomposition of representations $SO(D-1) \supset SO(D-2)$:

$$\square\square \rightarrow \square\square + \square + \bullet.$$

Reading this backwards we see that the light-cone states $\alpha^i_{-2}\lvert k\rangle$ and $\alpha^i_{-1}\alpha^j_{-1}\lvert k\rangle$ combine into a symmetric traceless tensor $\square\square$ of the massive little group $SO(D-1)$. In $D=4$, this state describes a massive spin 2 particle, corresponding to the five-dimensional representation of $SO(3)$. Staying in four dimensions for a bit longer, we make the following observation. In $D=4$, the rank n symmetric traceless tensor of $SO(3)$ corresponds to the spin n representation. In the open string spectrum states of the form $\alpha^{i_1}_{-1}\alpha^{i_2}_{-1}\cdots\alpha^{i_n}_{-1}\lvert k\rangle$ occur at level $N^\perp = n$ and are the highest spin states at a given mass. When drawing M^2 against spin, all string states organise into lines called *Regge trajectories*, and the maximal spin states are on the leading Regge trajectory:

$$\alpha'M^2 = N^\perp - 1 \xrightarrow{N^\perp = J_{\max}} \alpha'M^2 = J_{\max} - 1.$$

This explains why α' is called the *Regge slope*.

Historically, string theory started out as a candidate for a theory of strong interactions of hadrons, which, approximately, fit onto Regge trajectories. This was abandoned in favour of QCD, once partons were discovered, since the UV behaviour of string amplitudes is too soft. Ironically, through the *gauge/gravity correspondence*,[4] string theory has recently returned to its origins: at least supersymmetric versions of QCD seem to be 'holographically dual' to strings on a curved higher-dimensional space-time.

10.6 Lorentz Covariance in the Light-Cone Gauge

In the light-cone gauge unitarity is manifest while Lorentz covariance is not. In fact there is a potential violation of Lorentz covariance, for the following reason. Classically, the generators of Lorentz transformations take the form

$$J^{\mu\nu} = L^{\mu\nu} + S^{\mu\nu} = x^\mu p^\nu - x^\nu p^\mu - i\sum_{n=1}^{\infty}\frac{1}{n}\left(\alpha^\mu_{-n}\alpha^\nu_n - \alpha^\nu_{-n}\alpha^\mu_n\right), \tag{10.12}$$

---

[4] See Ammon and Erdmenger (2015) for an introduction. Chapter 1 of Green et al. (1987) and Chapter 18 of West (2012) review the historical relation between *dual resonance models* and string theory.

where we have separated the contributions of the centre of mass motion and excitations. We can transform the indices μ, ν to light-cone coordinates. In this frame the Lorentz Lie algebra requires in particular that $[J^{i-}, J^{j-}] = 0$ for $i, j = 1, \ldots, d$. When we promote the J^{i-} to operators on the light-cone Hilbert space, we need to take into account that they contain the operators α_m^-, which are given by infinite sums which are quadratic in the transverse oscillators α_m^i. This makes evaluating the commutators $[J^{i-}, J^{j-}]$ non-trivial.

The normal ordered transverse Virasoro operators

$$L_0^\perp = N^\perp + \frac{1}{2}\underline{p} \cdot \underline{p}, \quad L_m^\perp = \sum_{i=1}^{d} \sum_{n=1}^{\infty} \alpha_{m-n}^i \alpha_n^i, \quad m \neq 0, \tag{10.13}$$

satisfy the Virasoro algebra with central charge $c = d$:

$$[L_m^\perp, L_n^\perp] = (m - n)L_{m+n}^\perp + \frac{d}{12}(m^3 - m)\delta_{m+n}, \tag{10.14}$$

and

$$[L_m^\perp, \alpha_n^i] = -n\alpha_{m+n}^i.$$

At the classical level M^2, $N^\perp$ and $L_0^\perp$ are related by the classical mass formula (10.6) and by (10.13). One way of dealing with normal ordering effects is to use the ζ-function method. Here, we proceed differently and parametrise the ambiguity of the ground state energy by a constant a:

$$M^2 = 2\left(N^\perp - a\right).$$

Among the transverse Virasoro operators only $L_0^\perp$ has an ordering ambiguity. We take the operators $N^\perp$ and $L_0^\perp$ to be normal ordered and interpret the constant a as an ambiguity in the relation between $p^- = \alpha_0^-$ and $L_0^\perp$:[5]

$$p^- = \frac{1}{p^+}(L_0^\perp - a).$$

For $n \neq 0$ the operators α_n^- and $L_n^\perp$ are proportional,

$$\alpha_n^- = \frac{1}{p^+}L_n^\perp, \quad n \neq 0.$$

A straightforward but lengthy computation shows that

$$[J^{i-}, J^{j-}] = -\frac{1}{(p^+)^2} \sum_{m=1}^{\infty} A(m)(\alpha_{-m}^i \alpha_m^j - \alpha_{-m}^j \alpha_m^i), \tag{10.15}$$

where

$$A(m) = m\left(\frac{24 - d}{12}\right) + \frac{1}{m}\left(\frac{d - 24}{12} + 2(1 - a)\right), \tag{10.16}$$

(see Exercise 10.6.2). Lorentz covariance requires that $A(m) = 0$ for $m = 1, 2, \ldots$, which is equivalent to $d = 24$ and $a = 1$. This fixes the dimension of space time to be $D = d + 2 = 26$, and the value obtained for the shift of the ground state energy is consistent with the value obtained using ζ-function regularisation.

---

[5] Here, we make the *conventional* choice that the ambiguity in the ground state energy is absorbed into p^- rather than $L_0^\perp$.

While we have discussed the open string for definiteness, for the closed string one obtains the same values for the *critical dimension* $D = 26$ and for the shift in the operators $L_0^\perp$ and $\tilde{L}_0^\perp$, $a = \tilde{a} = 1$.

Exercise 10.6.1 Write down the states with $N^\perp = 3$ for the light-cone open string, determine the irreducible SO($D-2$) representations, and organise them into irreducible SO($D-1$) representations. This requires to generalise the branching rules given in Appendix E to include rank 3 symmetric tensors. The corresponding branching rule can be inferred by analysing the dimensions of the relevant representations. For $D = 4$, what is the spin of these massive string states?

Exercise 10.6.2 Determine the coefficient $A(m)$ in (10.15), thus completing the proof that open bosonic strings are Lorentz covariant if and only if $D = 26$ and $a = 1$. *Instructions:*

1. Explain why $[J^{i-}, J^{j-}]$ must take form (10.15) and show that (10.15) implies

$$\langle 0 | \alpha_m^k [J^{i-}, J^{j-}] \alpha_{-m}^l | 0 \rangle = -(p^+)^{-2} m^2 A(m) \left(\delta^{ik} \delta^{jl} - \delta^{jk} \delta^{il} \right), \quad \forall m > 0. \tag{10.17}$$

2. Show that

$$[J^{i-}, J^{j-}] = -(p^+)^{-2} \left(2ip^- p^+ S^{ij} - [S^i, S^j] - iS^i p^j + iS^j p^i \right), \tag{10.18}$$

where

$$S^i := p^+ S^{i-} \tag{10.19}$$

satisfies

$$[x^i, S^j] = -iS^{ij}. \tag{10.20}$$

3. Evaluate the r.h.s. of (10.18) between the same states as in (10.17) to extract $A(m)$, and show that $A(m) = 0$ for all m if and only if $D = 26$ and $a = 1$.

Remark: This exercise has been adapted from Section 2.3.1 of Green et al. (1987). Note that our conventions are different, in particular the normal ordering constant a is included in a different term. Our conventions are almost the same as in Zwiebach (2009), though our definition of the constant a differs by a minus sign. The computations for this exercise are quite lengthy, but straightforward.

10.7 Literature

A more detailed account on the light-cone quantisation of particles, fields and strings can be found in Zwiebach (2009). We have not discussed the *DDF construction* which relates covariant and light-cone quantisation and allows one to prove the no-ghost theorem for covariant quantisation. This is covered in Green et al. (1987), Chapter 2.3. String theory and gauge theory have a long common history, see, for example, Green et al. (1987), Chapter 1, Polyakov (1987), and West (2012), Chapter 18. More recently the relation between gauge theory and string theory has been reformulated in terms of the AdS/CFT, and, more generally, gauge/gravity correspondence, see Ammon and Erdmenger (2015) for an introduction.

11 Partition Functions II

We return to the topic of partition functions, this time emphasising the space-time perspective. Strings combine the degrees of freedom of a particle with those of an infinite set of harmonic oscillators, and their partition function is built accordingly. In Chapter 6, we have discussed the partition function for a two-dimensional chiral scalar field, and its relation to a string moving on a one-dimensional target space. We now extend this to open and closed strings in $D = 26$ dimensions. As a warm up and for later comparison, we start with the partition function of a massive scalar particle.

11.1 Partition Functions for Massive Particles

Particles can either be described from a word-line perspective or from a space-time perspective. From the world-line perspective we can define a partition function $Z(t)$ by taking the trace of the Euclidean time evolution operator for a finite interval of proper time t. The world-line Hamiltonian is $k^2 + m^2$, where k is the momentum of the particle. The trace is a Gaussian integral, of the type we have performed before when discussing the contribution of the centre of mass of a closed string to its partition function. Neglecting constant factors, we find for a particle propagating on a one-dimensional target space

$$Z(t) = \mathrm{Tr} e^{-tH} = \mathrm{Tr} e^{-t(k^2 + m^2)} = \int dk e^{-t(k^2 + m^2)} \cong t^{-1/2} e^{-tm^2}. \tag{11.1}$$

This is a *world-line partition function* for a *single* circular world-line of *fixed* Euclidean time t. To obtain the space-time partition function we need to sum over all circular world-lines modulo gauge symmetries. This will in particular include an integral over the proper time t, but one also needs to take into account world-line gauge invariance and gauge fixing. Also note that the partition function defined in quantum field theory will include a sum over the number disconnected circular world-lines, reflecting that quantum field theory is defined on the multiparticle Hilbert space.

We will now work out the partition function for a massive scalar field, and show that we recover (11.1) when we restrict to a single particle and fix the length of the loop. The Euclidean action for a free massive scalar field in D dimensions is

$$S[\phi] = \int d^D x \, \phi(-\Delta + m^2)\phi. \tag{11.2}$$

The associated partition function is

$$Z = \int D\phi e^{-S[\phi]}, \tag{11.3}$$

where $D\phi$ is a formal expression which indicates that we want to integrate over all configurations of the field ϕ. We leave aside the question how such an infinite-dimensional integral can be defined rigorously,[1] and evaluate it formally by generalising the standard formula for Gaussian integrals:

$$Z = \det^{-1/2}(-\Delta + m^2), \tag{11.4}$$

(see Appendix F.2). By taking the (natural) logarithm, we obtain the standard QFT formula for the logarithm of the one-loop partition function,

$$\log Z = \log \det^{-1/2}(-\Delta + m^2) = -\frac{1}{2}\operatorname{tr}\log(-\Delta + m^2). \tag{11.5}$$

The trace is understood as the sum over the eigenvalues of $\log(-\Delta + m^2)$. It can be evaluated by using the momentum representation (see F.3). The result is

$$\log Z = -\frac{1}{2}V_D \int \frac{d^D k}{(2\pi)^D} \log(k^2 + m^2), \tag{11.6}$$

where V_D is the volume of space-time. Unless we restrict to a finite region in space-time, this factor is infinite. The volume factor indicates that $\log Z$ is an extensive quantity. Finite quantities can be obtained by considering densities or ratios. For our purposes it is convenient to keep V_D as a formal constant. The infinite volume of space-time also has the effect that the eigenvalues of $-\Delta + m^2$ are continuous, and therefore the trace is an integral over momenta, rather than a sum. To compare our result to the fixed-time world-line partition function we employ the so-called *Schwinger proper time representation* of $\log(k^2 + m^2)$. This representation can be applied more broadly to recast a perturbative quantum field theory in D dimensions as a one-dimensional world-line quantum field theory living on its graphs. For our purposes we need the world-line expression for $\log(k^2 + m^2)$ given by

$$-\log(k^2 + m^2) \to \lim_{\epsilon \to 0} \int_\epsilon^\infty \frac{dt}{t} e^{-(k^2 + m^2)t}. \tag{11.7}$$

The integral is divergent for $\epsilon \to 0$, but can be assigned a finite value through its relation to the integral exponential function (see Appendix F.3 for details).

Using the integral representation we obtain

$$\log Z = \frac{1}{2}V_D \int \frac{d^D k}{(2\pi)^D} \int_0^\infty \frac{dt}{t} e^{-(k^2 + m^2)t}. \tag{11.8}$$

The Gaussian integral over momenta can be carried out, leaving us with

$$\log Z \cong V_D \int_0^\infty \frac{dt}{t} t^{-D/2} e^{-m^2 t}, \tag{11.9}$$

where we have omitted numerical constants. This expression can now be compared to the fixed-time world-line partition function $Z(t)$, and we find that

$$\log Z \cong V_D \int_0^\infty \frac{dt}{t} Z(t)^D. \tag{11.10}$$

Thus, up to the 'log' and the factor V_D, the D-dimensional space-time partition function is an integral over D copies of the word-line partition function for a one-dimensional target space. This can be interpreted as follows: the logarithm reflects

[1] See, e.g., Glimm and Jaffe (1981).

that Z is defined using the multiparticle Hilbert space of a quantum field theory. Put differently, Z is a generating functional for all graphs, including disconnected ones. The world-line partition function $Z(t)$, which is defined using the single particle Hilbert space, corresponds to the generating functional $\log Z$ of connected graphs. The factor V_D is interpreted as arising from integration over the possible positions of a single loop in space-time. This was not included in the world-line partition function, and appears as a multiplicative factor due to translation symmetry. Finally the integral over t is a sum over loops of different proper length.

With this interpretation it becomes clear that while the infinite constant V_D is an infrared (long distance, large volume) divergency, the divergence of the t-integral for short proper time $t \rightarrow 0$ is an ultraviolet (short distance, high energy) divergency. Thus a careful treatment involves to work on a finite volume (infrared cut-off) and with finite lower boundary ϵ for the t-integration (ultraviolet cut-off). After computing quantities corresponding to physical observable, one takes the thermodynamic limit (infinite volume) and removes the UV cut-off. At this step divergent terms have to be subtracted, while fixing any ambiguities by imposing normalisation conditions. While IR divergencies can be dealt with unambiguously by restricting oneself to suitable (observable) quantities, the removal of UV divergencies may require one to specify infinitely many normalisation conditions. In this case the theory is called *non-renormalisable*. Such theories have limited predictive power since they can only be used as *effective theories*, valid below a fixed energy scale (physical UV cut-off). A finite UV cut-off allows one to only keep finitely many terms, while neglecting terms which are small at energies below the cut-off. Theories where UV divergencies can be removed by imposing finitely many normalisation conditions are called *renormalisable* theories, and are potentially valid at arbitrarily high energies.

The example we have used above can be generalised in various directions. If we consider theories with more than one species of particles, the partition function will include a sum over their masses. If we include particles with spin and charges, this will introduce coefficients which reflect the degeneracy of states with a given mass, which belong to the same spin or internal symmetry multiplet. Finally, if we include fermions, these will enter with a sign relative to bosonic states, which reflects that their vacuum energy has the opposite sign.

We note that the method we have used to obtain the partition function needs to be modified when it is applied to massless particles, since the operator $-\Delta$ has zero modes which make its determinant vanish, and the trace of its logarithm diverge. A formula for the partition function can still be obtained by treating the zero-modes separately.[2]

11.2 Partition Functions for Open Strings

We now turn to open strings and define the analogue of (11.1). Instead of a circular word-line we now have a world-sheet annulus, or, equivalently, a cylinder, which

---

[2] See, for example, Polchinski (1998a), Weinberg (1987), and Ginsparg (1988).

arises from letting an open string propagate for a finite amount of time and then identifying world-sheet times $\sigma^0 \cong \sigma^0 + 2\pi t$. As Hamiltonian we take the light-cone Hamiltonian $L_0^\perp - 1$, and define the world-sheet partition function as

$$Z_{\text{LC}}^{\text{open}}(t) = \text{Tr}_{\mathcal{H}_{\text{LC}}} q^{L_0^\perp - 1}, \quad q = e^{-2\pi t}. \tag{11.11}$$

Since we know that the physical states of an open string correspond to an infinite tower of particles with increasing masses, we expect that (11.11) is an infinite sum of particle partition functions of the form (11.1), which thus encodes the number d_N of states at a given level N. To check this we first need to compute $Z_{\text{LC}}^{\text{open}}(t)$. One way is to use that the world-sheet Hamiltionian is the sum of a particle Hamiltonian (describing the centre of mass motion) and of an infinite sum of harmonic oscillator Hamiltonians. Computing the relevant harmonic oscillator partition function is the subject of Exercises 11.2.1–11.2.4.

Alternatively, we can adapt computations that we did in Chapter 6. There, we evaluated the chiral partition function for a closed string on a one-dimensional target space, that is for central charge $c = 1$:

$$Z_{(c=1),\text{osc}}^{\text{closed}}(\tau) = \text{Tr} q^{L_0 - \frac{1}{24}} = \eta(\tau)^{-1} = q^{-1/24} \prod_{k=1}^{\infty} \frac{1}{1-q^k}, \quad q = e^{2\pi\tau}, \tag{11.12}$$

where the complex parameter τ labels conformally inequivalent world-sheet tori. To obtain the oscillator part of $Z_{\text{LC}}^{\text{open}}(t)$, we proceed as follows. First, we have to take into account that the ends of an open string are not identified with one another, which prevents us to twist the σ^0-evolution by a σ^1-translation. Therefore, τ is purely imaginary, $\tau = it$. Second, the closed string computation was for a one-dimensional target space, whereas for the open string in the light-cone gauge the target space is effectively 24-dimensional. This is taken into account by raising the closed string result to the 24-th power. We have seen that for a conformal field theory with central charge c the operator L_0 is shifted by $-\frac{c}{24}$ when we map from the cylinder to the plane. For $c = 24$ this is the same shift $a = 1$ that we have obtained in chapter 10 using ζ-function regularisation, and, independently, by imposing closure of the Lorentz Lie algebra in the light cone gauge. This is one of many examples for the mutual consistency of world-sheet and space-time methods. Taking into account the contribution of the centre of mass motion, we obtain

$$Z_{\text{LC}}^{\text{open}}(t) = (it)^{-13} \eta(it)^{-24}. \tag{11.13}$$

Note that we have seen when discussing the relativistic particle that the correct power of the Schwinger parameter t is $t^{-D/2} = t^{-13}$. In contrast, for the oscillator degrees of freedom only light-cone directions count. The oscillator part

$$Z_{\text{osc}}^{\text{open}}(it) = \eta^{-24}(it) \tag{11.14}$$

which is new compared to the single particle partition function (11.1) encodes how many physical states an open string has at a given level. More precisely, if we expand

$$Z_{\text{osc}}^{\text{open}}(q) = \sum_{N=0}^{\infty} d_N q^{N-1}, \quad q = e^{2\pi it}, \tag{11.15}$$

then d_N is the number of states at level N.

For small N it is easy to obtain the coefficients d_N explicitly:

$$Z_{osc}^{open}(q) = q^{-1}\left(1 + 24q + 324q^2 + 3{,}200q^3 + \cdots\right), \qquad (11.16)$$

which indeed correctly encodes the number of open string states at levels $N = 0, 1, 2, 3$.

Evaluating d_N becomes more and more tedious for larger N. While no closed expression exists, one can obtain analytic expressions for the asymptotic behaviour of d_N for $N \to \infty$. If we interpret the partition function thermodynamically, this corresponds to performing a high temperature expansion. Mathematically, the problem of counting string states on a one-dimensional target space is the problem of counting the *partitions* of positive integers, that is, the inequivalent ways in which a positive integer can be written as a sum of positive integers. The asymptotic number of partitions can be computed by the *Hardy–Ramanujan formula*. The higher-dimensional case with indices $i, j, \ldots = 1, \ldots, d$ corresponds to a modified partitioning problem where numbers carry d distinct 'colours'. The generalised version of the Hardy–Ramanujan formula for this case is

$$d_N \approx \frac{1}{\sqrt{2}}\left(\frac{d}{24}\right)^{(d+1)/2} N^{-(d+3)/2} \exp\left(2\pi\sqrt{\frac{Nd}{6}}\right). \qquad (11.17)$$

We will obtain the leading term of this asymptotic formula using a high temperature expansion as an exercise (see Exercise 11.2.6). In this exercise we will compute the associated thermodynamic entropy, which is equal to the statistical entropy in the thermodynamic limit. The statistical entropy of a state with total excitation number N is the (natural) logarithm of the number of states with excitation number N:

$$S[N] = \log d_N \approx 2\pi\sqrt{\frac{Nd}{6}} - \frac{d+3}{4}\log N - \log\sqrt{2} + \frac{d+1}{4}\log\frac{d}{24} + \cdots \qquad (11.18)$$

In the exercise we will only be able to obtain the first term. As indicated, there are further contributions to $\log d_N$, which are not captured by the generalised Hardy–Ramanujan formula. These contributions vanish in the limit $N \to \infty$, but can be computed systematically using the so-called *Rademacher expansion* (see Section 11.4).

The space-time partition is obtained by summing over all possible one-loop world-sheet partition functions. Using the Schwinger proper time representation, we obtain the open string one-loop partition function

$$Z_1^{open} \cong V_{26}\int_0^\infty \frac{dt}{t}t^{-13}\eta(it)^{-24}. \qquad (11.19)$$

We have neglected numerical factors but display the volume factor. Compared to (11.9) we have dropped the log on the left hand side, to be consistent with conventions in the literature when working in a 'first quantised' formulation.

Exercise 11.2.1   Compute the partition function

$$Z_1 = \text{Tr}\,e^{-\beta H}$$

for a simple quantum harmonic oscillator

$$H = \hbar\omega\left(a^\dagger a + \frac{1}{2}\right), \quad [a, a^\dagger] = \mathbb{1}.$$

The trace over the Hilbert space $\mathcal{H}$ is defined as the sum over an orthonormal basis.

Exercise 11.2.2 Compute the partition function $Z_{(d)}$ for a system of d decoupled harmonic oscillators with the same frequency

$$H = \sum_{i=1}^{d} \hbar\omega \left((a^i)^\dagger a^i + \frac{1}{2} \right), \quad [a^i, (a^j)^\dagger] = \delta^{ij}\mathbb{1}.$$

Remark: these harmonic oscillators are distinguishable.

Exercise 11.2.3 Compute the partition function Z_N for N decoupled harmonic oscillators

$$H = \sum_{n=1}^{N} \hbar\omega_n \left(a_n^\dagger a_n + \frac{1}{2} \right), \quad [a_m, a_n^\dagger] = \delta_{mn}\mathbb{1}$$

with frequencies $\omega_n = n\omega_1$.

Exercise 11.2.4 Find the limit $Z_N \xrightarrow[N\to\infty]{} Z_\infty$ of the partition function computed in Exercise 11.2.3, using Appendices D.1 and D.2.

Exercise 11.2.5 Expand the partition function for low levels and show that it correctly encodes the number of physical states.

Exercise 11.2.6 Consider a system of infinitely many harmonic oscillators with Hamiltonian

$$H = \sum_{i=1}^{d} \sum_{n=1}^{\infty} n a_n^\dagger a_n.$$

For $d = 24$ this is the normal ordered light-cone Hamiltonian for the open string. We have neglected the shift in the ground state energy since we are interested in the asymptotic number of states for large energy. The energy of a state with occupation numbers $\{n_i\} = \{n_1, n_2, \ldots\}$ is

$$E_{\{n_i\}} = n_1 + 2n_2 + 3n_3 + \cdots = N,$$

where N is total occupation number. The associated partition function is

$$Z = \sum_{\{n_i\}} e^{-\beta E_{\{n_i\}}} = \sum_{N=0}^{\infty} d_N e^{-\beta N},$$

where $\beta = 1/T$ is the inverse temperature (we have set the Boltzmann constant to unity, $k_B = 1$), and where d_N is the number of states with energy $E = N$ that we want to find asymptotically for $N \to \infty$. The partition function Z for this system follows from the solution to the previous exercises. Given the partition function, we can compute the free energy F, entropy S and energy E of the corresonding canonical ensemble using the standard formulae

$$F = -\frac{1}{\beta} \log Z, \quad S = \beta^2 \frac{\partial F}{\partial \beta}, \quad E = -\frac{\partial \log Z}{\partial \beta}. \tag{11.20}$$

Show that

$$F = \frac{1}{\beta} \sum_{l=1}^{\infty} \log \left(1 - e^{\beta l} \right)^{-1}.$$

Approximate the sum by an integral, which you can solve in the limit of large temperature using the the Taylor expansion of $\log(1 - x)$, and thus show that

$$F = -\frac{d}{\beta^2}\frac{\pi^2}{6}.$$

Find S and E as functions of β and use your result to express S as a function of E. For $E = N$, using the relation between entropy and the number of states, you should find that the asymptotic number of states is

$$\log d_N \approx 2\pi\sqrt{\frac{Nd}{6}}.$$

11.3 Partition Functions for Closed Strings

For closed strings we can build upon our previous discussion in Chapter 6. The chiral partition function is defined as

$$Z_{\text{LC}}(\tau) = \text{Tr}_{\mathcal{H}_{\text{osc}}^{\text{LC}}} q^{L_0^\perp - 1}, \quad q = e^{2\pi i\tau},$$

where the trace is over the oscillator degrees of freedom. For $d = 24$ transverse degrees of freedom we obtain

$$Z_{\text{LC}}(\tau) = \eta^{-24}(\tau) = \frac{1}{q\prod_{k=1}^{\infty}(1 - q^k)}.$$

Combining left-movers and right-movers, the full oscillator partition function is

$$|Z(\tau)|^2 = \text{Tr}_{\mathcal{H}_{\text{osc}}^{\text{LC}}}\left(q^{L_0^\perp - 1}\bar{q}^{\tilde{L}_0^\perp - 1}\right) = |\eta(q)|^{-48}.$$

This does not yet take into account the centre of mass motion. As discussed in Chapter 6, each momentum component contributes a factor $\tau_2^{-1/2}$ to the partition function, where $\tau_2 = \text{Im}(\tau)$. The operators $L_0^\perp, \tilde{L}_0^\perp$ contain $n_T = 24$ transverse components of the momentum. Including them into the trace gives

$$Z_{\text{LC}}'(\tau, \bar{\tau}) = \tau_2^{-12}|\eta(q)|^{-48}.$$

This expression is modular invariant. However, from our discussion of the relativistic particle we know that the contribution of centre of mass degrees of freedom actually is $\tau_2^{-D/2} = \tau_2^{-13}$. This reflects that gauge symmetries work slightly different for the centre of mass and the oscillator degrees of freedom: oscillators are purely transverse, whereas momenta are not. Thus, the correct world-sheet partition function is

$$Z_{\text{LC}}(\tau, \bar{\tau}) = \tau_2^{-13}|\eta(q)|^{-48},$$

which is not modular invariant. However, the object that we need to be modular invariant is the space-time partition function that we obtain by summing over all inequivalent world-sheet tori, that is by integrating over $\tau = \tau_1 + i\tau_2$. From the previous discussion of the open string, we know that τ_2 corresponds to the Schwinger proper time t which we need to integrate over the interval $0 < \tau_2 < \infty$. For the closed string we also have to integrate over τ_1. Since tori related by $\tau_1 \rightarrow \tau_1 + 1$ are conformally equivalent, we restrict the integration to the interval $-\frac{1}{2} < \tau_1 < \frac{1}{2}$ and obtain

$$Z'_{\text{closed}} = V_{26} \int_{-1/2}^{1/2} d\tau_1 \int_0^\infty \frac{d\tau_2}{\tau_2} \tau_2^{-13} |\eta(q)|^{-48} = V_{26} \int_{-1/2}^{1/2} d\tau_1 \int_0^\infty \frac{d\tau_2}{\tau_2^2} \tau_2^{-12} |\eta(q)|^{-48}.$$

In the second step we have rearranged factors τ_2 so that the integrand is modular invariant. Modular invariance of Z'_{closed} requires that the integration measure $\frac{d\tau_1 d\tau_2}{\tau_2^2}$ is modular invariant. This is correct, since the measure is even invariant under the action

$$\tau \mapsto \frac{a\tau + b}{c\tau + d}, \quad \begin{pmatrix} a & b \\ c & d \end{pmatrix} \in \text{SL}(2, \mathbb{R}) \tag{11.21}$$

of the group $\text{SL}(2, \mathbb{R}) \supset \text{SL}(2, \mathbb{Z})$ on the upper half plane $\mathcal{H} = \{\tau \in \mathbb{C} | \tau_2 > 0\}$ (see Exercise 11.3.1). The action of $\text{SL}(2, \mathbb{R})$ on the upper half plane is transitive, with stabiliser group $\text{SO}(2)$. Therefore, $\mathcal{H}$ is a *homogeneous space*, $\mathcal{H} \cong \text{SL}(2, \mathbb{R})/\text{SO}(2)$. Moreover, the Riemannian space $(\mathcal{H}, \frac{d\tau_1 d\tau_2}{\tau_2^2})$ is a global Riemannian symmetric space.[3]

There is one last step we have to take. To count conformally inequivalent tori only once, we need to restrict to a fundamental domain $\mathcal{F} \subset \mathcal{H}$ of the action of the modular group on $\mathcal{H}$, as discussed in Chapter 6. By the restriction of the τ_1-integration we have already taken care of one of the generators of the modular group, $\tau \to \tau + 1$. The second generator is $\tau \to -\frac{1}{\tau}$, and, therefore, the natural choice of $\mathcal{F}$ is to remove the interior of the unit circle from the integration domain as well (see Figure 6.5). This is our final result:

$$Z_1^{\text{closed}} = V_{26} \int_{\mathcal{F}} d^2\tau \, \tau_2^{-14} |\eta(q)|^{-48}.$$

This gives rise to an important observation. For particles UV-divergencies arise from the limit $\tau_2 = t = \epsilon \to 0$. For closed strings this region is removed from the integration domain by the requirement of modular invariance. This illustrates that theories of closed strings have an intrinsic mechanism which prevents UV-divergencies.

Exercise 11.3.1 Show that the measure $\frac{d\tau_1 d\tau_2}{\tau_2^2}$ is invariant under the action $\tau \to (a\tau + b)(c\tau + d)^{-1}$ of $\text{SL}(2, \mathbb{R})$.

11.4 Further Remarks and Literature

The world-line formalism for quantum field theories is reviewed in Schubert (2001). The asymptotic expansion of partition functions has been studied in much detail in the mathematical and physical literature under the headings of *Rademacher expansion* and *Farey tail transform*. See, for example, Dijkgraaf et al. (2000) and de Boer et al. (2006). One important application is to show that the statistical entropy of certain D-brane configurations is equal to the thermodynamic entropy of black

---

[3] See Appendix G for a brief summary of homogeneous and symmetric spaces.

holes carrying the same quantum numbers, which suggests that D-branes can be used as models for the microstates of black holes, see Strominger and Vafa (1996), Maldacena (1996), Horowitz and Polchinski (1997). This description involves an interpolation in coupling space between a D-brane configuration in an on-shell background where backreaction is neglected but excited string states are included, and an effective field theory (supergravity) description of the black hole that forms once backreaction onto space-time is taken into account. Such an interpolation can be justified for BPS states, a concept that we will explain in Chapter 14.

PART IV

OUTLOOK

12   Interactions

At this point, we have obtained a solid understanding of the quantum theory of free strings. The resulting theory is bosonic, since it only contains tensor representations of the Poincaré group, requires $D = 26$ space-time dimensions, and has a tachyonic ground state. In Part IV, we address four immediate questions: how to introduce interactions, how to make string theory four-dimensional, how to add matter (fermions), and how to get rid of tachyons. In this chapter, we discuss interactions from two complementary perspectives: defining scattering amplitudes of string states in a fixed space-time background, and the dynamics of the background. Both approaches lead one to an effective field theory for the lightest modes of the string, which provides a bridge between the world-sheet formulation of string theory and standard quantum field theory, particle theory, and cosmology.

12.1 Amplitudes

In this section, we explain how scattering amplitudes of strings in Minkowski space-time are defined. We will restrict ourselves to closed strings, and explicitly compute the tree-level scattering between two ground state scalars (tachyons). Then, we discuss graviton scattering, characteristic properties of string amplitudes and how to obtain an effective action.

12.1.1 General Discussion of Amplitudes

Since we are building string theory from the world-sheet perspective, the following definition of scattering amplitudes will seem ad hoc. In short, we will postulate that space-time scattering amplitudes (S-matrix elements) can be defined using integrated correlation functions of the two-dimensional world-sheet field theory. In quantum field theory, there is a systematic way of deriving scattering amplitudes from fundamental principles, and the LSZ reduction formula expresses scattering amplitudes in terms of space-time correlation functions. The 'first quantised' world-sheet approach that we use in string theory does not allow such a derivation of scattering amplitudes. Historically string theory has evolved the other way round: certain amplitudes which were of interest for the theory of strong interactions in the 1970s were interpreted as scattering amplitudes of strings. We will understand how this came about once we have computed our first amplitude. For the time being, we postulate the following:

1. The total amplitude for a process with M external states is given by the sum of contributions from all gauge inequivalent world-sheets connecting the external states.
2. The contribution of a fixed world-sheet is given by the integrated correlation function of the vertex operators corresponding to the external states.

Let us fill in some more details. In a theory of closed oriented strings external states correspond to circles and the world-sheets representing scattering processes are connected oriented surfaces which have these circles as their boundaries. In the Polyakov approach, after imposing the conformal gauge, these surfaces are equipped with a conformal structure. Using conformal invariance, we can map the circles to points, and conformal field theory allows us to generate any physical state using a vertex operator inserted at the corresponding point. By trading boundaries for inserting vertex operators we now have world-sheets which are oriented and closed, that is compact and without boundary. The sum over inequivalent surfaces includes a sum over topologies. For closed oriented surfaces, the topological type is classified by a single invariant, the *genus g*. Starting from the two-sphere which has $g = 0$, surfaces with $g > 0$ are obtained by attaching g handles. For example a closed oriented surface with $g = 1$ is a torus. Instead of the genus g we can use the *Euler number* $\chi_g = 2 - 2g$, which is related to the integrated curvature of the surface.[1]

The basic interaction process for closed strings is the three-string vertex (see Figure 9.1), which describes the splitting of one string into two strings, or the fusion of two strings into one. To this vertex we associate a coupling constant, which, for the time being, we denote by κ. We note that in theories of point particles, which diagrammatically are described by graphs, we can have a variety of distinct vertices, each with its own coupling constant (see Figure 1.3). In string theory, the coupling constant is unique.[2]

For illustration, see Figure 12.1 for the first terms in the topological expansion (genus expansion) of a four-point amplitude. Since increasing the genus of the world-sheet by one introduces two closed string vertices, it follows that a genus g surface with M external states has $M + 2g - 2 = M - \chi(g)$ vertices and carries a factor $\kappa^{M-\chi(g)}$. Therefore, the topological expansion of an amplitude with M external states takes the form

$$A(1, 2, \ldots, M) = \sum_{g=0}^{\infty} \kappa^{M-\chi(g)} A(1, 2, \ldots)_g. \tag{12.1}$$

Apart from its topology and the vertex operators, the world-sheet carries a metric. In order for this metric to be globally well defined, we assume that the theory is defined in a Euclidean setting, with scattering amplitudes evaluated for physical momenta at the end.[3] Since we work with the Polyakov string in the conformal gauge, the metric has been gauge-fixed. While locally we can transform any world-sheet metric to the standard flat metric, this is not possible globally in general. For genus $g > 0$

---

[1] As this is a topological invariant, one can use any globally defined metric to compute the integrated curvature.

[2] This feature persists if we consider theories which include open strings, since the closed string coupling constant, open string coupling constant, and the coupling constants between closed and open strings are related to each other by unitarity.

[3] The Euclidean formulation was discussed in Section 4.1.

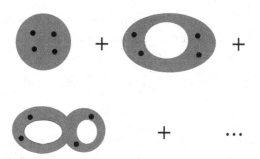

Fig. 12.1 The topological expansion of a four-point amplitude. Observe that these diagrams contain, two, four, and six closed string vertices, respectively.

there is a finite-dimensional family of conformally inequivalent metrics, and the global version of the conformal gauge is to choose a representative for each class. The fixed-genus amplitude $A(1, 2, \ldots)_g$ therefore includes a sum over conformally inequivalent surfaces of genus g. For an oriented surface a conformal structure is equivalent to a complex structure, and an oriented surface with a complex structure is called a *Riemann surface*. Thus closed string perturbation theory is a sum over Riemann surfaces which breaks up into a sum over topologies and an integral over conformal/complex structures. For $g = 0$, the conformal structure is unique and no integral needs to be performed. The conformal structure can be represented by the standard round metric on the two-sphere. The $g = 1$ surface is the torus, and we have seen in Section 6.1 that there is a complex one-dimensional family of complex structures labelled by the modular parameter τ which takes values in a fundamental domain $\mathcal{F} \subset \mathcal{H}$ in the upper half plane. Moreover, we know that the integral over complex structures is performed using the invariant measure $d^2\tau/\tau_2^2$. For $g > 1$ the so-called Riemann–Roch theorem implies that there is a complex $(3g - 3)$-dimensional family of conformally inequivalent surfaces. Schematically amplitudes take the form

$$A(1, 2, \ldots)_g = \int_{\mathcal{F}_g} d\mu_g(\vec{m}) A(1, 2, \ldots)_{g, \vec{m}}, \qquad (12.2)$$

where as indicated one has to integrate over the moduli space of conformal structures of a closed oriented surface Σ_g of genus g. Here, $d\mu_g(\vec{m})$ denotes the measure on this moduli space, and $\mathcal{F}_g$ the fundamental domain with respect to the action of the genus g modular group.

In the following, we focus on tree level amplitudes ($g = 0$), where the integration over world-sheet moduli is absent. Physical states are characterised by their momentum $k = (k^\mu)$ and polarisation tensor $\zeta = (\zeta_{\mu_1 \cdots \mu_n})$. To each such state we can associate a local vertex operator $V[k, \zeta](z, \bar{z})$, which is a conformal primary of weight $(1, 1)$. The CFT correlation function associated to a tree level process with M external string states is[4]

$$\langle 0 | V[k_1, \zeta_1](z_1, \bar{z}_1) \cdots V[k_M, \zeta_M](z_M, \bar{z}_M) | 0 \rangle. \qquad (12.3)$$

---

[4] On the sphere there is no global flat metric. But after performing a stereographic projection from the sphere to the complex plane, we can change the metric to the standard flat metric by performing a conformal transformation. As we will see in Section 12.1.2, the correlation function can be computed using methods from Part II.

From the space-time point of view, the location of vertex operators on the world-sheet has no invariant meaning. What enters into string amplitudes is the integrated world-sheet correlator

$$A(1, 2, \ldots)_{g=0} = \kappa^{M-2} \int_{\Sigma_0} d^2 z_1 \cdots d^2 z_M \langle 0 | V[k_1, \zeta_1](z_1, \bar{z}_1) \cdots$$
$$\cdots V[k_M, \zeta_M](z_M, \bar{z}_M) | 0 \rangle.$$

Note that since physical vertex operators have weight $(1, 1)$ this integral is conformally invariant. One point that still needs to be addressed is the existence of globally defined finite conformal transformations. Such transformations map configurations $z_1, z_2, \ldots z_M$ to equivalent configurations and, therefore, integrating over all positions amounts to an over-counting.

For $g = 0$ the conformal group $SL(2, \mathbb{C})$ is complex three-dimensional. To avoid over-counting we fix the positions of three vertex operators, the standard choice being $z_1, z_2, z_3 = \infty, 1, 0$.[5] Thus, the general form of the genus-0 closed string amplitude is

$$A(1, 2, \ldots)_{g=0} = \kappa^{M-2} \int_{\Sigma_0} d^2 z_4 \cdots d^2 z_M \langle 0 | V[k_1, \zeta_1](\infty) V[k_2, \zeta_2](1) V[k_3, \zeta_3](0)$$
$$V[k_4, \zeta_4](z_4, \bar{z}_4) \cdots V[k_M, \zeta_M](z_m, \bar{z}_M) | 0 \rangle. \tag{12.4}$$

As we can see, this prescription provides us with scattering amplitudes $A(1, 2, \ldots)$ which are functions of the momenta $k_1, k_2, \ldots$ and of the polarisation tensors $\zeta_1, \zeta_2 \ldots$ of the external states. Which states are initial states and which are final states depends on the choice of momenta. Scattering amplitudes are analytic functions of momenta and polarisations. Thus amplitudes where a particular particle is 'ingoing' are related by analytic continuation to other amplitudes where the same particle is 'outgoing'. Moreover one can relate momenta which are on-shell in Minkowski space-time to Euclidean momenta.

12.1.2 The Four Scalar Amplitude

For illustration we compute the simplest closed string scattering amplitude explicitly, the scattering of two scalar ground states (tachyons). As a first step we compute world-sheet correlation function

$$\langle -k_1 | : \exp(ik_2 \cdot X(z, \bar{z})) :: \exp(ik_3 \cdot X(w, \bar{w})) : | k_4 \rangle \tag{12.5}$$

of normal ordered exponentials between momentum eigenstates. Our conventions for momenta are

$$\langle 0 | e^{ikx} = \langle -k | , \quad \langle -k | l \rangle = \delta(k + l). \tag{12.6}$$

The product of two normal-ordered exponentials was computed before:

$$: \exp(ik_2 \cdot X(z, \bar{z})) :: \exp(ik_3 \cdot X(w, \bar{w})) :$$
$$= |z - w|^{\frac{k_2 \cdot k_3}{2}} \exp(ik_2 \cdot X(z, \bar{z}) + ik_3 \cdot X(w, \bar{w})).$$

---

[5] For $g = 1$, the only global conformal transformations are translations, which can be gauge-fixed by fixing the position of one vertex operator. For $g > 1$, there are no global conformal transformations.

We have seen that for this type of operator normal ordering by subtracting singular terms from the OPE is equivalent to normal ordering creation and annihilation operators. Therefore,

$$: \exp\left(ik_2 \cdot X(z,\bar{z}) + ik_3 \cdot X(w,\bar{w})\right):$$

$$= \exp\left(-\frac{1}{2} \cdot \sum_{n<0} \frac{1}{n} k_2 \cdot [\alpha_n z^{-n} + \tilde{\alpha}_n \bar{z}^{-n}] + \frac{1}{n} k_3 \cdot [\alpha_n w^{-n} + \tilde{\alpha}_n \bar{w}^{-n}]\right)$$

$$e^{i(k_2+k_3)\cdot x} |z|^{(k_2 \cdot p)/2} |w|^{(k_3 \cdot p)/2}$$

$$\exp\left(-\frac{1}{2} \sum_{n>0} \frac{1}{n} k_2 \cdot [\alpha_n z^{-n} + \tilde{\alpha}_n \bar{z}^{-n}] + \frac{1}{n} k_3 \cdot [\alpha_n w^{-n} + \tilde{\alpha}_n \bar{w}^{-n}]\right). \tag{12.7}$$

When evaluating this expression between momentum eigenstates, the terms involving oscillators do not contribute, due to normal ordering. Therefore, the correlation function is

$$|z - w|^{\frac{k_2 \cdot k_3}{2}} \langle -k_1 | e^{i(k_2+k_3)\cdot x} |z|^{(k_2 \cdot p)/2} |w|^{(k_3 \cdot p)/2} |k_4\rangle$$

$$= |z - w|^{\frac{k_2 \cdot k_3}{2}} |z|^{\frac{k_2 \cdot k_4}{2}} |w|^{\frac{k_3 \cdot k_4}{2}} \langle -k_1 - k_2 - k_3 | k_4\rangle$$

$$= |z - w|^{\frac{k_2 \cdot k_3}{2}} |z|^{\frac{k_2 \cdot k_4}{2}} |w|^{\frac{k_3 \cdot k_4}{2}} \delta(k_1 + k_2 + k_3 + k_4). \tag{12.8}$$

The δ-function imposes momentum conservation. Note that we have chosen a convention where all momenta are on the same footing and appear with the same sign. The vertex operator for particle 1 is located at ∞ and the one for particle 4 at 0 in the complex plane. To completely fix conformal invariance we set $z = 1$ for particle 2, and we relabel $w \to z$ for particle 3. While operator products are only defined if radially ordered, the resulting correlations functions can be extended analytically to meromorphic functions. The resulting the four scalar tree level amplitude is

$$A_4 = \kappa^2 \delta\left(\sum_{i=1}^{4} k_i\right) \int d^2z |1 - z|^{\alpha' k_2 \cdot k_3} |z|^{\alpha' k_3 \cdot k_4}, \tag{12.9}$$

where we have reconstructed the dimensionful Regge slope parameter α'. This integral can be evaluated using the integral representations of the Euler Γ- and Beta-functions (see Exercise 12.1.2). The result is

$$A_4 = \pi \kappa^2 \frac{\Gamma\left(-\frac{a}{2} - \frac{b}{2} - 1\right) \Gamma\left(\frac{a}{2} + 1\right) \Gamma\left(\frac{b}{2} + 1\right)}{\Gamma\left(-\frac{a}{2}\right) \Gamma\left(-\frac{b}{2}\right) \Gamma\left(\frac{a}{2} + \frac{b}{2} + 2\right)} \delta\left(\sum_{i=1}^{4} k_i\right), \tag{12.10}$$

where

$$a = \alpha' k_3 \cdot k_4, \quad b = \alpha' k_2 \cdot k_3. \tag{12.11}$$

To analyse the properties of this amplitude it is useful to introduce the *Mandelstam variables*

$$s = -(k_1 + k_2)^2 = -(k_3 + k_4)^2, \quad t = -(k_1 + k_3)^2 = -(k_2 + k_4)^2,$$
$$u = -(k_1 + k_4)^2 = -(k_2 + k_3)^2, \tag{12.12}$$

which are the Lorentz-invariant kinematical variables for a scattering process with two particles in the initial and in the final state. They are not independent (see

Exercise 12.1.1). Using the closed string mass formula, we can compute the tachyon mass

$$\alpha' M^2 = N + \tilde{N} - 4 \xrightarrow{N = \tilde{N} = 0} M^2 = -\frac{4}{\alpha'}. \tag{12.13}$$

Therefore,

$$s = -2k_3 \cdot k_4 - \frac{8}{\alpha'} \Rightarrow a = \alpha' k_3 \cdot k_4 = -\frac{\alpha'}{2} s - 4, \tag{12.14}$$

and similarly $b = -\frac{\alpha'}{2} u - 4$. The symmetries of the amplitude become manifest if we introduce

$$\alpha(x) := \frac{\alpha'}{2} x + 2, \quad x = s, t, u. \tag{12.15}$$

Note that $\alpha(s) + \alpha(t) + \alpha(u) = -2$ by momentum conservation. In terms of $\alpha(x)$ the amplitude takes the form

$$A_4 \cong \kappa^2 \frac{\Gamma\left(-\frac{1}{2}\alpha(s)\right) \Gamma\left(-\frac{1}{2}\alpha(t)\right) \Gamma\left(-\frac{1}{2}\alpha(u)\right)}{\Gamma\left(-\frac{1}{2}\alpha(t) - \frac{1}{2}\alpha(u)\right) \Gamma\left(-\frac{1}{2}\alpha(s) - \frac{1}{2}\alpha(u)\right) \Gamma\left(-\frac{1}{2}\alpha(s) - \frac{1}{2}\alpha(t)\right)}, \tag{12.16}$$

where we have suppressed numerical factors and the momentum conservation δ-function. The amplitude is now manifestly symmetric under permutations of the Mandelstam variables s, t, u. This property is referred as *duality*, which here refers to a duality between the kinematical channels where resonances occur. Historically, *dual amplitudes* and *dual resonance models* were the starting point of what we now know as string theory.

Poles in a scattering amplitude correspond to *resonances*, that is to the presence of intermediate on-shell states. At such points in momentum space the four-point amplitude splits into the product of the three-point amplitudes corresponding to the production and decay of the intermediate states. More precisely, the product of the two three-point amplitudes is the residue of a pole in $x - M^2$ in the four-point amplitude, where M is the mass of the intermediate particle. Poles in $x = s, t, u$ are said to occur in the x-channel.

Since poles in scattering amplitudes encode information about the particle spectrum of the theory, we can use this as a consistency test for our ad hoc definition of amplitudes. The Γ-function does not have zeros, but it has poles for $u = 0, -1, -2, \ldots$. Therefore the amplitude (12.10) has poles for

$$\alpha' x = -4, 0, 4, 8, \ldots = 2(N + \tilde{N} - 2), \tag{12.17}$$

which precisely matches the closed string mass formula. While the lowest resonance corresponds to an intermediate scalar, the amplitude also has an infinite series of further poles in all channels which correspond to intermediate excited closed string states. It is this property of dual amplitudes which lead to their re-interpretation as a theory of strings. String amplitudes have the special property that they can be written as infinite sums over resonances in one single channel. Thus the resonances in one channel encode the resonances in the other channels, which is only possible because there is an infinite number of resonances.[6]

---

[6] See, e.g., Green et al. (1987, Chapter 1); and West (2012, Chapter 18).

Exercise 12.1.1 Show that momentum conservation in a process involving four external particles with masses $m_i^2 = k_i^2$, $i = 1, 2, 3, 4$, implies that

$$s + t + u = m_1^2 + m_2^2 + m_3^2 + m_4^2. \tag{12.18}$$

Use this to show that $\alpha(s) + \alpha(t) + \alpha(u) = -2$ in a process involving the closed string scalar ground state.

Exercise 12.1.2 Show that

$$\int d^2z|z|^a|1 - z|^b = \pi \frac{\Gamma\left(-\frac{a}{2} - \frac{b}{2} - 1\right)\Gamma\left(\frac{a}{2} + 1\right)\Gamma\left(\frac{b}{2} + 1\right)}{\Gamma\left(-\frac{a}{2}\right)\Gamma\left(-\frac{b}{2}\right)\Gamma\left(\frac{a}{2} + \frac{b}{2} + 2\right)}.$$

Instruction: Use the integral representation of the Γ-function to derive the formula:

$$\frac{1}{x^{2w}} = \frac{1}{\Gamma(w)} \int_0^\infty t^{w-1} e^{-tx^2}.$$

Use this integral representation to convert the z-integral into a Gaussian integral. The remaining integrals can be performed using the integral representations of the Γ-function and of the Euler Beta-function (see Appendix D.3).

12.1.3 The Four Graviton Amplitude

It is instructive to consider one further example, the scattering of four massless closed string states. Suppressing numerical factors and the momentum conservation δ-function, the amplitude takes the form

$$A_4 \cong \kappa^2 \frac{\Gamma\left(-\frac{1}{4}\alpha's\right)\Gamma\left(-\frac{1}{4}\alpha't\right)\Gamma\left(-\frac{1}{4}\alpha'u\right)}{\Gamma\left(1 + \frac{\alpha'}{4}s\right)\Gamma\left(1 + \frac{\alpha'}{4}t\right)\Gamma\left(1 + \frac{\alpha'}{4}u\right)} K(\zeta_i, k_i), \tag{12.19}$$

where the factor $K(\zeta_i, k_i)$ contains the explicit dependence on polarisation tensors and momenta. The dependence on s, t, u through the Γ-functions is related to the one of the tachyon amplitude by shifting all arguments by $+1$.[7] The amplitude has poles for

$$\alpha'x = 0, 4, 8, \ldots, \quad x = s, t, u. \tag{12.20}$$

Thus, besides massless poles there is an infinite tower of massive poles corresponding to the massive excitations of a closed string. Note that there is no pole corresponding to the tachyon.

In general relativity graviton scattering amplitudes in Minkowski space-time are defined by splitting the metric into a background Minkowski metric and a dynamical graviton field, $g_{\mu\nu} = \eta_{\mu\nu} + \psi_{\mu\nu}$. In Section 9.1, we used this to argue that the stringy graviton has the same kinematical properties as a graviton in general relativity. Expanding to higher orders in $\psi_{\mu\nu}$ one can work out the Feynman rules and compute scattering amplitudes. When restricting to tree level diagrams, divergencies and renormalisation are not an issue. To compare the resulting amplitude A_4^{GR} to the string amplitude A_4 we take a low energy or particle limit of A_4 by sending the string tension to infinity, or, equivalently, the Regge slope parameter to zero, $\alpha' \to 0$. Using

---

[7] The arguments in the denominator have been rewritten using that for massless particles $s + t + u = 0$.

the functional equation of the Γ-function, we can separate the massless poles from the massive poles:

$$\frac{\Gamma\left(-\frac{\alpha'}{4}x\right)}{\Gamma\left(1+\frac{\alpha'}{4}x\right)} = \frac{1}{-\frac{\alpha'}{4}x}\frac{\Gamma\left(1-\frac{\alpha'}{4}x\right)}{\Gamma\left(1+\frac{\alpha'}{4}x\right)} = \frac{-4}{\alpha'x}\left(1+O(\alpha')\right). \tag{12.21}$$

It can be shown that the amplitude obtained in the limit $\alpha' \to 0$ agrees with the one obtained from general relativity:

$$A_4 \xrightarrow[\alpha'\to 0]{} A_4^{\text{GR}}. \tag{12.22}$$

Therefore, the identification of the symmetric traceless part of the massless closed string state with a graviton does not only make sense kinematically, but also within the interacting theory. Like for scalar scattering, string theory modifies gravity by adding an infinite tower of massive states in a very specific way.

12.1.4 Effective Actions from Amplitudes

The matching of tree level ($g = 0$) string amplitudes to those derived from a classical action can be used to derive an effective action for the particles which appear as external states in a process. In quantum field theory the effective action is the functional which generates the one-particle-irreducible Feynman graphs, that is connected graphs which remain connected when cutting a single line. This effective action is related to the generating functional $\log Z$ of connected graphs by a Legendre transform. At the classical level (tree level in terms of graphs), the effective action coincides with the classical action. Since the amplitudes for graviton-graviton scattering in Einstein gravity and string theory are the same in the low energy limit, the effective action for the string graviton must be the Einstein–Hilbert action, up to corrections which vanish for $\alpha' \to 0$. To be precise, the Ricci scalar R is an infinite series in the graviton field $\psi_{\mu\nu}$ and the graviton-graviton amplitude only tests finitely many of these directly. However, the underlying gauge symmetry of the tensor field implies that the terms which are tested explicitly must come from the expansion of the Ricci scalar. The full string theory tree level amplitude differs from the gravitational amplitude by an infinite series of α'-corrections, which are related to the infinite tower of resonances. Since α' has the dimension Length^2, these corrections involve higher derivatives, which due to gauge symmetry must organise into polynomials formed out of the Riemann tensor and its contractions. Such terms are negligible at low energies, but become relevant at energies close to the string scale.

Apart from the graviton there are two further massless particles in the closed string spectrum, the Kalb–Ramond field $B_{\mu\nu}$ and the dilaton ϕ. The leading order effective action takes the form

$$S_{\text{eff}} = \frac{1}{2\kappa^2}\int d^D x\sqrt{G}e^{-2\phi}\left(R + 4\partial_\mu\phi\partial^\mu\phi - \frac{1}{12}H_{\mu\nu\rho}H^{\mu\nu\rho}\right), \tag{12.23}$$

where $H_{\mu\nu\rho} = \partial_\mu B_{\nu\rho} + \text{cyclic}$ is the field strength of $B_{\mu\nu}$. While $D = 26$ is the critical dimension, we have left the dimension D general, since the following discussion also applies to situations where the theory has been *compactified* to $D < 26$ dimensions, a topic which is the subject of Chapter 13.[8] The way the action is parametrised is adapted to string perturbation theory, but somewhat unusual from the field theory or gravity point of view, because the gravitational term $\sqrt{G}e^{-2\phi}R$ does not take the standard Einstein–Hilbert form $\sqrt{g}R$. This is a conventional choice, because the gravitational term can be forced to take the standard from by making a field redefinition, more precisely a dilaton dependent conformal transformation of the metric such that

$$\sqrt{G}e^{-2\phi}R_G = \sqrt{g}R_g, \tag{12.24}$$

where R_G, R_g are the Ricci scalars of the metrics G and g, respectively. The two conformal frames are called the *string frame* and the *Einstein frame*, respectively. To find the conformal transformation which maps the string frame to the Einstein frame, we make the ansatz

$$G_{\mu\nu} = e^{2A\phi}g_{\mu\nu} \Rightarrow \sqrt{|G|} = e^{AD\phi}\sqrt{|g|}, \tag{12.25}$$

where A is a constant which is fixed by imposing (12.24). One can show that[9]

$$R_G = e^{-2A\phi}\left[R_g - 2A(D-1)(D-2)g^{\mu\nu}\nabla^{(g)}_\mu\partial_\nu\phi - A^2(D-1)(D-2)g^{\mu\nu}\partial_\mu\phi\partial_\nu\phi\right], \tag{12.26}$$

where $\nabla^{(g)}$ is the covariant derivative (Levi–Civita connection) with respect to the metric $g_{\mu\nu}$. Imposing (12.24) fixes A:

$$DA - 2 - 2A = 0 \Rightarrow A = \frac{2}{D-2}. \tag{12.27}$$

The Einstein frame action is obtained from the string frame action (12.23) through expressing the string frame metric $G_{\mu\nu}$ in terms of the Einstein frame metric $g_{\mu\nu}$. Note that one needs to keep track which metric is used to contract indices. In the string frame $\partial_\mu\phi\partial^\mu\phi = G^{\mu\nu}\partial_\mu\phi\partial_\nu\phi$, while in the Einstein frame one uses the metric $g^{\mu\nu}$. Also, note that $\sqrt{|g|}\nabla^{(g)}_\mu(\cdots)^\mu = \partial_\mu(\sqrt{|g|}(\cdots)^\mu)$ is a total derivative that we can discard (unless boundary terms are relevant) as it does not contribute to the equations of motion. Taking this into account, the Einstein frame effective tree level action is

$$S_{\text{eff}} = \frac{1}{2\kappa^2}\int d^Dx\sqrt{g}\left(R - \frac{4}{D-2}\partial_\mu\phi\partial^\mu\phi - \frac{1}{12}e^{-8/(D-2)\phi}H_{\mu\nu\rho}H^{\mu\nu\rho}\right). \tag{12.28}$$

The string frame has the disadvantage that standard formulae of general relativity, such as the ADM formula for the mass of an asymptotically flat space-time, need to be modified by dilaton dependent factors. Also note that, in the Einstein frame, we can see immediately that the dilaton has as a standard kinetic term, while in the string frame it is not obvious that is has positive kinetic energy. The advantage of the string frame is that the dependence of the tree level effective action on the dilaton is universal, that is, the only way the dilaton appears without a derivative is an overall factor $e^{-2\phi}$. It is, therefore, manifest that shifting the dilaton by a constant, $e^{-2\phi} \to e^{-2\phi-2\phi_0}$, is equivalent to rescaling the coupling constant $\kappa \to \kappa e^{\phi_0}$. The coupling constant κ is assigned to the three string vertex, and can be identified with

---

[8] We still need to assume that the underlying string theory is defined in 26 dimensions, because otherwise the dilaton term in the effective action is modified, as we will see in Section 12.2.3.
[9] See, e.g., Wald (1984, Appendix D).

the gravitational coupling. But, as long as we do not specify the constant part or vacuum expectation value of the dilaton, the constant κ has no physical meaning. At tree level, the equations of motion do not fix the constant part of ϕ. What we can do is to relate the dimensionful coupling κ, which defines the gravitational or Planck scale, to the dimensionful parameter α', which defines the string scale. Let κ_0 be the value where both scales are equal. By dimensional analysis this implies

$$\kappa_0 = (\alpha')^{(D-2)/4}.$$

Now, we can parametrise different values of the physical gravitational coupling κ using the vacuum expectation value ϕ_0 of the dilaton

$$\kappa = (\alpha')^{(D-2)/4} e^{\phi_0}.$$

As we have seen, an M-particle string amplitude carries a factor $\kappa^{M-\chi(g)}$. When eliminating κ in terms of α' this becomes a factor $e^{(M-\chi(g))\phi_0}$. Therefore, $g_S = e^{\phi_0}$ can be interpreted as a dimensionless coupling constant, which is associated to the closed string vertex. We remark that in the string frame effective action, g-loop corrections appear with a factor $e^{-\chi(g)\phi_0} = g_S^{-\chi(g)}$.

12.2 Curved Backgrounds

12.2.1 String Action for Curved Backgrounds and Background Fields

So far, we have defined string theory as a theory in Minkowski space-time. Like for a particle, we can couple the theory to a curved background metric, equivalently to a background gravitational field, by replacing the Minkowski metric by a general Riemannian metric. The resulting generalised Polyakov action is

$$S[X, h|G] = -\frac{1}{4\pi\alpha'} \int d^2\sigma \sqrt{h} h^{\alpha\beta} \partial_\alpha X^\mu \partial_\beta X^\nu G_{\mu\nu}(X). \tag{12.29}$$

The *background metric* $G_{\mu\nu}(X)$ is part of the definition of the action. Since the spectrum of the closed string contains a graviton, we expect that the space-time metric is fundamentally dynamical rather than a background. We will discuss later how the dynamics of the gravitational field enters into the picture. The generalised Polyakov action is a non-linear sigma model for the world-sheet fields X^μ, which are the components of a map Φ between two Riemannian manifolds, the world-sheet (Σ, h) and space-time (M, G):

$$\Phi : (\Sigma, h) \to (M, G). \tag{12.30}$$

Even after imposing the conformal gauge, this is in general an interacting theory with field-dependent couplings $G_{\mu\nu}(X)$. Therefore, it is non-trivial to show that the theory is renormalisable and conformal. Renormalisability can be phrased as a question about the behaviour of the theory under a change of scale. The renormalisation procedure which is needed to remove all ambiguities related to operator ordering introduces a scale dependence of the couplings. We can, for example, impose that the effective action of a theory is equal, at some energy scale Λ, to the classical action. The effective action at a lower scale $\mu < \Lambda$ is then computed by integrating over all momenta between μ and Λ. In general, this introduces infinitely many new terms into the action, but since most of these are higher derivative terms, they are irrelevant at low energies. In this way, one can define a theory at a high scale Λ and describe it effectively at a lower scale μ by discarding all but finitely many terms. In renormalisable theories one can do more, namely send the ultraviolet cut-off $\Lambda \to \infty$ while controlling the effective action exactly by imposing finitely many normalisation conditions. This can only work if under a change of scale the theory flows within a finite-dimensional subspace of the infinite-dimensional coupling space. For two-dimensional sigma models this concept of renormalisability needs to be generalised to a situation where the couplings are field-dependent. Renormalisability then means that the theory flows within a finite class of coupling functions. The class to which the generalised Polyakov action belongs contains two more terms, which can be interpreted as coupling the string to a background B-field and to a background dilaton field:

$$S[X, h|G, B, \phi] = S[X, h|G] + S[X, h|B] + S[X, h|\phi]. \tag{12.31}$$

The term involving the B-field is a two-dimensional Wess–Zumino term

$$S[X, h|B] = -\frac{1}{4\pi\alpha'} \int d^2\sigma \epsilon^{\alpha\beta} \partial_\alpha X^\mu \partial_\beta X^\nu B_{\mu\nu}(X). \tag{12.32}$$

The integrand is the pull-back of a two-form with components $B_{\mu\nu}$ from space-time to the world-sheet using the map Φ (12.30). This term is independent of both the world-sheet and the space-time metric. The dilaton term has the form

$$S[X, h|\phi] = +\frac{1}{4\pi} \int d^2\sigma \sqrt{h} R_h \phi(X), \tag{12.33}$$

where R_h is the Ricci scalar of the world-sheet metric $h_{\alpha\beta}$. The absence of a factor α' indicates that this term is generated at one-loop order in world-sheet perturbation theory. Two-dimensional sigma models have a perturbative expansion where the expansion parameter is the inverse curvature radius of the target space. This radius is measured in units of the string length $\sqrt{\alpha'}$. Therefore, world-sheet perturbation theory is an expansion in α' and quantum corrections to the world-sheet theory are referred to as α'-*corrections*. These corrections need to be distinguished from space-time quantum corrections which are encoded in the topology of the world-sheet and labelled by the coupling κ, or equivalently by the dimensionless coupling g_S. For constant dilaton fields, the action $S[X, h|\phi]$ is proportional to the Euler number

$$\chi(\Sigma) = 2 - 2g = \frac{1}{4\pi} \int_{\Sigma_g} d^2\sigma \sqrt{h} R_h \tag{12.34}$$

of the world-sheet. If we choose $\phi(X)$ to be constant, $\phi = \phi_0$, then the dilaton term contributes a weight factor

$$e^{-\frac{1}{4\pi} \int d^2\sigma \sqrt{h} R_h \phi_0} = e^{-(2-2g)\phi_0} \tag{12.35}$$

to the string world-sheet path integral. By counting vertices, we have seen before that an M-particle g-loop amplitude carries a weight factor

$$\kappa^{M-2+2g} \cong g_S^{M-2+2g} \cong e^{[M-(2-2g)]\phi_0}. \tag{12.36}$$

Thus, the weight factor $e^{-S[X,h|\phi_0]}$ precisely accounts for the g-dependence of amplitudes.[10] Again, this shows us that the perturbative expansion in the space-time sense is a topological expansion in g, equivalently in χ.

12.2.2 Conformal Invariance and the Equations of Motion

The two-dimensional quantum field theory defined by $S[X, h|G, B, \phi]$ is renormal-isable in the sense that its flow in coupling space (RG flow) is constrained to three types of couplings functions. While renormalisability means that α'-corrections do not introduce new types of couplings, they can still modify the coupling functions themselves. This is relevant for our next question, namely whether the resulting renormalisable quantum field theory is conformal. For string theory, this is an essential question since the conformal symmetry of the world-sheet action is a residual gauge symmetry, which is left after imposing the conformal gauge. Since the underlying symmetries, namely reparametrisation and Weyl invariance, are local symmetries, consistency requires that conformal symmetry is preserved.

In quantum field theories renormalisation introduces, in general, a dependence of the couplings constants g on the energy scale μ, that is, they become *running couplings* $g(\mu)$. Their scale dependence is encoded in the β-functions

$$\beta^{(g)} = \frac{\partial g}{\partial \log \mu} = \mu \frac{\partial g}{\partial \mu}. \qquad (12.37)$$

A quantum field theory is scale invariant (and under conditions discussed in Section 5.1 also conformally invariant) if the β-functions for all its couplings vanish. This concept can be generalised to the field-dependent couplings $G_{\mu\nu}(X), B_{\mu\nu}(X), \phi(X)$, where the β-functions are now functionals of the coupling functions. For string theory, a further tweak is required since the invariance one actually requires is the one under *local* Weyl transformations. We denote the corresponding modified β-functionals by $\bar\beta$. Thus, formally the condition for preserving local Weyl invariance is

$$\bar\beta[G_{\mu\nu}(X)] = \bar\beta[B_{\mu\nu}(X)] = \bar\beta[\phi(X)] = 0. \qquad (12.38)$$

These conditions can be evaluated in α'-perturbation theory. To lowest order the resulting conditions on $G_{\mu\nu}(X), B_{\mu\nu}(X), \phi(X)$ are exactly the Euler–Lagrange equations of the space-time effective action (12.23), assuming that we are in the critical dimension $D = 26$. This shows that imposing conformal invariance on the world-sheet action imposes the equations of motion on the background fields. In other words the background must be on-shell. Moreover, both string amplitudes and curved backgrounds allow one to deduce the equations of motion for the massless string modes. In particular, both approaches can be used to compute α'-corrections. One advantage of the β-function approach is that it allows to work directly with space-time tensors, such as the Riemann tensor and its contractions, while scattering amplitudes involve expanding such tensors around a fixed background with respect to the graviton field.

---

[10] When writing amplitudes in the path integral formalism, the factors κ^M associated with external states are absorbed into the vertex operators, see, e.g., Polchinski (1998a).

Apart from perturbative α'-corrections, the world-sheet theory may also receive non-perturbative corrections, in particular instanton corrections. Instantons are field configurations which extremise the Euclidean action functional and have a finite action. *World-sheet instantons* arise when the target space contains two-dimensional submanifolds S such that the map $\Phi : \Sigma \to S \subset M$ describing the embedding of the world-sheet into the target space extremises the world-sheet action. As an example, let us consider the case where both the world-sheet and S have the topology of a two-sphere: $\Sigma = \Sigma_0 \cong S^2$, $S \cong S^2$. Maps of the form $\Phi : S^2 \to S^2$ are classified by their winding number $k \in \mathbb{Z}$. Maps within the same class are equivalent under homotopy, that is, such maps can be deformed into each other continuously. The class $k = 0$ contains the constant map where S^2 is mapped to a point. In this case the world-sheet action takes the minimal value zero. This is the setting of standard perturbation theory where one expands around the vacuum. For $k \neq 0$ the map Φ covers $S^2 \cong S \subset M$ $|k|$-times, with the sign accounting for orientation. Since the Polyakov action has the same critical points as the Nambu–Goto action, which computes the area with respect to the pulled-back metric, the action is extremised for fixed S by the map Φ for which the pulled back volume is $|k|$-times the volume of $S \subset M$. However, we need to extremise Φ as a map $\Sigma \to M$, and it might be possible to deform S inside M such that its volume decreases. For example any two-sphere $S^2 \subset \mathbb{R}^n$, $n > 2$ can be deformed to a point with zero volume, which reduces the action to zero. Thus, for genus-0 world-sheet instantons to exist the target space M must have a submanifold $S \cong S^2$ which has minimal volume, that is, a volume that cannot be decreased by deformation. If M admits such a submanifold, there will be a series of world-sheet instanton corrections with weight factors

$$e^{-\frac{1}{4\pi\alpha'}S[X_0,h|G]} = e^{-\frac{1}{4\pi\alpha'}|k|\mathrm{vol}(S)}. \tag{12.39}$$

Note that the non-perturbative character of this correction is manifest in its dependence on α', since the function $e^{-1/\alpha'}$ is not analytic at $\alpha' = 0$.

We have only discussed the most simple example. Things get more complicated when including the B-field, when several submanifolds which give rise to extrema of Φ exist, and when considering world-sheets of higher genus $g > 0$. The theory of world-sheet instantons is highly developed for the case where the sigma-model target space is a Calabi–Yau manifold.

12.2.3 Consistent Backgrounds and Background Independence

Field configurations $G_{\mu\nu}, B_{\mu\nu}, \phi$ which solve the equations (12.38) are called *consistent string backgrounds*. To lowest order in α', these are the equations of motion of the effective action (12.23). The simplest solution is

$$G_{\mu\nu} = \eta_{\mu\nu}, \quad B_{\mu\nu} = 0, \quad \phi = \mathrm{const}, \tag{12.40}$$

which is, in fact, an exact solution in α' since we know that the standard Polyakov action defines a conformal field theory. If we only impose $B_{\mu\nu} = 0$, $\phi = 0$, then the α'-expansion of $\bar{\beta}[G_{\mu\nu}(X)] = 0$ starts as follows:

$$\alpha' R_{\mu\nu} + \mathrm{const}.\alpha'^2 R_{\mu\alpha\beta\gamma}R_\nu{}^{\alpha\beta\gamma} + O((\alpha')^3) = 0. \tag{12.41}$$

To leading order, this is the vacuum Einstein equation $R_{\mu\nu} = 0$, which is generated by the Einstein–Hilbert term in the effective action. At order α', there appears a four-derivative term quadratic in the Riemann tensor. This demonstrates the presence of a corresponding four-derivative term in the effective action, which is a one-loop correction in α'.

The leading order equations $R_{\mu\nu} = 0$ (for $B_{\mu\nu} = 0$, $\phi = 0$) leave us with a huge set of solutions besides Minkowski space, namely all *Ricci-flat* space-times. To cope with the problem that the critical space-time dimension $D = 26$ is larger than the actual dimension, we can look for space-times of the form $M_4 \times X$, where X is a compact space whose diameter is small compared to length scales that we can currently access. The simplest choice is a torus, where we keep a flat metric and just identify all extra dimensions periodically. In supersymmetric string theories one can impose that the lower-dimensional theory obtained by compactification is supersymmetric. In this case the Ricci flat manifold must in addition be a *Kähler manifold*. Ricci flat Kähler manifolds are called *Calabi–Yau manifolds*, and one knows much more about the resulting string compactifications than for generic Ricci-flat manifolds. We will discuss compactifications in some detail in the next chapter, where we also provide the definitions of Kähler and Calabi–Yau manifolds.

As an example of a different type of background, we consider the $\bar\beta$-equation for the dilaton:

$$\bar\beta^\phi = \frac{D - 26}{6} - \frac{\alpha'}{2}\nabla_\mu\nabla^\mu\phi + \alpha'\nabla_\mu\phi\nabla^\mu\phi - \frac{\alpha'}{24}H_{\mu\nu\rho}H^{\mu\nu\rho} + O(\alpha'^2) = 0. \quad (12.42)$$

The first term, which we have suppressed previously, shows how the conformal anomaly manifests itself when we try to formulate string theory in dimensions $D \neq 26$. In the effective action this induces an additional term

$$\Delta S = \frac{1}{2\kappa^2} \int d^D x \sqrt{G} e^{-2\phi} \frac{2(26 - D)}{3\alpha'}. \quad (12.43)$$

Solutions to the $\bar\beta$-equations with $D \neq 26$ exist once one allows a non-constant dilaton field. This illustrates that non-critical string theories with $D \neq 26$ are possible if we do not insist that the background is Minkowski space. Since a single string coordinate carries central charge $c = 1$, this corresponds to world-sheet theories with $c \neq 26$. This allows us to address a point which we have ignored so far: if conformal symmetry is a 'left-over' of reparametrisation and Weyl symmetry, which are *local* symmetries, shouldn't we insist on conformal symmetry without an anomaly, that is $c = 0$? In short, when gauge fixing a local symmetry we need to introduce Faddeev–Popov *ghost fields*, and it is the combined theory of string coordinates X^μ and ghosts which must have total charge $c_{\text{total}} = c + c_{\text{ghost}} = 0$ in order that the gauge-fixed theory preserves reparametrisation and Weyl invariance. The ghost fields carry $c_{\text{ghost}} = -26$, and one has two options: critical string theory, where the CFT of the string coordinates X^μ has $c = 26$ and cancels the conformal anomaly; or non-critical or Liouville string theory, where the X^μ have $c \neq 26$ and the conformal mode (overall scale) of the world-sheet metric becomes a dynamical field. In this case, the world-sheet action is extended by the Liouville action for the additional mode. Due to the absence of a Minkowski ground state, non-critical string theory is more complicated and less understood than critical string theory.

Let us next discuss how gravity is described by string theory. We have seen that this happens in two different ways: in a fixed background we have a state which in scattering amplitudes behaves like a graviton, up to higher derivative corrections. The background metric itself must satisfy Einstein's field equations, up to higher corrections, and enters into the effective action as a dynamical field. This twofold description of gravity reflects that the 'first quantised' world-sheet formulation of string theory lacks a feature that Einstein gravity has, namely *manifest background independence*. To define Einstein gravity we do not need to choose a particular background space-time, but instead we write down the Einstein–Hilbert action, which is a functional on the space of metrics.[11] Specific metrics arise as solutions to the variational problem. In absence of a cosmological constant, Minkowski space-time is distinguished by being a maximally symmetric solution, but we don't need to know this in advance to define the theory.

In contrast, our definition of string theory includes Minkowski space-time, or another consistent background, as part of the defining data. The way gravity enters is similar to the particle theorist's re-construction of Einstein gravity as the field theory of a 'massless spin-2 field'. This works by starting with a massless symmetric tensor field $\psi_{\mu\nu}$ on Minkowski space-time, then introducing self-interactions recursively while preserving its gauge symmetries, and ultimately recovering Einstein gravity by fusing the background Minkowski metric $\eta_{\mu\nu}$ with the graviton field to define a dynamical Riemannian metric $g_{\mu\nu} = \eta_{\mu\nu} + \psi_{\mu\nu}$.

In its current formulations, string theory is defined with respect to a specific reference background, the admissible backgrounds being identified by the $\bar{\beta}$-equations. Since the theory is not *manifestly* background independent, we need to show that this definition does not depend on which reference background we choose, since otherwise different backgrounds would define different theories. This can indeed be shown, at least locally. Here locally means that one can change the background data infinitesimally, and that if in addition certain integrability conditions are satisfied, this defines a finite deformation of the initial background in some neighbourhood in the space of all consistent backgrounds.[12]

Let us see how this works using an example. We take the Minkowski metric $\eta_{\mu\nu}$ as our reference background. Let $G_{\mu\nu}(X)$ be another consistent background. The difference $\psi_{\mu\nu}(X) = G_{\mu\nu}(X) - \eta_{\mu\nu}$ is what we call the graviton field, relative to the background $\eta_{\mu\nu}$. Now we substitute this decomposition into the world-sheet action:

$$S[X, h|G] = -\frac{1}{4\pi\alpha'} \int d^2\sigma \sqrt{h} G_{\mu\nu} \partial_\alpha X^\mu \partial^\alpha X^\nu$$
$$= S[X, h|\eta] - \frac{1}{4\pi\alpha'} \int d^2\sigma \sqrt{h} \psi_{\mu\nu}(X) \partial_\alpha X^\mu \partial^\alpha X^\nu. \tag{12.44}$$

Thus, the action on the background $G_{\mu\nu}$ splits into the action $S[X, h|\eta]$ on a Minkowski background plus a term involving the graviton field $\psi_{\mu\nu}(X)$. To understand this term, we remember that the graviton vertex operator is

---

[11] This is an oversimplification, since we need to obtain a globally well defined Riemannian manifold, which may require us to impose additional conditions. For example, unless the space-time manifold is closed, that is, a compact manifold without boundary, a well defined variational problem requires one to add a boundary term to the Einstein–Hilbert action.

[12] Globally, this space is quite complicated and in particular not a manifold, but a space which has components of different dimension. As a concrete example, we will study the space of one-dimensional compactifications in Chapter 13.

$$V[\zeta, k] = \zeta_{\mu\nu} \partial_\alpha X^\mu \partial^\alpha X^\nu e^{ik \cdot X}, \tag{12.45}$$

where the momentum k^μ and the polarisation tensor $\zeta_{\mu\nu}$ have to satisfy constraints, which are the linearised field equations in momentum space. To be precise, $V[\zeta, k]$ is the vertex operator for a momentum eigenstate, which is not normalisable. Normalisable states correspond to superpositions

$$V[\zeta(k)] = \int d^D k \zeta_{\mu\nu}(k) \partial_\alpha X^\mu \partial^\alpha X^\nu e^{ik \cdot X}. \tag{12.46}$$

This Fourier integral defines the position space representation $\psi_{\mu\nu}(X)$ of the graviton state:

$$V[\psi(X)] = \psi_{\mu\nu}(X) \partial_\alpha X^\mu \partial^\alpha X^\nu = \int d^D k \zeta_{\mu\nu}(k) \partial_\alpha X^\mu \partial^\alpha X^\nu e^{ikX}. \tag{12.47}$$

Thus, the difference between the two world-sheet actions is precisely the integrated vertex operator for a graviton state of the form $\psi_{\mu\nu}(X)$. Let us now compare the correlation functions of the world-sheet theories with background metrics $G_{\mu\nu}(X)$ and $\eta_{\mu\nu}$. In the path integral formulation field configurations are weighted with the negative exponential of the world sheet action. Therefore, the correlation functions differ by the insertion of the exponentiated graviton vertex operator

$$\langle \cdots \rangle_G = \langle \cdots e^{V[\psi]} \rangle_\eta. \tag{12.48}$$

Finally, what is the meaning of $e^{V[\psi]}$? Vertex operators create excitations from the ground state, and the exponentials of creation operators generate *coherent states*. Coherent states in a quantum theory are those states which saturate the Heisenberg uncertainty relation, and which are the closest approximation to classical states one can get in a quantum theory. The standard example is the harmonic oscillator, where the coherent states are Gaussian wave packages, for which the expectation values of position and momentum satisfy the classical equation of motion for a particle in a harmonic oscillator potential. These states involve contributions from all levels of the harmonic oscillator and are therefore complementary to the energy eigenstates, where the expectation values of both position and momentum are zero. Observe that energy eigenstates are very far from describing a classical particle in a potential. Coherent states can also be defined in quantum field theory, where they involve superpositions of states with any number of particles. These states are complementary to the usual particle-like states, but their field operators satisfy the classical field equations and in this sense they are the quantum states which are closest to classical field configurations. Applied to string theory we see that the two backgrounds differ by the insertion of the vertex operator for a coherent state. This shows that while we need to use a reference background to define the theory, the split between background and excitations around the background can be moved. The theory is background independent, but not manifestly background independent.

12.2.4 Marginal Deformations

To describe the deformations relating different consistent backgrounds more systematically, one uses the concept of a *marginal operator*. In a two-dimensional conformal field theory an operator V is called marginal if it has weight $(1, 1)$. The integrals of such operators are conformally invariant and can therefore be used to deform the action:

$$S \mapsto S + g \int d^2\sigma \sqrt{h} V. \tag{12.49}$$

However, in order for an operator to define a one-parameter family of conformal field theories it is not sufficient that V is marginal in the original theory, but it must remain marginal 'along the way'. Thus, the infinitesimal marginal deformation generated by V does not necessarily define a finite deformation. Marginal operators which generate finite deformations are called *exactly marginal*. If they exist one obtains families of conformal field theories, and thus of consistent string backgrounds. The global structure of the space of consistent string backgrounds adds another level of complication, since the number of exactly marginally operators can jump. As we will see in Chapter 13, this implies that the space of string backgrounds is not a manifold, but a 'stratified space' which consists of components which have different dimensions and which can intersect each other along lower-dimensional subspaces. Note that the concept of marginality used here is consistent with the terminology of the renormalisation group: marginal operators deform quantum field theories along special subspaces in the space of all couplings where theories are scale invariant and their β-functions vanish.

One might think that the transition to a second quantised, string field theory formalism should lead to a manifestly background independent formulation. To date this has not yet been achieved, since the definition of string field theory requires to specify a conformal field theory as reference background. However, it has been shown that superstring field theories are background independent in the sense used above. We also remark that if we talk about string theory using effective space-time actions, we are in the same position as in general relativity since we can start exploring the action without knowing its solutions first. However, this description is an approximation which is restricted to finitely many string modes and subject to α'- and g_S-corrections, and which is only valid locally in the sense that the space of consistent backgrounds has a complicated structure and can even change dimension. As we will see in Chapter 13, this is related to the fact that the separation into light modes, which are included in the effective action, and heavy modes, which are not, depends on the background, so that modes which are heavy in one background become massless once we deform it far enough. Thus, while effective actions are formally background independent, this only holds approximately and locally in the space of consistent string backgrounds.

12.3 Further Remarks and Literature

Computing amplitudes and effective actions quickly becomes computationally involved, and therefore we have restricted ourselves to one illustrative example, while otherwise we have summarised the key ideas and put them into context. Some textbooks, including Green et al. (1987) and Polchinski (1998a) discuss one-loop amplitudes in some detail. Polchinski (1998a, 1998b) also lays out the argument for the perturbative finiteness of string theory. For closed strings the most important ingredient for finiteness is modular invariance, which we have discussed using the one-loop partition function. What about open strings? While theories of closed strings can be consistent on their own, theories of open strings always also contain closed strings, because closed string states appear as intermediate states in open string loop diagrams.[13] The factorisation properties of amplitudes imply that while there are different types of couplings between open and closed strings, the corresponding coupling constants are related, so that a string theory always has a single independent coupling constant. For example, the coupling constant of a vertex with three closed strings is proportional to the square of the coupling constant of a vertex of three open strings. This can be understand intuitively by observing that one can cut a 'pair of pants' closed string vertex to obtain two vertices involving three open strings each. More generally, closed string amplitudes can be related to products of open strings amplitudes (Kawai et al., 1986). Since the massless sectors of open and closed strings contain Yang–Mills theories and gravity, respectively, this has given rise to the idea that gravitational theories can be constructed as squares or 'double copies' of Yang–Mills theories (see Bern et al., 2010a, 2010b).

Open string world-sheets do not have a modular group action which makes them UV finite, but finiteness follows from additional consistency conditions which are related to the factorisation properties of scattering amplitudes and the absence of tadpoles, that is of non-zero one-point functions. The absence of tadpoles determines the so-called Chan–Paton factors and the open string gauge group, which we will discuss briefly in Section 13.6.

Originally, string amplitudes were constructed in the framework of S-matrix theory, that is by directly imposing consistency conditions and further desirable properties on the amplitudes, see, for example, Scherk (1975), Green et al. (1987), and West (2012). While this allows one to construct loop amplitudes by seewing tree amplitudes, the Euclidean path integral approach of Polyakov (1981a), Polyakov (1981b) has turned out to be better suited for constructing higher loop amplitudes. However, the explicit computation of higher loop amplitudes ($g > 1$) still is very complicated. One technical complication is that one has to integrate over the moduli space of Riemann surfaces (and for superstrings, there are additional loop parameters to be integrated over). See, for example, D'Hoker and Phong (1988). The relation between curved backgrounds and world-sheet beta functions is explained in some more detail in Green et al. (1987) and Polchinski (1998a). For further reading, see, for example, Callan et al. (1985) and Tseytlin (1989). For non-critical string theories, see, for example, Polyakov (1981a), Polyakov (1981b), Polchinski (1998a), Polchinski (1998b), and Becker et al. (2007). The background independence of closed superstring field theory has been shown in Sen (2018), see there for earlier work on background independence. See also Erbin (2021) for a pedagogical discussion of background independence in string field theory.

---

[13] An intuitive explanation is set out in Section 7.7.

13 Dimensional Reduction and T-Duality

13.1 Overview

One of the characteristic features of string theory is that the number of space-time dimensions is fixed by imposing the absence of anomalies. Since the predicted number of space-time dimensions is higher than four, relating string theory to the real world requires one to explain why the predicted extra dimensions have not yet been observed. There are two different approaches to this:

1. The direct construction of string theories in four dimensions, with extra degrees of freedom described by an *internal conformal field theory*, which need not have a geometric interpretation. We will also refer to this as the *CFT-approach* or *algebraic approach*.

2. Splitting the higher-dimensional space-time into a four-dimensional space-time and a compact internal space. We will refer to this as *compactification* or the *geometric approach*.

String theory on 26-dimensional Minkowski space M_{26} corresponds to a specific conformal field theory with central charge $c = 26$, namely the sigma model defined by the Polyakov action with target space M_{26}. If we replace the target space by four-dimensional Minkowski space M_4, the theory will be inconsistent. Following the discussion in Section 12.2.3, the ultimate reason for the inconsistency is a 'central charge deficit' of the world-sheet CFT. Therefore, the most general solution is to add 'internal' degrees of freedom described by a conformal field theory with $c = 22$. Choosing this CFT to be a sigma model with a 22-dimensional target space X is a special choice. The simplest explanation for why the space X has not been detected so far is that X is compact with a diameter $l < 1/E$, where E is the energy scale that can currently be probed.[1] While the geometric approach is a subcase of the algebraic approach, it is by far the predominant approach, since conformal field theories with $c > 1$ are not solvable based on conformal symmetry alone, and since for studying compactifications powerful tools from differential and algebraic geometry are available. In this chapter, we will mostly focus on the simplest, one-dimensional case of compactification. This will already allow us to exhibit many typical features of string compactifications, such as their moduli spaces, T-duality and the emergence of non-abelian gauge symmetries and of extra light matter at special points of the moduli space. In Sections 13.7 and 13.8, we will give an outlook onto multi-dimensional compactifications.

---

[1] There are alternative scenarios. For example, in 'braneworld' models matter is realised by open strings with Dirichlet boundary conditions. Bounds on the size of extra dimensions are model dependent.

13.2 Closed Strings on S^1

Let us consider closed strings on a background of the form

$$M_{D-1} \times S_R^1, \tag{13.1}$$

where M_{D-1} is $(D-1)$-dimensional Minkowski space and where S_R^1 denotes a circle of radius R.[2] We split the coordinates accordingly into space-time coordinates X^μ and one internal coordinate X, which is subject to the identification

$$X \simeq X + 2\pi R. \tag{13.2}$$

Thus, we describe the circle as a quotient space of the real line:

$$S_R^1 \simeq \mathbb{R}/(2\pi R \mathbb{Z}).$$

With this identification strings along the X-direction are closed if

$$X(\sigma^0, \sigma^1 + \pi) = X(\sigma^0, \sigma^1) + 2\pi R m, \quad m \in \mathbb{Z}. \tag{13.3}$$

The integer parameter m is the *winding number* which tells us how many times the closed strings winds around the circle (modulo continuous deformations of the embedding, and with the sign accounting for direction). We also define a one-component *winding vector* $w = mR$, which depends on the radius R of the circle.[3]

In order for momentum eigenstates to be well defined, the eigenvalues k of the momentum operator p for the X-direction must be *quantised*:

$$e^{ikX} \simeq e^{ik(X+2\pi Rm)} \Rightarrow k \cdot 2\pi Rm \in 2\pi\mathbb{Z} \Rightarrow k = \frac{n}{R}, \quad n \in \mathbb{Z}. \tag{13.4}$$

We will call the integer $n \in \mathbb{Z}$ the *momentum number* and $k = \frac{n}{R}$ the (internal) *momentum vector*. The discreteness of momentum eigenvalues in systems of finite volume is a well-known feature of quantum particles. The new feature for strings is the existence of a second quantity with a discrete spectrum, the winding.

The presence of winding changes the zero mode part of the mode expansion:

$$X(\sigma^0, \sigma^1) = x + p\sigma^0 + 2w\sigma^1 + \cdots. \tag{13.5}$$

Compared to (2.64), we now have a term linear in σ^1. Defining *left-* and *right-moving momenta* by

$$p_L = \frac{1}{2}p + w, \quad p_R = \frac{1}{2}p - w, \tag{13.6}$$

the mode expansion becomes

$$X(\sigma^+, \sigma^-) = x + p_L \sigma^+ + p_R \sigma^- + \cdots, \tag{13.7}$$

where p_L and p_R are now independent. To quantise the zero mode sector we need to split the operator x into left- and right-moving parts x_L, x_R. We take x_L, x_R to be conjugate to the left- and right-moving momenta p_L, p_R, that is

---

[2]   We keep D generic, because much of what we say also applies to superstring theories, where the critical dimension is $D = 10$.

[3]   For multidimensional compactifications w becomes a multicomponent vector. Here, we use the term winding vector to distinguish w from the winding number m.

$$[x_L, p_L] = i = [x_R, p_R]. \tag{13.8}$$

The eigenvalues of p_L and p_R are denoted k_L and k_R and take the values

$$k_L = \frac{1}{2}k + w, \quad k_R = \frac{1}{2}k - w, \quad \text{where} \quad k = \frac{n}{R}, \quad w = mR. \tag{13.9}$$

We introduce the following notation for momentum eigenstates:

$$e^{ik_\mu x^\mu} e^{ik_L x_L} e^{ik_R x_R} |0\rangle = |k^\mu, (k_L, k_R)\rangle = |k^\mu, (k, w)\rangle = |k^\mu, (n, m), R\rangle. \tag{13.10}$$

States are labelled by the continuous space-time momentum k^μ and by their internal momentum and winding, or, equivalently, by their left- and right-moving momenta. We can also label the internal component by the integer quantum numbers m, n, together with the radius R. To obtain a basis for the Fock space, we need to include the oscillators, which now come in two types: those with a Lorentz index with respect to the space-time M_{D-1}, and internal oscillators which are Lorentz scalars:

$$\alpha^{\mu_1}_{-m_1} \cdots \alpha_{-k_1} \cdots \tilde{\alpha}^{\nu_1}_{-n_1} \cdots \tilde{\alpha}_{-l_1} \cdots |k^\mu, (k_L, k_R)\rangle, \quad m_i, n_i, k_j, l_j > 0. \tag{13.11}$$

We now have to re-evaluate the L_0 and $\tilde{L}_0$ constraints, taking into account that the mass M of a state is related to the space-time momentum p^μ, by $M^2 = -p^\mu p_\mu$. The result is, in string units:

$$M^2 = 4\left(N + \tilde{N} + \frac{1}{2}k_L^2 + \frac{1}{2}k_R^2 - 2\right), \tag{13.12}$$

$$N + \frac{1}{2}k_L^2 = \tilde{N} + \frac{1}{2}k_R^2. \tag{13.13}$$

The first formula is the mass formula where the left- and right-moving momenta now contribute together with the oscillators. The second formula is a modified level matching formula. It is useful to make the dependence on the radius R explicit by expressing k_L, k_R in terms of m, n, R:

$$M^2 = 4\left(N + \frac{1}{2}\left(\frac{n}{2R} + mR\right)^2 + \tilde{N} + \frac{1}{2}\left(\frac{n}{2R} - mR\right)^2 - 2\right), \tag{13.14}$$

$$N - \tilde{N} = mn. \tag{13.15}$$

This shows how the masses of states with $m \neq 0$ or $n \neq 0$ depend on R. Note that R drops out of the level matching condition, which tells us that the spectrum of physical states does not jump discontinuously when we vary R. We have seen before that Minkowski vacua depend on a continuous parameter, the value of the string coupling. For circle compactifications we now have found another deformation parameter or *modulus*, the radius R. Naively, the moduli space of S^1-compactifications is

$$\mathcal{M}_{S^1} = \{R \in \mathbb{R} | 0 < R < \infty\} = \mathbb{R}^{>0}. \tag{13.16}$$

In Section 13.4, we will see that not all these circles are distinct as string backgrounds. For the time being we focus on another question, namely the spectrum of massless states. As before, there will be tachyonic states, but we focus on features

Table 13.1. Generic massless states for circle compactification of a closed bosonic string

States	$\alpha_{-1}^{\mu}\tilde{\alpha}_{-1}^{\nu}|(0,0)\rangle$	$\alpha_{-1}^{\mu}\tilde{\alpha}_{-1}|(0,0)\rangle$	$\alpha_{-1}\tilde{\alpha}_{-1}^{\nu}|(0,0)\rangle$	$\alpha_{-1}\tilde{\alpha}_{-1}|(0,0)\rangle$
Space-time fields	$G_{\mu\nu}, B_{\mu\nu}, \phi$	$A_{\mu}^{(R)}$	$A_{\nu}^{(L)}$	σ

that bosonic strings share with superstrings. The most important states are therefore the massless states, like the graviton and photon, which tell us about the gauge symmetries of the theory. The conditions for a massless state are:

$$N + \tilde{N} + \frac{1}{2}k_L^2 + \frac{1}{2}k_R^2 = 0, \tag{13.17}$$

$$N + \frac{1}{2}k_L^2 = \tilde{N} + \frac{1}{2}k_R^2. \tag{13.18}$$

Since k_L, k_R depend on R, *generic massless states*, that is, states that are massless for all values of R must have $k_L = k_R = 0$. This leaves us with

$$k_L = k_R = 0 \Rightarrow N = \tilde{N} = 1. \tag{13.19}$$

The corresponding states are listed in Table 13.1. Apart from a lower-dimensional graviton $G_{\mu\nu}$, B-field $B_{\mu\nu}$, and dilaton ϕ, we obtain two vector fields $A_{\mu}^{(L/R)}$ and one scalar field σ.

Exercise 13.2.1   Derive the mass formula and level matching condition (13.13) for a circle compactification.

Exercise 13.2.2   Check that Table 13.1 corresponds to the decomposition of representations of the higher-dimensional Lorentz group with respect to the lower-dimensional Lorentz group.

13.2.1 Generic Massless States and Their Effective Field Theory

We now turn to the description of the generic massless states using effective field theory. The effective field theory for the graviton G_{MN}, B-field B_{MN}, and dilaton ϕ in D dimensions is

$$S_{\text{eff}} = \frac{1}{2\kappa_0^2} \int d^D x \sqrt{G} e^{-2\phi} \left(R_G + 4(\nabla\phi)^2 - \frac{1}{12}H_{MNP}H^{MNP} \right), \tag{13.20}$$

where R_G is the Ricci scalar of the string frame metric G_{MN}, and where $H_{MNP} = \partial_M B_{NP} + \partial_N B_{PM} + \partial_P B_{MN}$ is the field strength of the B-field. We split the coordinates as $x^M = (x^{\mu}, x)$ and decompose tensors accordingly. To perform a dimensional reduction from the D-dimensional string frame to the $(D-1)$-dimensional string frame, we use the following parametrisation:[4]

$$(G_{MN}) = \begin{pmatrix} \bar{G}_{\mu\nu} + e^{2\sigma}A_{\mu}A_{\nu} & e^{2\sigma}A_{\nu} \\ e^{2\sigma}A_{\mu} & e^{2\sigma} \end{pmatrix}. \tag{13.21}$$

---

[4] Different parametrisations lead to lower-dimensional metrics which differ from the string-frame metric by a conformal transformation.

Note that $\sqrt{G} = e^\sigma \sqrt{\bar{G}}$. While $\bar{G}_{\mu\nu}$ will turn out to be the $(D-1)$-dimensional string frame metric, we also obtain a *Kaluza–Klein vector* A_μ and a *Kaluza–Klein scalar* σ. By comparison to Table 13.1 we see that these fields account for part of the generic massless states. The remaining ones will arise from the reduction of the B-field, together with the dilaton.

To perform the reduction, we impose the identification $x \cong x + 2\pi R$ on the coordinate x and fix an on-shell background metric of the form

$$\langle (G_{MN}) \rangle = \begin{pmatrix} \eta_{\mu\nu} & 0 \\ 0 & e^{2\langle\sigma\rangle} \end{pmatrix}.$$

Here $\langle \cdots \rangle$ indicates that we choose a field configuration which satisfies the equations of motion. Specifically, we have chosen a constant background metric $\langle (G_{MN}) \rangle$, parametrised by an arbitrary constant value $\langle\sigma\rangle$ for the scalar σ, and an arbitrary constant value $\langle\phi\rangle$ for the dilaton. For the time being, we set the B-field to zero $\langle B_{MN} \rangle = 0$.

Let us analyse how we can parametrise inequivalent backgrounds. In the uncompactified theory, the physical gravitational coupling κ is given by

$$\frac{1}{2\kappa^2} = \frac{e^{-2\langle\phi\rangle}}{2\kappa_0^2}. \tag{13.22}$$

Inequivalent backgrounds are parametrised either by κ or by the dilaton vev $\langle\phi\rangle$. After compactification, we expect to have a second parameter, the radius of the circle. The radius of the circle appears twice: as the parametric radius R which is defined by the way we identify the coordinate x, $x \cong x + 2\pi R$, and through the vev of the scalar field σ, which controls the metric coefficient corresponding to the compactified direction. The quantity that can actually be measured is the physical radius (geodesic radius) ρ which we obtain by computing the length $l = 2\pi\rho$ of the circle using the background metric:

$$l = \int_0^{2\pi R} dx\, e^{\langle\sigma\rangle} = 2\pi R e^{\langle\sigma\rangle} = 2\pi\rho. \tag{13.23}$$

We can always compensate a change of the parametric radius R by a change of the vev $\langle\sigma\rangle$ and vice versa. To parametrise distinct circles we can either fix R and use $\langle\sigma\rangle$, or fix $\langle\sigma\rangle$ and use R. For the time being, we will fix $\langle\sigma\rangle$ and use R, though it is useful to keep in mind that changing R can be interpreted as changing the vacuum expectation value of the Kaluza–Klein scalar σ.

The next step is to impose the identification $x \cong x + 2\pi R$ on field configurations, which amounts to a Fourier decomposition. We illustrate this using a massless scalar field

$$\phi(x^\mu, x) = \sum_{n=-\infty}^{\infty} \phi_n(x^\mu) e^{inx/R}. \tag{13.24}$$

This is an infinite collection of lower-dimensional scalar fields. The masses of these fields are determined by the linearised field equations. Starting with a massless scalar in D dimensions, $\Box_D \phi = 0$, we obtain

$$\Box_{D-1} \phi_n = M_n^2 \phi_n, \quad M_n^2 = \frac{n^2}{R^2} e^{-2\langle\sigma\rangle} = \frac{n^2}{\rho^2}. \tag{13.25}$$

Thus, we obtain a massless field for $n = 0$, and an infinite *Kaluza–Klein tower* of massive fields for $n \neq 0$, which carry discrete momentum

$$p_n = \frac{n}{R} \tag{13.26}$$

along the circle. Note that reparametrisations of the circle

$$x \to x + \lambda(x^\mu) \tag{13.27}$$

act as gauge transformations on the fields ϕ_n, $n \neq 0$:

$$\phi_n(x^\mu)e^{inx/R} \to e^{in\lambda(x^\mu)/R}\phi_n(x^\mu)e^{inx/R}. \tag{13.28}$$

This shows that the Kaluza–Klein gauge group $U(1)_G$ is a remnant of the higher-dimensional reparametrisation invariance, and the $U(1)_G$ charge is given by the internal momentum k.

Let us now turn to the B-field. The simplest way to solve the higher-dimensional field equations is to set its field strength to zero, $\langle H_{MNP}\rangle = 0$. This equation can be solved by $\langle B_{MN}\rangle = 0$, and for circle compactifications this is the most general solution up to gauge transformations. A convenient decomposition of the B-field is

$$(B_{MN}) = \begin{pmatrix} \bar{B}_{\mu\nu} - A_{[\mu}A'_{\nu]} & A'_\nu \\ A'_\mu & 0 \end{pmatrix}, \tag{13.29}$$

where A_μ is the Kaluza–Klein gauge field and A'_μ the gauge field resulting from decomposing B_{MN} into a tensor field and a vector field. The vector field A'_μ is the gauge field of an abelian gauge symmetry $U(1)_B$, which descends from the gauge symmetry of the higher dimensional B-field. The corresponding charged states are the winding states of the closed string. Note that these winding states are not part of the Kaluza–Klein spectrum of the D-dimensional effective field theory, which only contains momentum modes through the Fourier decomposition. Winding states are a new, genuinely stringy feature. To see how the winding number and the $U(1)_B$ charge are related, we use world-sheet methods. First, remember that the minimal coupling of an electrically charge point particle to the electromagnetic field is described by

$$S_{\text{minimal}} = q \int_{\Sigma_1} d\tau A_M \frac{dx^M}{d\tau}. \tag{13.30}$$

The B-field term of the string world-sheet action can be viewed as a higher-dimensional analogue, describing the minimal coupling of a string to the B-field:

$$S[B] = T \int_{\Sigma_2} d^2\sigma \epsilon^{\alpha\beta} B_{MN}\partial_\alpha X^M \partial_\beta X^N. \tag{13.31}$$

Now consider a closed string winding around the S^1:

$$X^M = \begin{cases} X^\mu = x_0^\mu + p^\mu \sigma^0 + \cdots, \\ X = 2w\sigma^1 + \cdots, \quad w = mR. \end{cases} \tag{13.32}$$

We integrate over σ^1 and obtain a world-line action for the winding mode

$$S[B] = T(2\pi w) \int_{\Sigma_1} d\sigma^0 B_{\mu, D-1}\partial_0 X^\mu. \tag{13.33}$$

This shows how the $U(1)_B$ charge is related to the winding vector w and the string tension T. From (13.9) we see that the gauge fields A_μ, A'_μ with associated charges k, w are linear combinations of the gauge fields $A_\mu^{(L/R)}$ with associated charges k_L, k_R. Thus, we have two ways of viewing the gauge group of circle compactifications:

$$U(1)_L \times U(1)_R \cong U(1)_G \times U(1)_B. \tag{13.34}$$

To obtain the effective action for the lower-dimensional massless modes one drops all modes with a non-trivial dependence on the internal coordinate x. Evaluating the higher-dimensional action on such field configurations, and integrating over the coordinate x one can show that

$$\bar{S}_{\mathrm{eff}} = \frac{1}{\bar{\kappa}_0^2} \int d^{D-1}x \sqrt{\bar{G}} e^{-2\bar{\phi}} \Big[\bar{R} + 4(\nabla \bar{\phi})^2 - \frac{1}{12} \bar{H}_{\mu\nu\rho} \bar{H}^{\mu\nu\rho} - (\nabla \sigma)^2$$
$$- \frac{1}{4} \Big(e^{2\sigma} F_{\mu\nu} F^{\mu\nu} + e^{-2\sigma} F'_{\mu\nu} F'^{\mu\nu} \Big) \Big].$$

$\bar{H}_{\mu\nu\rho}$ differs from the field strength $H_{\mu\nu\rho}$ by Chern–Simons terms, which we do not give explicitly.[5] Note that in order to bring the effective action to standard form, we have to introduce a lower-dimensional dilaton

$$\bar{\phi} = \phi - \frac{1}{2}\sigma,$$

which is a linear combination of the higher-dimensional dilaton and the Kaluza–Klein scalar. The relation between the higher- and lower-dimensional gravitational coupling can be parametrised in various ways:

$$\frac{1}{2\bar{\kappa}^2} = \frac{1}{2\bar{\kappa}_0^2} e^{-2\langle\bar{\phi}\rangle} = \frac{2\pi R}{2\kappa_0^2} e^{-2\langle\bar{\phi}\rangle} = \frac{e^{-2\langle\phi\rangle}}{2\kappa_0^2} 2\pi R e^{\langle\sigma\rangle} = \frac{1}{2\kappa^2} 2\pi R e^{\langle\sigma\rangle} = \frac{1}{2\kappa^2} \mathrm{Vol}(S^1).$$
$$\tag{13.35}$$

Comparing the first and last expression, we see that the physical gravitational couplings $\kappa, \bar{\kappa}$ differ by a multiplicative factor involving the physical volume of the internal space.

13.2.2 U(1) Gauge Symmetry from the World-Sheet Point of View

From the space-time point of view, the $U(1)_L \times U(1)_R$ gauge symmetry descends from higher dimensional gauge symmetries. We now turn to the world-sheet perspective, where it is related to the presence of conserved chiral currents.

Let us collect the relevant formulae. The string coordinate along the compact direction is

$$X(z, \bar{z}) = \frac{1}{2} \Big(X(z) + \bar{X}(\bar{z}) \Big), \tag{13.36}$$

where

$$X(z) = x_L - ip_L \log z + i \sum_{m \neq 0} \frac{\alpha_m}{m} z^{-m}. \tag{13.37}$$

---

[5] See, e.g., Polchinski (1998a), Ortin (2004), or Kiritsis (2007) for details.

The left-moving momentum $p_L = \frac{p}{2} + w$ now contains the winding operator w. We focus on the holomorphic sector, where the conformal primaries are

$$H(z) = i(\partial_z X(z)) = \sum_m \alpha_m z^{-m}, \tag{13.38}$$

with $\alpha_0 = p_L$, and

$$E_{k_L}(z) =: \exp\left(ik_L X(z)\right) : . \tag{13.39}$$

These conformal fields have weights $(1, 0)$ and $\left(\frac{1}{2}k_L^2, 0\right)$, respectively. From the OPE $X(z)X(w) = -\log(z - w) + \cdots$ we can deduce:

$$H(z)H(w) = \frac{1}{(z - w)^2} + \cdots \tag{13.40}$$

$$H(z)e^{ik_L X(w)} = \frac{k_L}{z - w}e^{ik_L X(w)} + \cdots \tag{13.41}$$

$$e^{ik_L X(z)}e^{ik'_L X(w)} = (z - w)^{k_L k'_L}e^{i(k_L + k'_L)X(w)}\left(1 + \cdots\right) \tag{13.42}$$

$$T(z)e^{ik_L X(w)} = \left(\frac{\frac{k_L^2}{2}}{(z - w)^2} + \frac{\partial_w}{z - w}\right)e^{ik_L X(w)} + \cdots . \tag{13.43}$$

These formulas can also be obtained from our previous formulas (4.28), (4.51), (4.52), (4.53), by restricting them to one chiral sector, and replacing k by $k_L = \frac{1}{2}k + w$.

For generic R, the only conserved chiral current[6] is $H(z)$, and the associated conserved charge is the left-moving momentum:

$$H_0 := \oint \frac{dz}{2\pi i}H(z) = \alpha_0 = p_L. \tag{13.44}$$

Our basis (13.11) consists of p_L eigenstates. The operator H_0 generates the one-dimensional Lie algebra $\mathfrak{u}(1)$, and since H_0 commutes with the Virasoro operators L_m this is a symmetry algebra and physical states carry the discrete charge k_L under this algebra. The example of the energy-momentum tensor has taught us that with a conserved chiral current we can associate a larger, infinite-dimensional Lie algebra, which is generated by its Laurent modes

$$H_m = \oint \frac{dz}{2\pi i}H(z)z^m = \alpha_m. \tag{13.45}$$

These satisfy, as we know, the relations

$$[H_m, H_n] = m\delta_{m+n,0}. \tag{13.46}$$

The Lie algebra defined by the relations (13.46) is a $\mathbb{Z}$-graded Lie algebra, whose degree zero part is the $\mathfrak{u}(1)$ algebra related to the $U(1)$ gauge symmetry. It is an example of what is called a *current algebra*, which in our case is based on the group $G = U(1)$. The OPE (13.40) provides an equivalent formulation of this current algebra. We will use the notation $\hat{\mathfrak{u}}(1)$ for the algebra (13.46). The operators H_m, L_n form a closed algebra which extends the Virasoro algebra. Since $H(z)$ has weight

---

[6] Remember that this is an operator of weight $(1, 0)$ which does not depend on $\bar{z}$.

$(1, 0)$ we can write down the commutation relations between the operators H_m and the Virasoro operators L_n without computation:

$$[L_m, H_n] = -nH_{m+n}, \quad [L_m, L_n] = (m - n)L_{m+n} + \frac{c}{12}(m^3 - m). \tag{13.47}$$

The relations (13.45) and (13.47) show that $\hat{u}(1)$ and the Virasoro algebra Vir form a semi-direct sum Vir $\oplus_{s.d.}$ $\hat{u}(1)$, with the Virasoro algebra acting on $\hat{u}(1)$. This is our first example of an *extended chiral algebra*, that is of an algebra which enlarges the Virasoro algebra such that the number of chiral primaries which determine the CFT is reduced. Since only the degree zero part $u(1)$ of $\hat{u}(1)$ commutes with the Virasoro generators, the Lie algebra $\hat{u}(1)$ is not a symmetry algebra, and there are no further conserved charges in addition p_L. However, the Fock space organises into representations of the enlarged algebra Vir $\oplus_{s.d.}$ $\hat{u}(1)$, which is therefore called a *spectrum generating algebra*.

Mathematically, $\hat{u}(1)$ is a central extension of the U(1) loop algebra

$$\tilde{u}(1) = \mathbb{C}[z, z^{-1}] \oplus u(1), \tag{13.48}$$

where $\mathbb{C}[z, z^{-1}]$ is the algebra of complex-valued formal power series in z and z^{-1}. It admits a basis of the form $z^m \otimes p$, were $m \in \mathbb{Z}$ and where p is the $u(1)$ generator. The Lie bracket is defined by

$$[z^m \otimes p, z^n \otimes p] = z^{m+n} \otimes [p, p]. \tag{13.49}$$

This U(1) loop algebra is the Lie algebra of the U(1) loop group, that is of the infinite-dimensional group of maps $S^1 \rightarrow$ U(1). The algebra $\mathbb{C}[z, z^{-1}]$ appears through a Fourier (or Laurent) expansion of maps from the circle (or, by extension, from $\mathbb{C}^*$) into U(1). While the loop algebra is abelian, one can add a further generator k and define a Lie algebra structure on

$$\hat{u}(1) := \tilde{u}(1) \oplus \mathbb{C}k \tag{13.50}$$

by

$$[z^m \otimes p, z^n \otimes p] = m\delta_{m+n,0}k, \quad [k, z^m \otimes p] = 0. \tag{13.51}$$

The non-abelian $\mathbb{Z}$-graded Lie algebra $\hat{u}(1)$ is a central extension of the loop algebra $\tilde{u}(1)$. We recover the commutation relations (13.46) upon mapping

$$z^m \otimes p \mapsto H_m, \quad k \mapsto \mathbb{1}. \tag{13.52}$$

The representation that we have obtained through the conserved chiral current $H(z)$ is a 'level 1' or '$k = 1$' representation of the Lie algebra $\hat{u}(1)$ on the Fock space.[7]

Coming back to space-time aspects, we note that the $\hat{u}(1)$ symmetry realised on the world-sheet is a rigid symmetry, while the corresponding U(1) space-time symmetry is a local symmetry. We have seen that in the Fock space $\mathcal{F}$ used in covariant quantisation the constraints impose gauge fixing conditions, and thus local gauge symmetries manifest themselves in 'already gauge fixed form', that is through the decoupling of unphysical and null states. We can summarise the relation between

---

[7] So-called higher level representations where $k > 1$ can be realised using other types of CFTs.

world-sheet and space-time symmetries as follows:[8] 'The local $U(1)_L \times U(1)_R$ space-time gauge symmetry of circle compactifications of closed strings arises from a rigid infinite-dimensional $\hat{u}(1)_L \times \hat{u}(1)_R$ symmetry on the world-sheet.'

In Section 13.2.4, we will see that this holds more generally, and in particular extends to non-abelian gauge symmetries. Therefore, we can note more generally: *Local space-time gauge symmetries for closed strings arise from rigid world-sheet symmetries and the corresponding infinite-dimensional chiral current algebras.*

13.2.3 Enhanced Symmetries and Extra Massless States

So far, we have looked at states which are massless for any value of R. In addition, there are states which are only massless for special values of R. These states can be determined by analyzing the mass formula (13.13). By inspection, there are three types of extra massless states.

1. States with $N = 0, k_L^2 = 2, \tilde{N} = 1, k_R = 0$:

$$\tilde{\alpha}_{-1}^{\mu}|(\pm\sqrt{2}, 0)\rangle, \quad \tilde{\alpha}_{-1}|(\pm\sqrt{2}, 0)\rangle, \tag{13.53}$$

where $k_L = \pm\sqrt{2}$, and where we have suppressed the space-time momentum k^{μ} for simplicity. Depending on whether the remaining right-moving oscillator is a space-time oscillator or an internal oscillator, the resulting states are massless vectors $A_{\mu}^{\pm(L)}$ or massless scalars $\sigma_{\pm,0}$ which are charged under the gauge group $U(1)_L$. For convenience we will normalise the $U(1)_L$ charges such that states with $k_L = \pm\sqrt{2}$ carry charge $q_L = \pm 1$. These states are massless if and only if $R = \frac{1}{2}\sqrt{2}$. By restoring α', we see that these extra states occur when the radius R of the circle is equal to the string length:

$$R = \sqrt{\alpha'} = l_S. \tag{13.54}$$

This is called the *critical radius*, or for reasons that will become clear when we discuss T-duality, the *self-dual radius*.

2. States with $\tilde{N} = 1, k_L = 0, \tilde{N} = 0, k_R^2 = 2$:

$$\alpha_{-1}^{\mu}|(0, \pm\sqrt{2})\rangle, \quad \alpha_{-1}|(0, \pm\sqrt{2})\rangle. \tag{13.55}$$

This is similar to the previous case with $U(1)_L$ replaced by $U(1)_R$, and again happens at the critical radius $R = \sqrt{\alpha'}$:

3. States with $N = \tilde{N} = 0, k_L^2 = k_R^2 = 2$:

$$|(\pm\sqrt{2}, \pm\sqrt{2})\rangle, \tag{13.56}$$

where all four combinations of signs are allowed. These four states are scalars $\sigma_{\pm\pm}$ with charges $(\pm 1, \pm 1)$ under $U(1)_L \times U(1)_R$. Massless states of the form σ_{+-}, σ_{-+} exist at radii

$$R = \frac{2}{|m|}\sqrt{\alpha'}, \quad \text{where} \quad m = \pm 1, \pm 2, \dots, \tag{13.57}$$

---

[8] Here, and occasionally in the following, we use 'symmetry' in the sense of extended chiral algebras rather than in the strict sense.

while massless states of the form σ_{++}, σ_{--} exist at radii

$$R = \frac{|n|}{2}\sqrt{\alpha'}, \quad \text{where} \quad n = \pm 1, \pm 2, \ldots . \tag{13.58}$$

Looking at the properties of the extra massless states we can interpret what happens at the special values of R.

1. For $R = \sqrt{\alpha'}$ we have four extra massless charged vector bosons with charges $(\pm 1, 0)$ and $(0, \pm 1)$ under $U(1)_L \times U(1)_R$. These fit into the adjoint representation $(\mathbf{3}, \mathbf{0}) \oplus (\mathbf{0}, \mathbf{3})$ of the group $SU(2)_L \times SU(2)_R$, which indicates an enhanced, non-abelian gauge symmetry at this point. Therefore there needs to be a Higgs mechanism which gives a mass to the charged vector bosons for $R \neq \sqrt{\alpha'}$, and this mass should be controlled by the vev of a field which is proportional to $\delta R = R - \sqrt{\alpha'}$. Using the effective field theory analysis of Section 13.2.1 we know that this is the vev of the Kaluza–Klein scalar σ. To see that the light spectrum is consistent with a Higgs mechanism we note that the charges of the nine scalars $\sigma, \sigma_{\pm,0}, \sigma_{0,\pm}, \sigma_{\pm\pm}, \sigma_{\pm\mp}$ fill out the 'bi-adjoint' $(\mathbf{3}, \mathbf{3})$ representation of the gauge group $SU(2)_L \times SU(2)_R$. If our interpretation is correct then away from $R = \sqrt{\alpha'}$ the scalars $\sigma_{\pm,0}, \sigma_{0,\pm}$ should have the same masses as the vector fields $A_\mu^{\pm(L/R)}$ since they will provide them with their longitudinal degree of freedom. See Exercise 13.2.4.

2. For $R = 2\sqrt{\alpha'}$ and $R = \frac{1}{2}\sqrt{\alpha'}$ there are four massless scalars $\sigma_{\pm\pm}$, which are charged under $U(1)_L \times U(1)_R$. We will see later that at these points we can turn on a vacuum expectation value for one of the charged scalars, which then breaks the $U(1)_L \times U(1)_R$ symmetry and parametrises a new branch of our moduli space.

3. Finally, there are infinitely many further points where two charged scalars become massless, either σ_{++} and σ_{--}, or σ_{+-} and σ_{-+}. At these points there is no gauge symmetry enhancement or further symmetry breaking.

Special points in a moduli space of consistent string backgrounds, where additional massless states appear, are often referred to as ESPs. Depending on context this acronym stands for *enhanced symmetry point* (gauge symmetry enhanced) or *extra species point* (extra massless matter, no gauge symmetry enhancement).

Exercise 13.2.3 Work out the complete list of states which become massless for some value of R, thus confirming the statements made in the preceding section.

Exercise 13.2.4 Work out the dependence on $\delta R = R - \sqrt{\alpha'}$ of the masses of the states $A_\mu^{\pm(L/R)}$, $\sigma_{\pm 0}, \sigma_{0,\pm} \sigma_{\pm\pm}$, for $|\delta R| \ll 1$. Show that the masses change smoothly with R, and that the spectrum is consistent with a Higgs mechanism. Show that some of the scalar fields are tachyonic for $\delta R \neq 0$.

Exercise 13.2.5 Work out the light spectrum around the point where $R = 2\sqrt{\alpha'}$. How do the masses of the extra light states depend on $\delta R = R - 2\sqrt{\alpha'}$?

13.2.4 SU(2) Gauge Symmetry from the World-Sheet Point of View

In this section, we will identify the conserved chiral currents and extended chiral algebras corresponding to the enhanced $SU(2)_L \times SU(2)_R$ gauge symmetry at the critical radius $R = \sqrt{\alpha'}$. For simplicity we focus on one chiral sector of the world-sheet theory. To identify the additional conserved chiral currents we note that chiral operators of the form : $\exp(k_L X(z))$: have conformal weight $(\frac{1}{2}k_L^2, 0)$, and thus are conserved chiral currents if $k_L^2 = 1$. Also note that these operators are precisely the zero (space-time) momentum parts of the vertex operators for the charged vector bosons $A_\mu^{\pm(L)}$. Therefore the world-sheet CFT has additional conserved chiral currents if and only if there are vertex operators for massless charged vector fields.

We need to analyse the OPEs between the conformal fields

$$H(z) = i\partial_z X(z), \quad E(z) =: \exp(\sqrt{2}X(z)) :, \quad F(z) =: \exp(-\sqrt{2}X(z)) : . \quad (13.59)$$

Using (13.40), (13.41), (13.42) for $k_L = \sqrt{2}, k_L' = -\sqrt{2}$ we find

$$E(z)F(w) = \frac{1}{(z-w)^2} + \frac{\sqrt{2}H(w)}{z-w} + \cdots , \quad (13.60)$$

$$H(z)E(w) = \frac{\sqrt{2}}{z-w}E(w) + \cdots , \quad (13.61)$$

$$H(z)F(w) = \frac{-\sqrt{2}}{(z-w)}F(w) + \cdots . \quad (13.62)$$

Note that when applying (13.42) we need to include all terms which come with negative powers of $(z - w)$. Introducing the mode operators

$$H_m = \oint \frac{dz}{2\pi i}H(z)z^m, \quad E_m = \oint \frac{dz}{2\pi i}E(z)z^m, \quad F_m = \oint \frac{dz}{2\pi i}F(z)z^m, \quad (13.63)$$

we have the commutation relations

$$[H_m, H_n] = m\delta_{m+n,0}, \quad [E_m, F_n] = m\delta_{m+n,0} + \sqrt{2}H_{m+n}, \quad (13.64)$$

$$[H_m, E_n] = \sqrt{2}E_{m+n}, \quad [H_m, F_n] = -\sqrt{2}F_{m+n}. \quad (13.65)$$

To understand this $\mathbb{Z}$-graded Lie algebra, we first look at its degree zero subalgebra, which has the non-trivial relations

$$[E_0, F_0] = \sqrt{2}H_0, \quad [H_0, E_0] = \sqrt{2}E_0, \quad [H_0, F_0] = -\sqrt{2}F_0. \quad (13.66)$$

This is the Lie algebra $\mathfrak{su}(2)$, written in terms of a Cartan generator H_0 and two ladder operators E_0, F_0. Up to normalisation H_0, E_0, F_0 are *Chevalley generators* for $\mathfrak{su}(2)$, or more precisely for its complexification, the complex simple Lie algebra A_1. See Appendix G for a brief summary of the relevant concepts. The relations for standard Chevalley generators h, e, f are

$$[e, f] = h, \quad [h, e] = 2e, \quad [h, f] = -2f, \quad (13.67)$$

and, therefore, $h = \sqrt{2}H_0, e = E_0, f = F_0$. By taking suitable complex linear combinations we can obtain generators T^1, T^2, T^3 which satisfy the standard $\mathfrak{su}(2)$ relations

$$[T^a, T^b] = i\epsilon_{abc}T^c, \quad a, b, c = 1, 2, 3 \quad (13.68)$$

for Hermitian generators.[9] Similarly, we can find complex linear combinations of H_m, E_m, F_m so that the full algebra takes the form

$$[T_m^a, T_n^b] = i\epsilon_{abc}T_{m+n}^c + m\delta^{ab}\delta_{m+n,0}. \quad (13.69)$$

---

[9] To obtain real structure constants one needs to use the corresponding anti-Hermitian generators $t_a = -iT_a$, which form a real basis for $\mathfrak{su}(2)$. In the physics literature one usually prefers the Hermitian generators, because in quantum theory observables are represented by Hermitian operators.

This is an infinite-dimensional $\mathbb{Z}$-graded Lie algebra which contains $\mathfrak{su}(2)$ as its degree zero part. It provides another example of a current algebra which generalises $\mathfrak{su}(2)$ in the same way as $\hat{\mathfrak{u}}(1)$ extends $\mathfrak{u}(1)$. We denote this algebra by $\hat{\mathfrak{su}}(2)$.

Mathematically, $\hat{\mathfrak{su}}(2)$ is an *untwisted affine Kac–Moody algebra*. Affine Kac–Moody algebras are infinite-dimensional $\mathbb{Z}$-graded Lie algebras which are similar to finite-dimensional simple Lie algebras but have generalised Cartan matrices where one of the eigenvalues is zero. Untwisted affine Kac–Moody algebras are in one-to-one correspondence with simple Lie algebras and their Cartan matrices are encoded by the extended Dynkin diagrams of the corresponding simple Lie algebras.[10] Untwisted affine Kac–Moody algebras can be realised as follows. Starting from a compact, real, finite-dimensional Lie algebra $\mathfrak{g}$ with generators T^a, normalised to have scalar product $(T^a, T^b) = \delta^{ab}$, and with structure constants f_c^{ab}, we go to the corresponding loop algebra $\tilde{\mathfrak{g}}$, that is to the Lie algebra of the group of maps $S^1 \rightarrow G$, where G is a Lie group with Lie algebra $\mathfrak{g}$. Then we add two further generators k and d to obtain the Lie algebra $\hat{\mathfrak{g}}$. As a vector space

$$\hat{\mathfrak{g}} = \tilde{\mathfrak{g}} \oplus \langle k \rangle \oplus \langle d \rangle, \tag{13.70}$$

and the Lie algebra structure is defined by

$$[z^m \otimes T^a, z^n \otimes T^b] = z^{m+n} f_c^{ab} T^c + k m \delta_{m+n,0} \delta^{ab}, \quad [d, z^m \otimes T^a] = m z^m \otimes T^a. \tag{13.71}$$

This shows that k is central and that $\tilde{\mathfrak{g}} \oplus \langle k \rangle$ is a central extension of $\tilde{\mathfrak{g}}$. The operator d operates as a derivation on $\tilde{\mathfrak{g}} \oplus \langle k \rangle$, which reads out the degree of homogeneous elements. This implies that $\hat{\mathfrak{g}}$ is a semi-direct extension of $\tilde{\mathfrak{g}} \oplus \langle k \rangle$. It can be shown that all untwisted affine Kac–Moody algebras can be realised in this way.

Comparing (13.69) to (13.71), we see that we can match the first set of relations in (13.71) by representing the central element k by the unit $\mathbb{1}$. Thus we obtain a level 1 or $k = 1$ representation. We also need to identify the derivation d. To do this we note that the conformal fields $T^a(z)$ have been obtained by taking linear combinations of the fields $H(z), E(z), F(z)$, and therefore they have same conformal weights $(1, 0)$. This fixes their commutation relations with the Virasoro generators to be

$$[L_m, T_n^a] = -n T_{m+n}^a. \tag{13.72}$$

Therefore, the derivation d is the negative of the chiral world-sheet Hamiltonian, $d = -L_0$. Note that besides d there is an infinite set of further derivations given by the action of the L_m, $m \neq 0$. As for $\hat{\mathfrak{u}}(1)$, the Virasoro algebra and $\hat{\mathfrak{su}}(2)$ form a semi-direct sum.[11] Since only the degree zero subalgebra $\mathfrak{su}(2)$ commutes with the Virasoro generators, only $\mathfrak{su}(2)$ is a symmetry in the strict sense, while the extended chiral algebra $\hat{\mathfrak{su}}(2)$ is a spectrum generating algebra.

Throughout this section we have not been careful in distinguishing between the complex Lie algebra A_1 and the real Lie algebra $\mathfrak{su}(2)$.[12] In particular, we have have freely taken complex linear combinations of generators. Symmetry algebras in quantum theory are realised by *unitary* representations. A_1 is the complex simple

---

[10] Twisted affine Kac–Moody algebras are then obtained by acting in all possible ways with outer automorphisms.

[11] Using the so-called *Sugawara construction* one can write the energy-momentum tensor as a quadratic expression in the $\hat{\mathfrak{su}}(2)$ currents.

[12] See Appendix G for some background on Lie algebras.

Lie algebra generated by H_0, E_0, F_0. It has two real forms, that is, there are two non-isomorphic real Lie algebras with complexification A_1. The so-called normal form or split form is $\mathfrak{sl}(2,\mathbb{R})$, which is the real Lie algebra generated by H_0, E_0, F_0. The other is the compact real form $\mathfrak{su}(2)$, where compact refers to the compactness of the corresponding Lie group. This is the real Lie algebra generated by $-iT_a$, $a = 1, 2, 3$. Among the (semi-)simple Lie algebras, only the compact real forms have finite-dimensional unitary representations. Therefore the symmetry algebra of our theory is $\mathfrak{su}(2)$.

Another question is whether the infinitesimal (Lie algebra) action we have found exponentiates to a finite group action. For infinite-dimensional Lie algebras this is not automatic. However, the $\widehat{\mathfrak{su}}(2)$ representations constructed above are so-called *integrable highest weight representation*, which means that all $\mathfrak{su}(2)$ sub-representations can be exponentiated ('integrated') to obtain SU(2) representations. This is the property that we need for physics: the particle spectrum organises itself into SU(2) multiplets.

Exercise 13.2.6   Derive the commutation relations (13.64), (13.65) and (13.65) from the OPE (13.60), (13.61), and (13.62).

Exercise 13.2.7   Find linear combinations T_0^1, T_0^2, T_0^3 of E_0, F_0, H_0 which satisfy the standard $\mathfrak{su}(2)$ commutation relations

$$[T_0^a, T_0^b] = i\epsilon_{abc} T_0^c. \tag{13.73}$$

Write down the corresponding chiral currents.

13.2.5   U(1) Gauge Symmetry, Higgs Effect, and Massive Vector Fields

We now turn to the space-time perspective and explore how additional light states and the Higgs mechanism can be described using effective field theory. To start, we review how the Higgs mechanism works for a single vector field. Apart from serving as a toy example, this will have an application later when we show that at the special radii $R = 2\sqrt{\alpha'}$ and $R = \frac{1}{2}\sqrt{\alpha'}$ there is a Higgs mechanism which allows to make the $U(1)_L \times U(1)_R$ gauge bosons massive.

The Abelian Higgs model is the simplest example for the Higgs mechanism, with Lagrangian

$$\mathcal{L} = -\frac{1}{4} F_{\mu\nu} F^{\mu\nu} - |D_\mu \phi|^2 - V(\phi). \tag{13.74}$$

The covariant derivative of the complex scalar field ϕ is

$$D_\mu \phi = (\partial_\mu + ig A_\mu)\phi, \tag{13.75}$$

where g is the gauge coupling. As scalar potential we choose the 'Mexican hat' potential

$$V = \lambda(2|\phi|^2 - c^2)^2, \tag{13.76}$$

where $c > 0$ is a constant which parametrises the shape of the potential.

This theory contains a vector field and a complex scalar field. To identify its particle content, we need to expand around a ground state, and take into account the local U(1) gauge symmetry, which acts by

$$\phi \to e^{i\alpha(x)}\phi, \quad A_\mu(x) \to A_\mu(x) - \frac{1}{g}\partial_\mu\alpha(x). \tag{13.77}$$

It can be shown that the energy density T^{00} satisfies $T^{00} \geq 0$ and is minimised if

$$F_{\mu\nu} = 0, \quad D_\mu\phi = 0, \quad V(\phi) = 0. \tag{13.78}$$

These conditions characterise the ground state(s) of the theory and can be solved by[13]

$$A_\mu = 0, \quad |\phi|^2 = c^2. \tag{13.79}$$

For $c \neq 0$ we have a family of gauge equivalent ground states which are parametrised by a circle $S^1 \subset \mathbb{R}^2$ of radius c. To proceed we pick one ground state $\phi_0 = c$ and expand $\phi(x)$ around it:

$$\phi = \frac{1}{\sqrt{2}}e^{-ia(x)}(\varphi(x) + c). \tag{13.80}$$

Then we impose the gauge condition $a(x) = 0$ which eliminates the angular part of $\phi(x)$:

$$\phi = \frac{1}{\sqrt{2}}(\varphi(x) + c). \tag{13.81}$$

This completely fixes our freedom of making local $U(1)$ transformations, and therefore all remaining degrees of freedom must be physical. To determine their masses and spins, we substitute (13.81) back into the Lagrangian and obtain

$$\mathcal{L} = -\frac{1}{4}F_{\mu\nu}F^{\mu\nu} - \frac{1}{2}m_A^2 A_\mu A^\mu - \frac{1}{2}\partial_\mu\varphi\partial^\mu\varphi - \frac{1}{2}m_H^2\varphi^2 + \cdots, \tag{13.82}$$

where we omitted interaction terms, and where

$$m_A^2 = g^2 c^2, \quad m_H^2 = 8\lambda c^2, \quad c^2 = |\langle\phi\rangle|^2. \tag{13.83}$$

This Lagrangian contains mass terms for the vector field A_μ and for the real scalar field φ, which are proportional to the vacuum expectation value $|c|$. The first two terms comprise the Proca Lagrangian for a massive vector field, see Exercise 13.2.8. The physical degrees of freedom are a massive vector field and massive real scalar.

The second case is $c = 0$, where the potential has a unique minimum at $\phi_0 = 0$. Since ϕ does not have a vacuum expectation value, no mass term for the vector field is generated. As we have seen in our discussion of the Maxwell field, a massless vector field has $D - 2$ independent components. The theory with $c = 0$ contains one massless vector field and one massless complex scalar. Comparing the two cases we see that the Higgs effect redistributes one degree of freedom from the scalars to the vector, in order that the vector field can become massive.

In the Abelian Higgs model the vacuum expectation value of ϕ is fixed by the scalar potential. This is different for circle compactifications of the closed string, where the vacuum expectation value of the scalar σ is a free parameter, indicating that the scalar potential is flat in the σ-direction. A similar situation can be obtained in the Abelian Higgs model by taking the limit $\lambda \to 0$. Then $\langle\phi\rangle$ becomes a free parameter, and the masses are

$$m_A^2 = g^2 c^2, \quad m_H = 8\lambda c^2 \to 0, \quad c^2 = |\langle\phi\rangle|^2. \tag{13.84}$$

---

[13] Note that on spaces with non-trivial topology, like a torus, we might have non-trivial Wilson lines, and hence solutions of $F_{\mu\nu} = 0$ which are not gauge equivalent to $A_\mu = 0$.

The vector field retains its mass, while the physical Higgs scalar is now massless and has an arbitrary vacuum expectation value.

Exercise 13.2.8   The *Proca Lagrangian* for a massive vector field is given by

$$\mathcal{L} = -\frac{1}{4}\left(\partial_\mu A_\nu - \partial_\nu A_\mu\right)\left(\partial^\mu A^\nu - \partial^\nu A^\mu\right) - \frac{1}{2}m_A^2 A_\mu A^\mu. \tag{13.85}$$

Show that this theory has $D - 1$ propagating degrees of freedom, which transform as a vector under the massive little group $SO(D-1)$.

Exercise 13.2.9   The *free Abelian Higgs model* is defined by taking the limit $\lambda \to \infty$. This enforces the constraint $2|\phi|^2 = c^2$ and fixes the radial component of ϕ. Show that the resulting Lagrangian is equivalent to the $U(1)$ gauge invariant *Stückelberg Lagrangian*

$$\mathcal{L} = -\frac{1}{4}\left(\partial_\mu A_\nu - \partial_\nu A_\mu\right)\left(\partial^\mu A^\nu - \partial^\nu A^\mu\right) - \frac{m_A^2}{2}\left(A_\mu + \partial_\mu B\right)\left(A^\mu + \partial^\mu B\right) \tag{13.86}$$

for a massive vector field. How is the Stückelberg Lagrangian related to the Proca Lagrangian?

13.2.6 $SU(2)_L \times SU(2)_R$ Gauge Symmetry from the Space-Time Point of View

To prepare ourselves for describing the Higgs mechanism for the $SU(2)_L \times SU(2)_R$ gauge symmetry of circle compactifications, we first review a simpler example, the adjoint Higgs mechanism for a single $SU(2)$. This theory contains three vector fields A_μ^a and three real scalars ϕ^a, where $a = 1, 2, 3$. Since the adjoint representation of $SU(2)$ is the fundamental or vector representation **3** of $SO(3)$, gauge transformations act through multiplication of vectors by $SO(3)$ matrices. If we fix the subgroup $U(1) \cong SO(2) \subset SO(3)$ which acts by rotation around the 3-axis, we can assign $U(1)$ charges to fields. The fields ϕ^3, A_μ^3 are neutral, while $\phi^\pm = \phi^1 \pm i\phi^2$ and $A_\mu^\pm = A_\mu^1 \pm i A_\mu^2$ carry charges ± 1.

To write down a Lagrangian which is invariant under local gauge transformations, we need to replace partial derivatives by $SU(2)$-covariant derivatives. For the scalars these take the form

$$D_\mu \phi^a = \partial_\mu \phi^a - g\epsilon^{abc} A_\mu^b \phi^c, \tag{13.87}$$

while the covariant field strength for the vector fields takes the form

$$F_{\mu\nu}^a = \partial_\mu A_\nu^a - \partial_\nu A_\mu^a - g\epsilon^{abc} A_\mu^b A_\nu^c. \tag{13.88}$$

To induce the Higgs effect in a gauge theory we add a 'Mexican hat' type scalar potential. The resulting Lagrangian (often called the Georgi–Glashow model) is

$$\mathcal{L} = -\frac{1}{4}F_{\mu\nu}^a F^{a\mu\nu} - \frac{1}{2}(D_\mu \phi^a)^2 - \frac{\lambda}{4}(\phi^a \phi^a - c^2)^2, \tag{13.89}$$

where $c \geq 0$ is a parameter which controls the shape of the potential. As in the Abelian case we have to expand around a ground state and distinguish physical from gauge degrees of freedom.

It can be shown that the energy density associated to this Lagrangian is positive definite and that it takes its minimal value zero if

$$F^a_{\mu\nu} = 0, \quad D_\mu \phi^a = 0, \quad V(\phi^a) = 0. \tag{13.90}$$

Locally we can solve the equation $F^a_{\mu\nu} = 0$ by $A^a_\mu = 0$ without loss of generality.[14] Then the scalars are constant, and to obtain the ground state it remains to minimise the potential:

$$V(\phi^a) = 0 \Rightarrow \phi^a \phi^a = c^2. \tag{13.91}$$

Thus, for $c \neq 0$ the ground state manifold is a two-sphere $S^2 \subset \mathbb{R}^3$ of radius c. Since the gauge group acts on the ϕ^a by rotations, all points on this sphere are equivalent. The constant $c > 0$ is the gauge invariant overall radial part of the vacuum expectation value of the scalar fields ϕ^a, $c = \sqrt{\langle \phi^a \phi^a \rangle}$. We can pick one point on the sphere to represent the ground state, for example $\phi^a_0 = c\delta^{a,3}$. This choice is still invariant under the subgroup SO(2) $\subset$ SO(3) which acts by rotations around the 3-axis. Fluctuations around the ground state are parametrised as

$$\phi^a = c\delta^{a,3} + \varphi^a. \tag{13.92}$$

We now impose the gauge conditions $\varphi^1 = \varphi^2 = 0$. Note that this is only a partial gauge fixing, since the gauge condition is invariant under U(1) $\cong$ SO(2) $\subset$ SO(3) $\cong$ SU(2). By substituting the decomposition (13.92) and the gauge condition into the Lagrangian, we obtain

$$\mathcal{L} = \sum_{i=1,2} \left(-\frac{1}{4}(F^i_{\mu\nu}F^{i|\mu\nu})_{\text{lin}} - \frac{1}{2}g^2 c^2 A^i_\mu A^{i|\mu} \right) \tag{13.93}$$

$$- \frac{1}{4}(F^3_{\mu\nu}F^{3|\mu\nu})_{\text{lin}} - \frac{1}{2}\partial_\mu \varphi^3 \partial^\mu \varphi^3 - \lambda c^2 (\varphi^3)^2 + \cdots, \tag{13.94}$$

where we omit interaction terms. The notation $(F^i_{\mu\nu})_{\text{lin}}$ stands for the linearised fields strength tensor $(F^i_{\mu\nu})_{\text{lin}} = \partial_\mu A^i_\nu - \partial_\nu A^i_\mu$. The first line (13.93) is the sum of two Proca Lagrangians for massive vector fields. This shows that the two charged vector fields $A^\pm_\mu = A^1_\mu \pm iA^2_\mu$ acquire a mass. The first term in (13.94) is the standard Maxwell Langrangian for the neutral vector field A^3_μ, which therefore remains massless. This is consistent with the residual U(1) gauge symmetry. Finally we have one real massive physical Higgs field φ^3. The masses of the fields are

$$m^2_\pm = m^2_{1,2} = g^2 c^2, \quad m_3 = 0, \quad m^2_H = 2\lambda c^2, \quad c^2 = \langle \phi^a \phi^a \rangle. \tag{13.95}$$

This shows how the adjoint Higgs mechanism works. Two of the real scalars are traded for the longitudinal degrees of freedoms of the two charged vector fields which become massive. The third, neutral vector field remains massless, and there is one massive real Higgs field. For $c = 0$ no mass terms are generated and there are three massless vector fields and three massless real scalars. As in the Abelian Higgs model we can consider the limit $\lambda \to 0$ where the potential is eliminated. Then the vacuum expectation value c becomes a free parameter. In this limit the Higgs field φ^3 is massless, but the charged vector fields still have a mass proportional to the vacuum expectation value.

---

[14] See Section 13.6 for the effect of Wilson loops.

Now we turn to the situation realised in circle compactifications of closed strings, where the gauge symmetry is $SU(2)_L \times SU(2)_R$ and the nine scalars $\sigma, \sigma_{\pm,0}, \sigma_{0,\pm}$, $\sigma_{\pm,\pm}$ fill out the bi-adjoint representation $(\mathbf{3,3})$. Note that we do not simply have two copies of the adjoint $SU(2)$ Higgs model. From our previous analysis of the spectrum we know (compare Exercise (13.2.4)): (i) the potential has one flat direction, parametrised by the vacuum expectation value of the KK-scalar σ which is proportional to $\delta R = R - \sqrt{\alpha'}$, (ii) $M^2 \propto (\delta R)^2$ for the gauge bosons, while $M^2 \propto \pm \delta R$ for $\sigma_{\pm\pm}$. I.p. half of the scalars $\sigma_{\pm\pm}$ become tachyonic away from the critical radius.

These features suggest a potential which is cubic in the σ-scalars. Let us sketch how to set up an effective field theory for the six $SU(2)_L \times SU(2)_R$ gauge bosons and nine scalars $\sigma, \sigma_{\pm,0}, \sigma_{0,\pm}, \sigma_{\pm\pm}$. We will neglect all other string states and only want to show that one can reproduce the qualitative dependence of the masses of string states on the vacuum expectation value $\langle\sigma\rangle \propto \delta R$. We can choose linear combinations σ_{ia}, $i, a = 1, 2, 3$ of the nine scalars such that σ_{ia} transform under $SU(2)_L \times SU(2)_R$ as

$$\sigma_{ia} \to R_i{}^j R'^{b}_a \sigma_{jb} \tag{13.96}$$

where $R, R' \in SO(3)$. For definiteness, we choose σ_{ia} such that $\sigma, \sigma_{\pm,0}, \sigma_{0,\pm}, \sigma_{\pm\pm}$ are the charge eigenstates with respect to the maximal abelian subgroup $SO(2) \times SO(2)$, where the $SO(2)$ subgroups act by rotation around the 3-axis. The natural candidate for an $SU(2)_L \times SU(2)_R$ invariant potential is

$$V = \det \Sigma, \quad \Sigma = (\sigma_{ia})_{i,a=1,2,3}. \tag{13.97}$$

Mass terms for the charged gauge bosons $A_\mu^{\pm,0}$ and $A_\mu^{0,\pm}$ should arise through the Higgs mechanism from the gauge covariantised scalar kinetic term

$$\mathrm{Tr}\left(D_\mu \Sigma D^\mu \Sigma^T\right) = D_\mu \sigma_{ia} D^\mu \sigma_{ai}, \tag{13.98}$$

where

$$D_\mu \sigma_{ia} = \partial_\mu \sigma_{ia} - g_L \epsilon_{ijk} A_\mu^{j,0} \sigma_{ka} - g_R \epsilon_{abc} A_\mu^{0,b} \sigma_{ic} \tag{13.99}$$

is the $SU(2)_L \times SU(2)_R$ covariant derivative.

Exercise 13.2.10   Show that the potential (13.97) is gauge invariant and has a single flat direction. For this it is helpful to first show that Σ can be diagonalised by an $SU(2)_L \times SU(2)_R$ transformation. Check that the covariantised scalar kinetic term is gauge invariant. Work out the dependence of the masses of vector and scalar fields on the vacuum expectation value, and show that this dependence is the same as for circle compactifications of closed strings, evaluated to leading order in δR. It is helpful to find the explicit relation between the scalars σ_{ia}, $i, a = 1, 2$ and the charge eigenstates $\sigma_{\pm\pm}$.

13.2.7 The Higgs Effect from the World-Sheet Point of View

From the space-time point of view, the moduli space of circle compactifications is parametrised by the vacuum expectation value of the KK scalar σ. We now discuss the Higgs effect from the world-sheet point of view. In Section 12.2.4, it was mentioned that conformal field theories with central charge $c \geq 1$ can form continuous families which are related to each other by marginal deformations. The world-sheet action for the compactified dimension is

$$S[X] = -\frac{1}{4\pi\alpha'} \int d^2\sigma \partial_\alpha X \partial^\alpha X G_{D-1,D-1}, \tag{13.100}$$

where we have imposed the conformal gauge and used that the space-time metric G_{MN} factorises. It is convenient to identify the coordinate X according to $X \cong X + 2\pi$. Then the radius R of the circle is encoded by the metric coefficient

$$G_{D-1,D-1} = R^2. \tag{13.101}$$

The marginal deformation corresponds to changing R. The corresponding marginal operator is the vertex operator of the KK-scalar σ, evaluated at zero space-time momentum, $\partial_\alpha X \partial^\alpha X$. This is obviously an exactly marginal operator, since it is proportional to the world-sheet Lagrangian. The infinitesimal change of the action is

$$\delta S = -\frac{1}{4\pi\alpha'} \int d^2\sigma \partial_\alpha X \partial^\alpha X \delta G_{D-1,D-1}, \tag{13.102}$$

where

$$\delta G_{D-1,D-1} = 2R\delta R. \tag{13.103}$$

We can express this in terms of the constant part (vacuum expectation value) of the Kaluza–Klein scalar σ:

$$G_{D-1,D-1} = R^2 = \alpha' e^{2\langle\sigma\rangle} \Rightarrow \delta\langle\sigma\rangle = \frac{\delta R}{R}. \tag{13.104}$$

Here we have chosen a normalisation where the critical radius $R = \sqrt{\alpha'}$ corresponds to $\langle\sigma\rangle = 0$.

In order to find other exactly marginal operators, it is useful to note that the marginal operator $\partial_\alpha X \partial^\alpha X \cong \partial_z X \partial_{\bar{z}} X$ is the product of two conserved chiral currents, $J_1 = H = i\partial_z X(z)$ and $\bar{J}_1 = \bar{H} = i\partial_{\bar{z}} \bar{X}(\bar{z})$. For generic R, these are the only conserved chiral currents. We have seen that at $R = \sqrt{\alpha'}$ there are the additional conserved chiral currents $J_{2,3} \cong E \pm F$, and $\bar{J}_{2,3} \cong \bar{E} \pm \bar{F}$. This allows us to construct $3 \times 3 = 9$ marginal operators. However, these marginal operators do not create additional deformations, because we also have the enhanced SU(2) × SU(2) symmetry. This acts as SO(3) × SO(3) on the triples (J_1, J_2, J_3) and $(\bar{J}_1, \bar{J}_2, \bar{J}_3)$ of currents and allows one to map all marginal operators to the generic one.

There are other points in the moduli space where extra massless states occur. The corresponding vertex operators, taken at zero space-time momentum, are marginal operators which can be used to deform the action. In particular, we have seen that at the points where $R = 2\sqrt{\alpha'}$ and $R = \frac{1}{2}\sqrt{\alpha'}$ there are four additional massless scalars $\sigma_{++}, \sigma_{+-}, \sigma_{-+}, \sigma_{--}$. Note that the currents $J_i, \bar{J}_i$ $i = 1, 2$ are not conserved at these points in the moduli space. However their products $J_i\bar{J}_j$, which are the (zero momentum) vertex operators of the extra four massless states can be shown to be exactly marginal. Together with $J_3\bar{J}_3$ there are five exactly marginal operators. The CFT has a continuous O(2) × O(2) symmetry which relates the marginal operators $J_i\bar{J}_j$ with $i, j = 1, 2$ to one another, but not to $J_3\bar{J}_3$.[15] Thus, there are two independent

---

[15] The U(1) × U(1) $\cong$ SO(2) × SO(2) symmetry gets enhanced by an additional $\mathbb{Z}_2 \times \mathbb{Z}_2$ symmetry, see Dijkgraaf et al. (1988).

marginal operators. The second marginal operator creates a new branch of the moduli space, and in a later section we will see that this corresponds to the compactification on an *orbifold* of the circle.

The other points with extra massless states are $R = \frac{|n|}{2}\sqrt{\alpha'}, \frac{2}{|m|}\sqrt{\alpha'}$, where $|n|, |m| = 3, 5, 6, 7, \ldots$ At these points there are two charged massless scalars, either σ_{++}, σ_{--}, or σ_{+-}, σ_{-+}, but there is no symmetry enhancement or breaking, and there are no extra directions in moduli space.

13.2.8 Partition Functions for Winding States

Through compactification the continuous momentum eigenstates have been replaced by discrete momentum and winding eigenstates along the S^1-direction. In this section, we work out the resulting modification of the partition function, and verify that it is still modular invariant. We consider the $c = 1$ CFT of modes associated with the S^1. The standard basis for these states is

$$\alpha_{-k_1} \cdots \tilde{\alpha}_{-l_1} \cdots |(m, n), R\rangle, \quad 0 < k_1 \le k_2 \le \cdots, \quad 0 < l_1 \le l_2 \le \cdots, \quad (13.105)$$

where $k_i, l_j, m, n \in \mathbb{Z}$. The trace over the oscillators works as before in Section 6.2, so that

$$Z_R(\tau, \bar{\tau}) = \frac{1}{\eta(\tau)\eta(\bar{\tau})} \Theta_{\Gamma_{1,1}(R)}(\tau, \bar{\tau}), \quad (13.106)$$

where

$$\Theta_{\Gamma_{1,1}(R)}(\tau, \bar{\tau}) = \sum_{m,n \in \mathbb{Z}} q^{\frac{1}{2}\left(\frac{n}{2R}+mR\right)^2} \bar{q}^{\frac{1}{2}\left(\frac{n}{2R}-mR\right)^2} = \sum_{(k_L, k_R) \in \Gamma_{1,1}(R)} q^{\frac{1}{2}k_L^2} \bar{q}^{\frac{1}{2}k_R^2}. \quad (13.107)$$

We introduced the notation

$$\Gamma_{1,1}(R) = \left\{ (k_L, k_R) = \left(\frac{n}{2R} + mR, \frac{n}{2R} - mR \right) \in \mathbb{R}^2 \,\middle|\, m, n \in \mathbb{Z} \right\} \quad (13.108)$$

for the lattice of left- and right-moving momenta, or lattice of momenta and windings. Such lattices are also known as *Narain lattices* or *momentum lattices*. We will refer to $\Theta_{\Gamma_{1,1}(R)}(\tau, \bar{\tau})$ as the *lattice partition function*. On $\Gamma_{1,1}(R)$ we define the indefinite scalar product

$$(k_L, k_R) \cdot (k_L', k_R') = k_L \cdot k_L' - k_R \cdot k_R', \quad (13.109)$$

which has the remarkable property of being independent of R and integer-valued:

$$(k_L, k_R) \cdot (k_L', k_R') = nm' + mn' \in \mathbb{Z}. \quad (13.110)$$

To check modular invariance we need to work out the modular transformation properties of the lattice partition function $\Theta_{\Gamma_{1,1}(R)}$. We will do this for a more general class of Narain lattices, namely lattices of the form $\Gamma_{p,q}$. Therefore our results will apply to two important generalisations: (i) compactifications on tori $T^d = \mathbb{R}^d/\Lambda$, where Λ is a d-dimensional lattice, so that k_L, k_R are replaced by d-component vectors, and (ii) heterotic string theories where $p \ne q$ (see Section 14.7).

First, we introduce a few definitions about lattices.

Let Γ be a lattice of rank n equipped with a scalar product of signature (p, q).

- The *dual lattice* Γ^* contains all vectors which have integer scalar products with all vectors of Γ:

$$\Gamma^* = \{W \in \mathbb{R}^n | W \cdot V \in \mathbb{Z}, \quad \forall V \in \Gamma\}. \tag{13.111}$$

- Γ is *integral* if all scalar products are integer valued, equivalently, if Γ is contained in its dual, $\Gamma \subset \Gamma^*$.
- Γ is *unimodular* if the volume of its fundamental cell is unity.
- Γ is *self-dual* if it is integral and unimodular, equivalently, if it is equal to its dual, $\Gamma = \Gamma^*$.
- Γ is *even* if the square norm of any lattice vector is an even integer.

For a rank n lattice $\Gamma \subset \mathbb{R}^n$ we can choose a lattice basis e_i, $i = 1 \ldots, n$, that is a basis of $\mathbb{R}^n$ such that

$$\Lambda = \{v \in \mathbb{R}^n | v = m^i e_i, \ m^i \in \mathbb{Z}\}. \tag{13.112}$$

The dual lattice has a natural dual basis e^{*i} defined by $e^{*i} \cdot e_j = \delta^i_j$. The Gram matrices of the scalar product with respect to these bases are $g_{ij} = e_i \cdot e_j$ and $g^{ij} = e^{*i} \cdot e^{*j}$, where $(g^{ij}) = (g_{ij})^{-1}$. Note that the volumes of the fundamental cells of Γ and Γ^* are $\text{vol}(\Gamma) = (\det \Gamma)^{1/2} := |\det(g_{ij})|^{1/2}$ and $\text{Vol}(\Gamma^*) = (\det \Gamma^*)^{1/2} := |\det(g^{ij})|^{1/2}$, respectively. This shows that integral lattices must be unimodular in order to be self-dual.

Exercise 13.2.11   Show that for circle compactifications the one-dimensional lattices Λ of windings and Λ^* of momenta are dual to each other. What are the one-dimensional basis vectors?

Exercise 13.2.12   Find a lattice basis for $\Gamma_{1,1}(R)$. Express the basis vectors in terms of the basis vectors of Λ, Λ^*.

Exercise 13.2.13   Show that $\Gamma_{1,1}(R)$ is even and self-dual.

To obtain the modular transformation behaviour of a lattice partition function $\Theta_{\Gamma_{p,q}}$ it is sufficient to study the two generators $\tau \to \tau + 1$ and $\tau \to -\frac{1}{\tau}$ of the modular group $\text{SL}(2, \mathbb{Z})$. We observe that

$$\Theta_{\Gamma_{p,q}}(\tau + 1, \bar{\tau} + 1) = \sum_{(k_L, k_R) \in \Gamma_{p,q}} q^{\frac{1}{2}k_L^2} \bar{q}^{\frac{1}{2}k_R^2} e^{2\pi i \frac{1}{2}(k_L^2 - k_R^2)} = \Theta_{\Gamma_{p,q}}(\tau, \bar{\tau}), \tag{13.113}$$

provided that the lattice $\Gamma_{p,q}$ is even.

To work out the transformation $\tau \to -\frac{1}{\tau}$ is more complicated, but can be done using the *Poisson resummation formula*. To start with, we consider the simpler problem of a one-dimensional lattice partition function

$$\Theta(\tau) = \sum_{m \in \mathbb{Z}} q^{\frac{1}{2}m^2}. \tag{13.114}$$

This corresponds to a chiral partition function for winding modes only. The one-dimensional version of the Poisson resummation formula states that

$$f(a) := \sum_{m \in \mathbb{Z}} e^{-\pi a m^2} = \frac{1}{\sqrt{a}} \sum_{n \in \mathbb{Z}} e^{-\pi n^2/a} = \frac{1}{\sqrt{a}} f(1/a), \quad \text{where} \quad a > 0. \tag{13.115}$$

Exercise 13.2.14   Compute the Fourier transform of the periodic function,

$$f(x, a) = \sum_{m \in \mathbb{Z}} e^{-\pi a(x+m)^2}, \quad a > 0. \tag{13.116}$$

Evaluate your result at $x = 0$ and show that you obtain the Poisson resummation formula (13.115).

To apply (13.115) to $\Theta(\tau)$ we need to extend this formula to the case where $a = -i\tau = (\tau_2 - i\tau_1)$ is complex, with $\text{Re}(a) > 0$. Since τ takes values in the upper half plane, we have $\tau_2 > 0$ and the formula applies for $\tau_1 = 0$. One can then extend it to $-\frac{1}{2} \leq \tau_1 \leq \frac{1}{2}$ by analytic continuation, with the result:

$$\Theta(-1/\tau) = \frac{1}{\sqrt{-i\tau}} \sum_{n \in \mathbb{Z}} q^{\frac{1}{2}n^2} = \frac{1}{\sqrt{-i\tau}} \Theta(\tau). \tag{13.117}$$

We now generalise this to higher-dimensional lattice partition functions $\Theta_{\Gamma_{p,q}}(\tau, \bar{\tau})$. This is straightforward, since we only need to replace one-dimensional by higher-dimensional Fourier series, see Appendix C for a summary.

We extend the partition function to a function on $\mathbb{R}^{p+q}$ which is periodic with respect to the lattice $\Gamma_{p,q}$:

$$\Theta_{\Gamma_{p,q}}(\tau, \bar{\tau}, x) = \sum_{(k_L, k_R) \in \Gamma_{p,q}} e^{i\pi\tau(k_L + x_L)^2 - i\pi\bar{\tau}(k_R + x_R)^2}. \tag{13.118}$$

This function has a Fourier expansion

$$\Theta_{\Gamma_{p,q}}(\tau, \bar{\tau}, x) = \sum_{k \in \Gamma_{p,q}^*} c_k(\tau, \bar{\tau}) e^{2\pi i k \cdot x}, \tag{13.119}$$

where $\Gamma_{p,q}^*$ is the dual lattice, with Fourier coefficients

$$c_k(\tau, \bar{\tau}) = \frac{1}{\text{vol}(\Gamma_{p,q})} \int_{F.C.} d^{p+q}x \Theta_{\Gamma}(\tau, \bar{\tau}, x) e^{-2\pi i k \cdot x}, \tag{13.120}$$

where the integral is over a fundamental cell F.C. of $\Gamma_{p,q}$. The prefactor $\text{vol}(\Gamma_{p,q})$ generalises the usual normalisation factor of a Fourier series, compare Appendix C.

By repeating the same steps as for the one-dimensional lattice partition function, we obtain:

$$\Theta_{\Gamma_{p,q}}(\tau, \bar{\tau}) = \sum_{k \in \Gamma_{p,q}^*} c_k(\tau, \bar{\tau}) = \frac{1}{\text{vol}(\Gamma_{p,q})} \sum_{k \in \Gamma_{p,q}^*} (-i\tau)^{-p/2} (i\bar{\tau})^{-q/2} e^{-\frac{i\pi}{\tau} k_L^2} e^{\frac{i\pi}{\bar{\tau}} k_R^2}. \tag{13.121}$$

Exercise 13.2.15   Fill out the missing steps in deriving (13.121).

The only natural way in which this expression can be related to $\Theta_{\Gamma_{p,q}}(-1/\tau, -1/\bar{\tau})$ is if the lattice $\Gamma_{p,q}$ is self-dual, $\Gamma_{p,q} = \Gamma_{p,q}^*$, which in particular makes it unimodular, $\text{vol}(\Gamma_{p,q}) = 1$. For self-dual lattices we have

$$\Theta_{\Gamma_{p,q}}(\tau, \bar{\tau}) = (-i\tau)^{-p/2} (i\bar{\tau})^{-q/2} \Theta_{\Gamma_{p,q}}(-1/\tau, -1/\bar{\tau}). \tag{13.122}$$

To summarise, we have shown that for even self-dual lattices:

$$\Theta_{\Gamma_{p,q}}(\tau + 1, \bar{\tau} + 1) = \Theta_{\Gamma_{p,q}}(\tau, \bar{\tau}), \quad \Theta_{\Gamma_{p,q}}(-1/\tau, -1/\bar{\tau}) = (-i\tau)^{p/2}(i\bar{\tau})^{q/2}\Theta_{\Gamma_{p,q}}(\tau, \bar{\tau}).$$
$$(13.123)$$

If we supplement these expressions with the contributions of p left-moving and q right-moving oscillators, the resulting partition functions

$$Z(\tau, \bar{\tau}) = \frac{1}{\eta^p \bar{\eta}^q} \Theta_{\Gamma_{p,q}}(\tau, \bar{\tau}) \tag{13.124}$$

have the following modular transformation properties:

$$Z(\tau + 1, \bar{\tau} + 1) = e^{-2\pi i \frac{p-q}{24}} Z(\tau, \bar{\tau}), \quad Z(-1/\tau, -1/\bar{\tau}) = Z(\tau, \bar{\tau}). \tag{13.125}$$

For even self-dual lattices with $p = q$, such as $\Gamma_{1,1}(R)$, the combined partition functions for oscillators, momentum, and winding modes are modular invariant. In particular, the partition function $Z_R(\tau, \bar{\tau}) = (\eta\bar{\eta})^{-1}Z_{\Gamma_{1,1}(R)}(\tau, \bar{\tau})$ for string modes on a circle of radius R is modular invariant. Our derivation has been more general and applies to higher-dimensional, toroidal compactifications and to left-/right-asymmetric ('heterotic') constructions. In the latter case one needs to make sure that non-trivial phases cancel against other contributions to the partition functions. Even self-dual lattice can be used to construct modular invariant lower-dimensional string theories. They only exist for signature (p, q) where $p - q = 0$ modulo 8. In non-Euclidean signatures even self-dual lattices are unique up to isometries, that is up to $O(p, q)$ transformations. Since string partition functions are invariant under $O(p) \times O(q) \subset O(p, q)$, the resulting theories are locally parametrised by the symmetric space $O(p, q)/O(p) \times O(q)$.[16] In Euclidean signature there is a finite number of even self-dual lattices which are not related to one another by orthogonal transformations. Their classification is known for dimensions 8, 16, 24.

13.3 Closed Strings on $S^1/\mathbb{Z}_2$ and Orbifolds

If a world-sheet conformal field theory has a discrete symmetry, one can construct a new CFT by projecting onto the states which are invariant under the symmetry. We have seen an example of such a construction when we defined non-oriented strings by projecting onto states which are invariant under world-sheet parity in Section 2.2.9. Another example is replacing a non-compact direction in space-time by a circle, as we did in the last section. In this case, the symmetry is the translational or shift

---

[16] $O(p, q)$ transformations which map a given lattice basis to another lattice basis preserve the lattice $\Gamma_{p,q}$ and are symmetries. The corresponding discrete subgroup $O(p, q, \Gamma) \cong O(p, q, \mathbb{Z})$ is the so-called T-duality group associated with the lattice $\Gamma_{p,q}$. While this symmetry looks obvious from the point of view of the Narain lattice $\Gamma_{p,q}$, some of these transformations have a highly non-trivial and interesting action on space-time, whose geometry is encoded in the lattice Λ. We will discuss T-duality in Section 13.4.

Table 13.2. Transformation of generic massless states under the orbifold twist

States	Fields	$\mathbb{Z}_2$-action
$\alpha^\mu_{-1}\tilde{\alpha}^\nu_{-1}|(0,0)\rangle$	$G_{\mu\nu}, B_{\mu\nu}, \phi$	$+1$
$\alpha^\mu_{-1}\tilde{\alpha}_{-1}|(0,0)\rangle$	$A^{(R)}_\nu$	-1
$\tilde{\alpha}_{-1}\alpha^\nu_{-1}|(0,0)\rangle$	$A^{(L)}_\nu$	-1
$\alpha_{-1}\tilde{\alpha}_{-1}|(0,0)\rangle$	σ	$+1$

symmetry of Minkowksi space. The projection onto invariant states initially breaks modular invariance, which, however, can be restored by including strings which are only closed up to the action of the symmetry group. For circle compactifications we have seen that we have to include winding states, and that the combined set of winding and momentum vectors must define an even and self-dual lattice. The additional sectors in the Hilbert space are called *twised sectors*. A CFT obtained by dividing out a symmetry is called an *orbifold* of the initial CFT. Orbifold means 'space of orbits' and refers to the identification of states related by the symmetry. In the context of geometric compactifications, the term orbifold is often reserved for constructions where the symmetries one divides out include reflections or discrete rotations. These symmetries are distinguished from translation symmetries by having *fixed points*. We illustrate this procedure using the simplest case, which also allows us to continue our study of one-dimensional compactifications.

The real line $\mathbb{R}$ and the circle $S^1_R = \mathbb{R}/(2\pi R\mathbb{Z})$ are invariant under reflections around an arbitrary point. Let us choose this point to be the origin, so that the reflection acts by $X \mapsto -X$. We call the resulting space $S^1_R/\mathbb{Z}_2$ the *orbicircle*. Combining the actions of the two identifications $X \cong X + 2\pi R \cong -X$, any point on the real line can be mapped to the interval $0 \le X \le \pi R$, whose endpoints are the two independent fixed points of the *orbifold twist* $X \mapsto -X$ (modulo translations).

The action of the twist is determined by this geometric action, since all conformal fields are constructed out of X. The mode operators transform as follows:

$$\alpha_m \mapsto -\alpha_m, \quad \tilde{\alpha}_m \mapsto -\tilde{\alpha}_m, \quad p_L \mapsto -p_L, \quad p_R \mapsto -p_R. \tag{13.126}$$

States with $k_L = k_R = 0$ are eigenstates of the twist with eigenvalues ± 1. The transformations of the generic massless states are summarised in Table 13.2, where $\mathbb{Z}_2$ refers to the group generated by the twist. Since in the orbicircle theory we only keep twist-invariant states, the $U(1)_L \times U(1)_R$ gauge bosons are projected out and the remaining states are the graviton, B-field, dilaton and Kaluza–Klein scalar. More generally, all states with an odd number of internal oscillators $\alpha_{-m}, \tilde{\alpha}_{-m}$ are projected out.

On the winding and momentum quantum numbers m, n the reflection acts as $(m, n) \mapsto (-m, -n)$. Twist eigenstates are obtained by taking suitable linear combinations:

$$|(m, n), R\rangle + |(-m, -n), R\rangle \mapsto |(-m, -n), R\rangle + |(m, n), R\rangle,$$
$$|(m, n), R\rangle - |(-m, -n), R\rangle \mapsto -|(-m, -n), R\rangle + |(m, n), R\rangle.$$

Only the invariant combinations remain in the spectrum.

The additional identification $X \cong -X$ allows us to admit more general closed string boundary conditions. Besides strings which are already closed on $\mathbb{R}$, and winding states that close on $\mathbb{R}$ up to a translation by $2\pi m R$, we now have *twisted states*, which close up to a reflection (modulo translations). In the *twisted sector*, the mode expansion is determined by the condition

$$X(e^{2\pi i}z, e^{-2\pi i}\bar{z}) = -X(z,\bar{z}) + 2\pi m R. \tag{13.127}$$

Imposing this on the mode expansion

$$X(z,\bar{z}) = x - \frac{i}{2}p_L \log z - \frac{i}{2}p_R \log \bar{z} + \frac{i}{2}\sum_{k \neq 0}\frac{\alpha_k}{k}z^m + \frac{i}{2}\sum_{l \neq 0}\frac{\tilde{\alpha}_l}{l}\bar{z}^n, \tag{13.128}$$

we obtain the following conditions

$$x + \pi(p_L - p_R) = -x + 2\pi m R \,, \quad p_L = -p_L \,, \quad p_R = -p_R, \tag{13.129}$$

$$\alpha_k e^{2\pi k} = -\alpha_k, \quad \tilde{\alpha}_l e^{2\pi l} = -\tilde{\alpha}_l. \tag{13.130}$$

The conditions on the momenta $p_{L,R}$, which result from comparing the logarithmic terms, imply that $p_L = 0, p_R = 0$, that is, in the twisted sector there are no internal momenta and windings. The first condition, which comes from matching the constant parts, then simplifies to $x = -x + 2\pi m R$. Since we have already imposed $x \cong x + 2\pi R$, this equation has two inequivalent solutions:

$$x_{(1)} = 0, \quad x_{(2)} = \pi R. \tag{13.131}$$

These are the two fixed points of the transformation $X \mapsto -X$ in the interval $0 \leq X < 2\pi R$, which correspond to the endpoints of the interval $S^1_R/\mathbb{Z}_2$. Twisted strings have their centre of mass located at the fixed points, and they do not have momentum and winding along the X-direction. Finally, the condition imposed on the oscillators can be solved by taking k, l to be strictly half-integer valued,

$$k, l = \pm\frac{1}{2}, \pm\frac{3}{2}, \cdots . \tag{13.132}$$

Half integer powers z^k transform into $-z^k$ under $z \mapsto e^{2\pi i}z$, and therefore the corresponding configurations only describe closed strings upon using the orbifold identification $X \cong -X$. Due to the half-integer mode expansion the twisted fields $X(z,\bar{z})$ are no longer single-valued in the complex plane.[17] We now have enough information to write down a basis for the twisted states. The twisted sector has two ground states $|x_{(i)}\rangle$, $i = 1, 2$ corresponding to the two fixed points. Adding oscillators and space-time momentum, we obtain the basis vectors

$$\alpha_{-m_1}^{\mu_1} \cdots \tilde{\alpha}_{-n_1}^{\nu_1} \cdots \alpha_{-k_1} \cdots \tilde{\alpha}_{-l_1} \cdots |k^\mu, x_{(i)}\rangle. \tag{13.133}$$

The contribution of the internal oscillators to the mass follows from the commutation relation

$$[N, \alpha_{-k}] = k\alpha_{-k}, \quad k \in \mathbb{Z} + \frac{1}{2}, \tag{13.134}$$

---

[17] One can still build a meaningful conformal field theory where all physical correlation functions are single valued, see Dixon et al. (1987); Hamidi and Vafa (1987).

which is the 'analytic extension' of the one for integer mode number. The mass formula takes the form

$$\frac{1}{8}M^2 = N + a = \tilde{N} + \tilde{a}. \tag{13.135}$$

To compute the masses of twisted sector states, we need to know the masses of the twisted sector ground states. Here we can adapt the method that we used previously to compute the effect on normal ordering on CFT ground states. The operator L_0 of the twisted sector of the orbicircle CFT has classically the form

$$L_0 = \sum_{k \in \mathbb{Z} + \frac{1}{2}} \alpha_{-k}\alpha_k. \tag{13.136}$$

If we bring this to normal ordered form we obtain

$$L_0 = N + a, \quad N = \sum_{k=\frac{1}{2},\frac{3}{2},\dots} \alpha_{-k}\alpha_k, \quad a = \frac{1}{2}\sum_{k=\frac{1}{2},\frac{3}{2},\dots} k. \tag{13.137}$$

The divergent normal ordering constant

$$a = \frac{1}{2}\sum_{k=\frac{1}{2},\frac{3}{2},\dots} k = \frac{1}{2}\sum_{n=0}^{\infty}\left(n + \frac{1}{2}\right) \tag{13.138}$$

can be regularised using the Hurwitz ζ-function[18]

$$\zeta(s, b) := \sum_{n=0}^{\infty}\frac{1}{(n+b)^s}, \quad \text{Re}(s) > 1, \text{Re}(b) > 0, \tag{13.139}$$

which can be extended to a meromorphic function on the complex plane. In particular, this function is holomorphic for the values $s = -1, q = \frac{1}{2}$. Using that

$$a = \zeta(-n, b) = -\frac{B_{n+1}(b)}{n+1}, \tag{13.140}$$

for $n = 0, 1, 2, \dots$, where $B_n(b)$ are the Bernoulli polynomials, one obtains

$$\zeta(-1, b) = -\frac{1}{2}B_2(b) = -\frac{1}{2}\left(\frac{1}{6} - b + b^2\right) \tag{13.141}$$

and

$$a = \frac{1}{2}\zeta\left(-1, \frac{1}{2}\right) = -\frac{1}{24} + \frac{1}{16} = \frac{1}{48} > 0. \tag{13.142}$$

Thus replacing integer by half-integer oscillators shifts the normal ordering constant from $-\frac{1}{24}$ by $\frac{1}{16}$ to $\frac{1}{48}$. This tells us two things: firstly, the ground state energy is positive, and the twisted sectors do not contribute to the massless spectrum for the orbicircle. Secondly, the *twist operators* τ_i which map the SL$(2, \mathbb{C})$ ground state $|0\rangle$ to the two twisted sector ground states $|x_{(i)}\rangle = \tau_i(0)|0\rangle$ have conformal weights $\left(\frac{1}{16}, \frac{1}{16}\right)$. These *twist operators* can be constructed explicitly.

Like for the winding states in circle compactifications, including the twisted sectors is not optional, but required for maintaining modular invariance. In general, modular transformations map different (twised/untwisted) boundary conditions to

---

[18] See Appendix D.2.2 for infomation about the Hurwitz ζ-function and the Bernoulli numbers.

each other, and a modular invariant partition functions requires one to combine sectors in such a way that they form an *orbit* under the action of the modular group. Since the modular transformation $\tau \rightarrow -1/\tau$ exchanges world-sheet space with world-sheet time, there is a relation between imposing twisted boundary conditions, which affects world-sheet space, and imposing projections, which affects world-sheet time. The most obvious way to obtain a modular invariant partition function is to sum over all boundary conditions in both world-sheet space and time. For the orbicircle theory this works as follows. Let us denote the chiral partition function of the original theory by $Z_1^1 = \text{Tr}_{(1)}(q^{L_0})$. To project onto twist-invariant states we need to insert a projector $\Pi = \frac{1}{2}(1 + g)$, where g with $g^2 = 1$ is the generator of the twist group $\mathbb{Z}_2$. The projected partition function is $\text{Tr}_{(1)}\left(\frac{1}{2}(1 + g)q^{L_0}\right) = \frac{1}{2}\left(Z_1^1 + Z_1^g\right)$. Our notation indicates that the insertion of g can be interpreted as choosing g-twisted boundary conditions in world-sheet time. Similarly we denote the partition function for g-twisted strings by $Z_g^1 = \text{Tr}_{(g)}(q^{L_0})$ and its projection onto twist invariant states by $\text{Tr}_{(g)}(q^{L_0}) = \frac{1}{2}\left(Z_g^1 + Z_g^g\right)$. The modular invariant partition function of the orbicircle model is the sum of the projected untwisted and twisted sector partition functions,

$$Z_{\text{orbifold}} = \frac{1}{|G|} \sum_{g,h \in G} Z_g^h, \tag{13.143}$$

where $G = \mathbb{Z}_2$ is the twist group and $|G| = |\mathbb{Z}_2| = 2$ its order. In this formula it is manifest that we sum over boundary conditions in both world-sheet directions. Formula (13.143) can be applied for any abelian twist group G, while for non-abelian groups the summation needs to be restricted to pairs (g, h) of group elements which commute with each other. If the twist group operates with more than one orbit on the boundary conditions, one may have the option to introduce relative phases between these orbits ('discrete torsion'). To obtain all necessary conditions for modular invariance at arbitrary genus of the world-sheet, it is sufficient to have a modular covariant one-point function on the world-sheet torus, together with crossing symmetry for the four-point function on the world-sheet sphere.[19]

We now return to the specific case of the orbicircle. As for circle compactifications, there is a moduli space, which we parametrise by the radius R of the underlying circle.

Exercise 13.3.1   Work out the massless spectrum of the orbifold theory at $R = \sqrt{\alpha'}$ and show that it agrees with the massless spectrum of a circle compactification at $R = 2\sqrt{\alpha'}$. Work out the masses of the extra light states to leading order in $\delta R = R - \sqrt{\alpha'}$, where R is the orbicircle radius.

Exercise 13.3.1 shows that at $R = \sqrt{\alpha'}$ there is an enhanced $U(1)_L \times U(1)_R$ gauge symmetry, which is Higgsed for $\delta R = R - \sqrt{\alpha'} \neq 0$. Moreover, the spectrum matches with the spectrum of the circle compactifcation at $R = 2\sqrt{\alpha'}$, and we have seen that the circle theory has a second independent marginal operator at this point, indicating that this is an intersection point of the circle and orbicircle moduli spaces.

---

[19] See Sonoda (1988a,b). Crossing symmetry means that when evaluating a four-point correlation function it does not matter in which way one reduces it to three-point functions using OPEs.

13.4 T-Duality for Closed Strings

We return to circle compactifications. As already remarked in passing, automorphisms of the Narain lattice $\Gamma_{1,1}(R)$ map the states of the theory onto themselves and are therefore symmetries of the CFT. While this is completely obvious from the Hilbert space point of view, the geometric implications are far reaching, and lead to a new, stringy symmetry called *T-duality*, which we will start to explore in this section.

By inspection of the mass formula

$$M^2 = 4\left(N + \frac{1}{2}\left(\frac{n}{2R} + mR\right)^2 + \tilde{N} + \frac{1}{2}\left(\frac{n}{2R} - mR\right)^2 - 2\right), \tag{13.144}$$

we find that it is symmetric under the exchange of momenta and windings

$$m \leftrightarrow n \Leftrightarrow \frac{1}{2}k \leftrightarrow w, \tag{13.145}$$

provided that we invert the radius of the circle, measured in string units,

$$R \to \frac{1}{2R} \Leftrightarrow \frac{R}{\sqrt{\alpha'}} \to \frac{\sqrt{\alpha'}}{R}, \tag{13.146}$$

where in the second relation we have restored α'. This implies that in order to decide whether the radius of the circle is R or $R' = \frac{\alpha'}{R}$, we need to specify which modes are momentum modes and which are winding modes. Our initial identification is certainly correct for $R \gg \sqrt{\alpha'}$ where classical geometry applies. However, for radii smaller than the string scale $R \ll \sqrt{\alpha'}$ it is more natural to reinterpret winding modes as momentum modes, and the other way round. The limit $R \to 0$ can then be understood as an alternative decompactification limit $R' = \frac{\alpha'}{R} \to \infty$. In this limit the original winding modes form a continuum, and therefore should be re-interpreted as momentum modes, while the original momentum modes becoming infinitely massive and decouple. In the intermediate, stringy regime $R \approx \sqrt{\alpha'}$ momenta and winding modes have comparable masses, and the geometric interpretation is ambiguous. Since large and small radii are related by a symmetry, we can interpret the string length $\sqrt{\alpha'}$ as a minimal length scale. Note that $R = \sqrt{\alpha'}$ is precisely the point in moduli space where the gauge symmetry is enhanced to $SU(2)_L \times SU(2)_R$. At this point the operators $\exp(\pm i\sqrt{2}X(z))$, $\exp(\pm i\sqrt{2}\bar{X}(\bar{z}))$ have the same properties as the derivatives $\partial_z X(z), \partial_{\bar{z}}\bar{X}(\bar{z})$ of the string coordinates, which suggests to look for an interpretation in terms of a 'stringy, non-commutative' geometry with $SU(2)_L \times SU(2)_R$ valued coordinates.

In terms of left- and right-moving momenta, T-duality acts as

$$(k_L, k_R) \to (k_L, -k_R). \tag{13.147}$$

As this leaves k_L^2 and k_R^2 invariant, it is clear that this is a symmetry of the lattice partition function. The corresponding automorphism of the lattice $\Gamma_{1,1}(R)$ acts on the basis vectors as[20]

$$(R, -R) \to (R, R) = \left(\frac{1}{2R'}, \frac{1}{2R'}\right), \quad \left(\frac{1}{2R}, \frac{1}{2R}\right) \to \left(\frac{1}{2R}, -\frac{1}{2R}\right) = (R', -R'), \tag{13.148}$$

---

[20] Here we use a result which is obtained in Exercise 13.2.12.

where the relabelling $R' = \frac{1}{2R}$ brings the lattice basis back to standard form. We can also interpret T-duality as exchanging the lattice Λ of winding vectors and the lattice Λ^* of momentum vectors.

All even self-dual lattices of signature $(1,1)$ are related by $O(1,1)$ transformations, and the automorphism group is the subgroup $O(1,1) \cap GL(\Gamma_{1,1}) \cong O(1,1) \cap GL(2,\mathbb{Z}) = O(1,1,\mathbb{Z})$ which acts cristallographically, that is, which maps lattice bases to lattice bases. The transformation (13.147) is a $O(1,1)$ transformation with determinant -1, which therefore does not belong to the connected component of the unit of $O(1,1)$.

In position space, T-duality acts as a 'chiral reflection'

$$(X(z), \bar{X}(\bar{z})) \rightarrow (X(z), -\bar{X}(\bar{z})) , \tag{13.149}$$

which only acts on the right-moving part of the string coordinate. This fixes the action on the world-sheet CFT by determining its action on oscillators,

$$\alpha_m \rightarrow \alpha_m , \quad \tilde{\alpha}_m \rightarrow -\tilde{\alpha}_m . \tag{13.150}$$

It can be shown that the CFTs related by this transformation are equivalent. In particular, the partition function $Z_R(\tau, \bar{\tau}) = (\eta, \bar{\eta})^{-1} \Theta_{\Gamma_{1,1}(R)}(\tau, \bar{\tau})$ is invariant.

The chiral reflection (13.149) does not have an interpretation in 'classical' geometry, and has motivated the idea to describe closed strings on toroidal backgrounds (and some generalisations thereof) using generalisations of standard Riemannian geometry. These come in two flavours: in *generalised geometry* one works with tensor fields which have a doubled index range, while in *doubled geometry* one doubles the dimension of space itself. At the level of the effective field theory the idea is to treat the winding modes on the same footing as the momentum modes, while a conventional Kaluza–Klein reduction of the effective field theory only includes the momentum modes.

In both approaches one works with two natural metrics, which are derived from the two natural scalar products $k_L \cdot k_L' \pm k_R \cdot k_R'$. The corresponding Gram matrices with respect to our standard basis of $\Gamma_{1,1}(R)$ are

$$\mathcal{H} = \begin{pmatrix} 2R^2 & 0 \\ 0 & \frac{1}{2R^2} \end{pmatrix} \quad \text{and} \quad \eta = \begin{pmatrix} 0 & 1 \\ 1 & 0 \end{pmatrix} . \tag{13.151}$$

We have already encountered the indefinite scalar product η when discussing level matching and modular invariance. It defines a flat metric of neutral signature. The positive definite scalar product is R-dependent and enters into the mass formula:

$$M^2 = 4 \left(N + \tilde{N} + \frac{1}{2} M^i \mathcal{H}_{ij} M^j - 2 \right) , \quad (M^i) = (m, n) , \quad i,j = 1, 2 . \tag{13.152}$$

The positive definite, moduli dependend metric $\mathcal{H}$ is is called the *generalised metric*.[21] In generalised and doubled geometry it is interpreted as a generalisation of the standard Riemannian metric of space-time.

T-duality is also a symmetry for orbicircle theories and identifies radii which are inversely related in string units. As for the circle compactification the self-dual radius

---

[21] In the conventions used later for toroidal compactifications, this is actually the inverse of the generalised metric, compare Section 13.7.

$R = \sqrt{\alpha'}$ coincides with gauge symmetry enhancement. In the following section, we explain the relation between T-duality and gauge symmetries.

13.5 T-duality and Target Space Gauge Symmetries

At the self-dual point $R = \sqrt{\alpha'}$, the generic $U(1)_L \times U(1)_R$ gauge symmetry of circle compactifications is enhanced to $SU(2)_L \times SU(2)_R$. In this section, we discuss the relation of T-duality with this $SU(2)_L \times SU(2)_R$ symmetry and obtain a complete, global description of the moduli space of one-dimensional string compactifications.

As briefly reviewed in Appendix G, simple Lie algebras are determined by their system of root vectors, which can be written as non-negative integer linear combinations of a set of *simple roots*. The *Weyl group* is the group generated by the reflections at the hypersurfaces orthogonal to the simple roots, and the actions of elements of the Weyl group are called *Weyl twists*. The group $SU(2)$ has just one simple root, which we can take to be the one-component vector $\alpha = (\sqrt{2})$. The simple roots of a simple Lie algebra generate the so called root lattice Λ_R which for $SU(2)$ is $\Lambda_R = \sqrt{2}\mathbb{Z}$. Observe that the $SU(2)$ enhancement happens precisely when the $SU(2)$ root lattice Λ_R is contained in the momentum lattice $\Gamma_{1,1}(R)$ as a sublattice of the form $(\Lambda_R, 0)$ or $(0, \Lambda_R)$. The $SU(2)$ Weyl twist is the reflection $\alpha \to -\alpha$ and the Weyl group is isomorphic to $\mathbb{Z}_2$. The action of the Weyl group on the roots induces an inner automorphism of the corresponding Lie algebra:

$$E \mapsto F, \quad F \mapsto E, \quad H \mapsto -H. \tag{13.153}$$

The ladder operators E, F are permuted according to the action on the corresponding roots, which for $SU(2)$ means that $E \leftrightarrow F$. The Cartan subalgebra is dual (with respect to the natural scalar product, the Killing form) to the vector space generated by the root lattice (by passing from integer to real or complex linear combinations). Therefore the Weyl twist acts by $H \mapsto -H$. In the world-sheet CFT, where $H = i(\partial_z X(z))$, $E =: \exp(i\sqrt{2}X(z)):$ and $F =: \exp(-i\sqrt{2}X(z)):$, the Weyl twist is induced by $X(z) \to -X(z)$ which is the chiral reflection (13.149), composed with an overall reflection.

This shows that T-duality can be identified with a particular $SU(2)$ gauge transformation. Away from the self-dual point the $SU(2)$ symmetry is Higgsed, but as a local gauge symmetry it is not literally broken. The marginal operator $J_3 \bar{J}_3 = H\bar{H} = \partial_z X(z) \partial_{\bar{z}} \bar{X}(\bar{z})$, is sent to its negative $-J_3 \bar{J}_3$ under the Weyl twist of either $SU(2)$. If we take $\delta R = R - \sqrt{\alpha'}$ to be the order parameter (proportional to the vacuum expectation value of the Higgs field) then T-duality identifies δR with $-\delta R$. Therefore the field theoretic moduli space, that is the range of δR, is not the real axis $\mathbb{R}$ but the half-axis $\mathbb{R}^{\geq 0}$. This space is in one-to-one correspondence with the geometrical moduli space $\{R \in \mathbb{R} | R \geq \sqrt{\alpha'}\}$, that is the space of inequivalent circles. Also note that at the point $R = \sqrt{\alpha'}$ the operator $J_3 \bar{J}_3$ is only 'half of a marginal operator', since deformations to smaller and to larger R are equivalent.

This connection between T-duality and gauge symmetries extends to higher-dimensional compactifications. It suggests a picture where the enhancement of target space gauge symmetries, which happens at fixed points of T-duality transformations, 'un-Higgses' part of the full gauge symmetry of string theory. This symmetry group is infinite-dimensional, but Higgsed to a finite-dimensional gauge group once one chooses a background.

The SU(2) Weyl twist can be diagonalised by a basis transformation, to the effect that in the new basis its action is $H' \mapsto H'$, $E' \mapsto E'$ and $F' \mapsto -F'$. The basis transformation mixes $\partial_z X(z)$ with $\exp(\pm i\sqrt{2}X(z))$, so that the new string coordinate $X'(z)$ is a combination of the old coordinate $X(z)$ and SU(2) ladder operators. This illustrates that at the self-dual point classical geometry, where X is distinguished as a coordinate, fails to account for the full picture. The action on X' which induces the Weyl twist is not a reflection, but a shift,

$$X' \rightarrow X' + \sqrt{\alpha'}\pi = X' + \pi\frac{1}{2}\sqrt{2}. \tag{13.154}$$

This is called the *twist-shift equivalence*: Weyl twists can be realised equivalently as reflections (or rotations) and as translations of the string coordinate X.

This provides another way of understanding the global geometry of the moduli space. While orbifolding with respect to the reflection $X \mapsto -X$ brings us from circle to orbicircle compactifications with the same radius, orbifolding with respect to the corresponding shift changes the radius of the circle (or of the orbicircle) by $R \rightarrow \frac{1}{2}R$. By T-duality this is equivalent to doubling the radius, $R \rightarrow 2R$. Thus we see that the circle compactifications at $R_{S^1} = 2\sqrt{\alpha'}$ and $R_{S^1} = \frac{1}{2}\sqrt{\alpha'}$, which are equivalent to each other by T-duality, are also equivalent, by twist-shift equivalence, to the orbicircle compactification at $R_{S^1/\mathbb{Z}_2} = \sqrt{\alpha'}$. This means that the moduli space of circle and orbicircle compactifications consists of the circle line $\sqrt{\alpha'} \leq R_{S^1} < \infty$ and the orbicircle line $\sqrt{\alpha'} \leq R_{S^1/\mathbb{Z}_2} < \infty$, which branches of at the point $R_{S^1} = 2\sqrt{\alpha'}$ (see Figure 13.1).

Let us describe the effective field theory around the branch point. At the branch point we have a $U(1)_L \times U(1)_R$ gauge symmetry and the five massless scalars $\sigma, \sigma_{\pm\pm}$, as well as the graviton, B-field and dilaton which we discard in the following discussion. There are two one-dimensional branches connected to this point: a *Coulomb branch* where the abelian gauge group remains unbroken while four charged scalars become massive upon giving a vacuum expectation value to the fifth, neutral scalar, and a *Higgs branch* where one of the charged scalars acquires a vacuum expectation value, and the $U(1)_L \times U(1)_R$ symmetry gets Higgsed. The Higgs mechanism absorbs two scalars, while two of the remaining scalars become massive, and the scalar acquiring the vacuum expectation value remains massless. From our previous analysis of the string spectrum we know that the scalar masses are proportional to the vacuum expectation values δR_{S^1}, $\delta R_{S^1/\mathbb{Z}_2}$ along each branch, and that half of the scalars $\sigma_{\pm\pm}$ become tachyons away from the special point. The scalar potential must be $O(2) \times O(2)$ symmetric, and the vacuum expectation values $\pm\delta R_{S^1/\mathbb{Z}_2}$ must be equivalent along the Higgs branch, in order that the theory is T-duality invariant along the orbicircle line. The simplest scalar potential satisfying these criteria is

$$V = \sigma \det \Sigma, \quad \Sigma = (\sigma_{ij})_{i,j=1,2}, \tag{13.155}$$

where σ is the KK scalar of the circle compactification and σ_{ij} are linear combinations of the other four other scalars, which are chosen such that σ_{ij} transforms as a vector under the first and second $O(2)$ group with respect to the first and

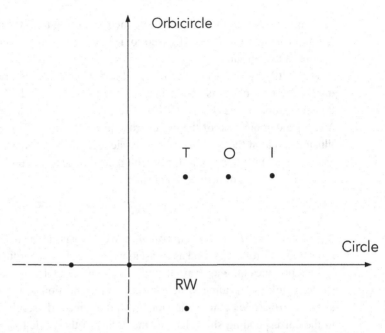

Fig. 13.1 The moduli space of known $c = 1$ confomal field theories. By T-duality the points of the circle and orbicircle line with $R < \sqrt{\alpha'}$ (dashed) are equivalent to points of inverse radius, in string units (straight). The two lines intersect at $R_{\text{circle}} = 2\sqrt{\alpha'}$, which is equivalent to $R_{\text{orbicircle}} = \sqrt{\alpha'}$. The space-time gauge groups are $\text{SU}(2)^2$ for $R_{\text{circle}} = \sqrt{\alpha'}$, $\text{U}(1)^2$ for other points on the circle line and no gauge group for all other points. The three isolated points result from orbifolding the $\text{SU}(2)^2$ theory by the three non-abelian discrete subgroups $T, O, I \subset \text{SU}(2)$. The fourth isolated theory denoted RW has been constructed as a limit of the minimal series.

second index, respectively. This is the basis corresponding to the marginal operators $J_3 \bar{J}_3, J_i \bar{J}_j$ of the circle line.

Exercise 13.5.1   Find the critical points of the potential (13.155) and show that they form two lines or 'branches' which intersect at a point. Find the dependence of the scalar masses on the vacuum expectation values δR_{S^1}, $\delta R_{S^1/\mathbb{Z}_2}$ along each branch and show that it is qualitatively the same as for the string spectrum around the branch point. Explain how, upon coupling the scalars to two vector fields, you can implement the Higgs mechanism along one of the branches.

The orbicircle theory at $R_{S^1/\mathbb{Z}_2} = \sqrt{\alpha'}$ is an orbifold of the circle theory at $R_{S^1} = \sqrt{\alpha'}$ by the Weyl group $\mathbb{Z}_2 \subset \text{SU}(2)$. At the $\text{SU}(2)_L \times \text{SU}(2)_R$ enhancement point we can construct further orbifold theories using discrete subgroups of $\text{SU}(2)$. Discrete subgroups of $\text{SU}(2)$ organise into two infinite families C_n, D_n and three exceptional subgroups. The subgroups C_n are cyclic of order n and relate the $\text{SU}(2)^2$ enhancement point to points on the circle line with $R_{S^1} = n\sqrt{\alpha'}$. The elements of the subgroups D_n contain in addition the Weyl twist and relate the $\text{SU}(2)^2$ enhancement point to points on the orbicircle line with radius $R_{S^1/\mathbb{Z}_2} = n\sqrt{\alpha'}$. The three remaining subgroups T (tetrahedral), O (octahedral), and I (icosahedral) are non-abelian, and their action is more general than changing the radius, possibly combined with a Weyl twist. Orbifolding by them leads to three new $c = 1$ conformal field theories which are isolated, that is, which do not have any marginal deformations. These three theories are rational CFTs, and the same is true for all CFTs on the circle and orbicircle lines for which $(R/\sqrt{\alpha'})^2$ takes rational values. For these CFTs the partition function is a finite sum of Jacobi θ-functions. For the circle and orbicircle line it makes sense to say that the rational theories form a dense subset. The full moduli space of one-dimensional string compactifications is not a manifold but a more general *stratified space* which has components of different dimensions which can be disjoint or intersect each other over lower-dimensional subspaces. There is at least one further, non-rational CFT with $c = 1$, which has been constructed as a limit of the the minimal series (see Runkel and Watts (2002)). This theory can be interpreted as a more complicated type of orbifold (see Gaberdiel and Suchanek (2012)). Figure 13.1 provides a summary.

13.6 T-duality for Open Strings

Let us now consider what happens if we apply T-duality to open strings. If we compactify open strings with Neumann boundary conditions on a circle, the momentum becomes discrete, but we do not have winding states. In contrast, if we impose Dirichlet boundary conditions along a circular direction, we can create winding states along the circle since the ends of the string are kept fixed, but there is no momentum. Since we know that for closed strings T-duality exchanges momentum and winding states, we expect that for open strings it exchanges Neumann and Dirichlet boundary conditions. Using the interpretation of T-duality as a chiral reflection,

$$X(\sigma^0, \sigma^1) = X_L(\sigma^0 + \sigma^1) + X_R(\sigma^0 - \sigma^1) \rightarrow X'(\sigma^0, \sigma^1) = X_L(\sigma^0 + \sigma^1) - X_R(\sigma^0 - \sigma^1),$$
(13.156)

we immediately see that

$$\partial_0 X = \partial_1 X', \tag{13.157}$$

which confirms our expectation. There is however one puzzle: for Dirichlet boundary conditions we can fix both ends of the string at different points, and the mass spectrum will change if we change the distance between the D-branes along the circle, similar to (2.81). In particular, there will be additional massless states if we put the D-branes on top of each other. This feature seems to be missing for Neumann boundary conditions so far.

To investigate this we consider the mode expansions. Starting with a single string coordinate with Neumann boundary conditions

$$X(\sigma^0, \sigma^1) = x + p\sigma^0 + i \sum_{n \neq 0} \frac{1}{n} \alpha_n e^{-in\sigma^0} \cos(n\sigma^1), \tag{13.158}$$

where $p = \frac{m}{R}, m \in \mathbb{Z}$ is now discrete, we apply T-duality and obtain

$$X'(\sigma^0, \sigma^1) = 2w'\sigma^1 + \sum_{n \neq 0} \frac{1}{n} \alpha_n e^{-in\sigma^0} \sin(n\sigma^1), \quad w' = \frac{1}{2} p. \tag{13.159}$$

T-duality maps $R \rightarrow \frac{1}{2R} = R'$, and momenta to windings,

$$w' = \frac{m}{2R} = mR'. \tag{13.160}$$

Since

$$X'(\sigma^0, \sigma^1 + \pi) = X'(\sigma^0, \sigma^1) + 2\pi m R', \tag{13.161}$$

we see that the wound open strings start and end at the same point, $x' \cong x' + 2\pi R'$.

As a generalisation, we impose generic Dirichlet boundary conditions where the positions of the endpoints are $x_i = 2\pi \theta_i R'$. This shifts the winding spectrum by $\Delta w' = \Delta \theta R'$, where $0 \leq \Delta \theta = \theta_2 - \theta_1 < 1$ covers all possible relative positions, up to integer windings:

$$X'(\sigma^0, \sigma^1 + \pi) = X'(\sigma^0, \sigma^1) + 2\pi (m + \Delta \theta) R', \tag{13.162}$$

and the winding spectrum is

$$w' = (m + \theta_2 - \theta_1) R'. \tag{13.163}$$

Going back to Neumann boundary conditions, this corresponds to a modified momentum spectrum

$$p = \frac{m + \theta_2 - \theta_1}{R}. \tag{13.164}$$

From quantum mechanics we know that shifts in the momentum spectrum occur when a charged particle is moving in a magnetic field. For an open string we can assign charges to the endpoints. This means that we extend the Hilbert space by the product of two representation spaces for a gauge group G.

We can infer which group G we have to introduce by going back to the T-dual picture with Dirichlet boundary conditions. Here, the Hilbert space is extended by an additional factor which encodes on which D-brane a string starts and ends. In a background with two D-branes (see Figure 2.2), we denote the ground states of the four sectors as $|1, 1\rangle, |1, 2\rangle, |2, 1\rangle, |2, 2\rangle$. From Section 2.2.8, we known that we have the normal spectrum of oscillators, which will now be extended by winding states, and that for strings ending on two different branes the mass spectrum is shifted by a term proportional to the distance of the branes. Assuming Neumann boundary conditions along the non-compact directions, the lightest vector particles in such a background are

$$\alpha^\mu_{-1} |1, 1\rangle, \quad \alpha^\mu_{-1} |1, 2\rangle, \quad \alpha^\mu_{-1} |2, 1\rangle, \quad \alpha^\mu_{-1} |2, 2\rangle. \tag{13.165}$$

The two states where the string starts and ends on the same D-brane are massless irrespective of the D-brane positions, which shows that the generic gauge symmetry is U(1)×U(1). The other two vector fields are massive for generic D-brane positions, but become massless if we put the two D-branes on top of each other, which indicates a Higgs mechanism. It can be shown that strings in the $|i, j\rangle$ sector carry charge $+1$ under U(1)$_i$ and charge -1 under U(1)$_j$. Thus, $|i, i\rangle$ strings are neutral whereas $|1, 2\rangle$ and $|2, 1\rangle$ strings carry charges $(1, -1)$ and $(-1, 1)$, respectively. By taking linear combinations of the U(1) factors, one sees that the four vector bosons fit into the adjoint representation of U(2) $\cong$ U(1) × SU(2). The scalar fields required for an adjoint Higgs mechanism breaking U(2) to U(1) × U(1) are provided by the four states $\alpha_{-1}|i,j\rangle$. The scalars $\alpha_{-1}|1,1\rangle$, $\alpha_{-1}|2,2\rangle$ are massless for all configurations and encode the position of the D-brane along the compact direction. The two charged scalars, which provide the longitudinal components of the charged vector bosons if the D-branes are separated, become massless when the D-branes coincide and are then on the same footing as the D-brane coordinates. Thus, one can regard the four scalars as a non-abelian, U(2)-valued coordinate, providing another indication that at the string scale classical geometry should be replaced by a generalised, 'stringy' geometry.

Returning to the T-dual description with Neumann boundary conditions we see that when dualising a configuration with two D-branes, the extra data $|i,j\rangle$ associate $U(1) \times U(1)$ charges to the two endpoints, which together encode a representation of U(2). The labels i,j associated with the endpoints of a Neumann open strings are called *Chan-Paton factors*. What remains to be clarified is what corresponds to the D-brane positions in the Neumann picture, and how the breaking of the U(2) gauge symmetry can be understood. We already noted that the shift in the momentum spectrum resembles an electrical particle in a magnetic field. To review this consider a particle of charge $q \in \mathbb{Z}$ moving on a one-dimensional space in the presence of a gauge field A, with world-line action[22]

$$S = \int dt \left(\frac{1}{2}\dot{x}^2 - \frac{1}{2}M^2 + q\dot{x}A \right). \tag{13.166}$$

We take the gauge potential A to be constant,

$$A = -\frac{\theta}{2\pi R}. \tag{13.167}$$

While this implies that the field strength is zero, this gauge potential still has a non-trivial effect if space is a circle. In particular, A can only be set to zero by a gauge transformation locally, but not globally.

Exercise 13.6.1 Show that the gauge potential (13.167) is locally of pure gauge form $A = -i\lambda^{-1}(x)\partial_x\lambda(x)$. Show that the Wilson line (or Wilson loop)

$$W_q = \exp\left(iq \oint_{S_R^1} A dx \right) \tag{13.168}$$

is gauge invariant and compute its value for A given by (13.167). Use your results to explain why we can set $A = 0$ locally, but not globally.

In the canonical formalism, we can see the effect of a Wilson line by computing the canonical momentum

$$p = \frac{\partial L}{\partial \dot{x}} = \dot{x} - q\frac{\theta}{2\pi R}. \tag{13.169}$$

This shows that the canonical momentum p is different from $\dot{x}$ in the presence of a Wilson line. Since p is the operator conjugate to the position x, its spectrum must take the form $k = \frac{m}{R}$ to ensure that momentum eigenstates generated by operators e^{ikx} are well defined. However, the term entering into the Hamiltonian is $\dot{x}^2$, and $\dot{x}$ has the shifted spectrum

---

[22] We use (1.54) with $K = M^2$ and impose the gauge $e = 1$. Here, we use the notation M for the mass, since m is used for the winding or momentum quantum number.

$$\dot{x} = \frac{m}{R} + q\frac{\theta}{2\pi R}. \tag{13.170}$$

In our example of a Neumann string obtained by T-dualising a configuration with two D-branes, we have two point-like ends and a U(2) gauge theory. The condition on the matrix valued U(2) gauge field $A_\mu = A_\mu^a T^a$ to have vanishing field strength is

$$F_{\mu\nu}^a = \partial_\mu A_\nu^a - \partial_\nu A_\mu^a + f^{abc} A_\mu^b A_\nu^c = 0. \tag{13.171}$$

This equation is solved by restricting A_μ^a to a Cartan subalgebra, which eliminates the commutator term, and then choosing it to be constant. Along a circle such configurations cannot be gauged away because they imply non-zero Wilson lines for the maximal Abelian subgroup $U(1) \times U(1) \subset U(2)$. We can parametrise the gauge fields as

$$A^a T^a = -\frac{1}{2\pi R}\mathrm{diag}(\theta_1, \theta_2). \tag{13.172}$$

The parameters θ_1, θ_2 enter into the open string mass formula through the shift in the spectrum of the discrete momentum along the circle

$$\alpha' M^2 = N - 1 + \left(\frac{m}{R} - \frac{\Delta\theta}{2\pi R}\right)^2, \quad \Delta\theta = \theta_2 - \theta_1. \tag{13.173}$$

We have restricted ourselves to configurations with two D-branes, equivalently to an open string gauge group $G = U(2)$, and used T-duality to relate purely Neumann boundary conditions (a space-filling D-25 brane) to strings with Dirichlet boundary conditions in one direction (D-24 branes). By continuing with T-duality transformations we can introduce Dirichlet boundary conditions in more directions and move on to D-23 branes, D-22 branes, and so on. We can also have $N > 2$ D-branes, which results in a $U(N)$ gauge theory. If we divide out by world-sheet parity we obtain a theory of non-oriented strings, which changes the gauge group. There are two options: if we keep states which are invariant under world-sheet parity, then the gauge group is reduced to the symplectic group $USp(N)$ whose Lie algebra consists of symmetric matrices. If we keep states which are odd under world-sheet parity, then the gauge group is $O(N)$, whose Lie algebra consists of antisymmetric matrices.

We note that T-duality shows that including Dirichlet boundary conditions in open string theory is mandatory rather than optional. Moreover, theories which are initially defined as theories of closed strings, such as the Type-II superstring theories to be discussed in Chapter 14, contain open string sectors where the open strings are localised on D-branes. Note that the existence of open string sectors supported by D-branes does not imply that there also exist open strings which have Neumann boundary in all directions, and which therefore can move freely in Minkowski space-time. One might think that this should be possible, because T-duality exchanges Dirichlet and Neumann boundary conditions. However, while this is true as long as we have compactified the theory on a circle, a suitable decompactification limit may not exist. There is an asymmetry between Dirichlet and Neumann boundary conditions when we decompactify: with Dirichlet boundary conditions we can break the gauge group by moving D-branes apart, while there is no such mechanism for Neumann boundary conditions, since Wilson loops require a non-contractible circle to support them.

13.7 Toroidal Compactification of Closed Strings

13.7.1 The Narain Lattice

As an example for higher-dimensional compactifications we now turn to compactifications on tori $T^d \cong \mathbb{R}^d/\Lambda$, where Λ is a d-dimensional lattice. To parametrise different tori we use a constant symmetric $d \times d$ matrix G_{ij}, $i,j = 1,\ldots,d$. There are two natural non-redundant parametrisations. The first is to choose a constant metric G_{ij} on $\mathbb{R}^d$ and to fix the lattice to be the standard lattice $\Lambda = \mathbb{Z}^d$. Alternatively, we can fix a flat metric on $\mathbb{R}^d$ and parametrise lattices up to rotations by their *lattice metric* $G_{ij} = e_i \cdot e_j$, where e_i is a lattice basis, and the scalar product is the one associated with our fixed metric on $\mathbb{R}^d$.[23] To check the parameter count, let us fix the standard metric on $\mathbb{R}^d$. Then all choices of lattice bases are related by the action of $GL(d, \mathbb{R})$, and since bases related by $O(d)$ transformations define the same torus, the space of inequivalent flat tori is locally parametrised by $GL(d, \mathbb{R})/O(d)$. The corresponding $d^2 - \frac{1}{2}d(d-1) = \frac{1}{2}d(d+1)$ parameters fit into a symmetric matrix G_{ij}. In fact, symmetric matrices provide a standard coordinate system on the symmetric space $GL(d, \mathbb{R})/O(d)$. This space comes with a natural Riemannian metric, which appears in the dimensionally reduced effective action in the form of a non-linear sigma model for the Kaluza–Klein scalars (see Section 13.7.4).

So far the discussion is the same as for a point particle. For string compactifications we can choose a B-field background with vanishing field strength, $H_{ijk} = 0$. For tori of dimension $d \geq 2$ the integration of the B-field over a two-subtorus $C \subset T^d$, $C \cong T^2$ defines the analogue of a Wilson line, based on the gauge-invariant quantity

$$\oint_C B = \int_C B_{ij} dX^i dX^j.$$

Inequivalent flat B-fields ($H_{ijk} = 6\partial_{[i}B_{jk]} = 0$) can be parametrised by constant B-fields, which involves $\frac{1}{2}d(d-1)$ parameters. Together with the $\frac{1}{2}d(d+1)$ 'metric' moduli G_{ij}, this brings the moduli count up to d^2. When working out the spectrum we will identify the corresponding d^2 massless scalar fields.

We consider d string coordinates X^i subject to the toroidal closed string boundary conditions

$$X^i(\sigma^0, \sigma^1 + \pi) = X^i(\sigma^0, \sigma^1) + 2\pi w^i, \quad \mathbf{w} = (w^i) \in \Lambda. \tag{13.174}$$

This introduces the winding vector w^i into the mode expansion,

$$X^i(\sigma^0, \sigma^1) = x^i + p^i\sigma^0 + 2w^i\sigma^1 + \cdots. \tag{13.175}$$

We then define left- and right-moving momenta:

$$p^i_{L/R} = \frac{1}{2}p^i \pm w^i \Leftrightarrow p^i = p^i_L + p^i_R, \quad w^i = \frac{1}{2}(p^i_L - p^i_R). \tag{13.176}$$

It is straightforward to write down a basis for the Fock space:

$$\alpha^{\mu_1}_{-m_1} \cdots \alpha^{i_1}_{-k_1} \cdots \tilde{\alpha}^{\nu_1}_{-m_1} \cdots \tilde{\alpha}^{j_1}_{-l_1} \cdots |k^\mu, (\mathbf{k}_L, \mathbf{k}_R)\rangle, \tag{13.177}$$

---

[23] A more invariant characterisation is to note that G_{ij} encodes the lengths and intersection angles between closed curves on the torus which represent the generators of its first homology group by geodesics.

Table 13.3. Generic massless states in a d-dimensional toroidal compactification

State	Fields	Interpretation
$\alpha^{\mu}_{-1}\tilde{\alpha}^{\nu}_{-1}\|k^{p}(0,0)\rangle$	$G_{\mu\nu}, B_{\mu\nu}, \phi$	Graviton, B-field, dilaton
$\alpha^{\mu}_{-1}\tilde{\alpha}^{j}_{-1}\|k^{p}(0,0)\rangle$	$G_{\mu j}, B_{\mu j}$	$2d$ vector fields
$\alpha^{i}_{-1}\tilde{\alpha}^{\nu}_{-1}\|k^{p}(0,0)\rangle$	$G_{i\nu}, B_{i\nu}$	
$\alpha^{i}_{-1}\tilde{\alpha}^{j}_{-1}\|k^{p}(0,0)\rangle$	G_{ij}, B_{ij}	d^2 scalar fields

where $\mu = 0, \ldots D - d - 1$ are space-time indices, where $i = 1, \ldots, d$ are internal indices, and k^{μ}, $(\mathbf{k}_L, \mathbf{k}_R)$ are the eigenvalues of the space-time momentum p^{μ} and of the internal left- and right-moving momenta $(\mathbf{p}_L, \mathbf{p}_R)$. The resulting mass and level matching formulae are

$$\alpha' M^2 = 2\left(N + \frac{1}{2}\mathbf{k}_L^2 + \tilde{N} + \frac{1}{2}\mathbf{k}_R^2 - 2\right), \tag{13.178}$$

$$N - \tilde{N} = \frac{1}{2}(\mathbf{k}_L^2 - \mathbf{k}_R^2).$$

The states with $N = \tilde{N} = 1$, $\mathbf{k}_L = \mathbf{k}_R = 0$ which are massless for generic values of the moduli G_{ij}, B_{ij} are listed in Table 13.3.

Besides the lower-dimensional graviton, B-field and dilaton we have $2d$ abelian vector fields. Half of them are Kaluza–Klein vector fields, corresponding to the mixed components $G_{\mu i} = G_{i\mu}$ of the metric, the other half result from decomposing the B-field. The corresponding charges are the internal momenta and windings. The generic gauge group is

$$U(1)^{2d} \cong U(1)_G^d \times U(1)_B^d \cong U(1)_L^d \times U(1)_R^d. \tag{13.179}$$

Depending on which linear combinations of the abelian vector fields we take, the corresponding charges are either the momenta and windings or the left- and right-moving momenta. Finally, we have d^2 scalars which are moduli and parametrise the deformations of the metric and B-field on the torus.

To work out the quantisation condition for the internal momenta, we need to distinguish between the kinetic and the canonical momenta. The eigenvalues $\mathbf{k} = (k^i)$ of the kinetic momentum $\mathbf{p} = (p^i)$ enter into the mass formula (13.178). The canonical momentum $\mathbf{p}_{\text{can}}$ satisfies the canonical commutation relation $[x^i, p^j_{\text{can}}] = iG^{ij}$, where $G^{ij} = e^{*i} \cdot e^{*j}$ is the lattice metric of the dual lattice,

$$e^{*i} \cdot e^{*j} = G^{ij}, \quad e^{*i} \cdot e_j = \delta^i_j. \tag{13.180}$$

The eigenvalues of $\mathbf{p}_{\text{can}}$ take values in the dual lattice Λ^*. The (total, integrated) canonical momentum of the string is

$$p_i^{\text{can}} = \int_0^{\pi} d\sigma^1 \frac{\partial \mathcal{L}}{\partial \partial_0 X^i}. \tag{13.181}$$

If a constant B-field is present, then the action is, in the conformal gauge,[24]

$$S = \int d^2\sigma \, \mathcal{L} = -\frac{1}{2\pi} \int d^2\sigma \left(G_{ij}\partial_\alpha X^i \partial^\alpha X^j - B_{ij}\partial_\alpha X^i \partial_\beta X^j \epsilon^{\alpha\beta} \right), \qquad (13.182)$$

where $\epsilon^{\alpha\beta}$ is the two-dimensional ϵ-tensor. While for constant B the second term is a total derivative and does not contribute to the equations of motion, it induces a shift between the kinetic momentum and the canonical momentum:

$$p_i^{\text{can}} = G_{ij}p^j + 2B_{ij}w^j. \qquad (13.183)$$

Exercise 13.7.1   Verify the relation (13.183) between the canonical and kinetic momentum.

If we express the eigenvalues $k^i_{L/R}$ of the left- and right-moving momenta in terms of $(w^i) \in \Lambda$ and $(p_i^{\text{can}}) \in \Lambda^*$,

$$k^i_{L/R} = \frac{1}{2}k^i \pm w^i = \frac{1}{2}G^{ij}k_j^{\text{can}} - G^{ij}B_{jk}w^k \pm w^i, \qquad (13.184)$$

we see explicitly how the mass formula depends on the d^2 parameters G_{ij}, B_{ij}. The left- and right-moving momenta form the momentum lattice or *Narain lattice* $\Gamma_{d,d}$. By using the quantisation conditions satisfied by $\mathbf{w} = m^i e_i$ and $\mathbf{p}^{\text{can}} = n_i e^{*i}$, we can read off a standard basis for $\Gamma_{d,d}$ which is expressed in terms of the lattice bases e_i, e^{*i} and depends on the moduli G_{ij}, B_{ij}:

$$(\mathbf{k}_L, \mathbf{k}_R) = m^i k_i + n_j \bar{k}^j, \qquad (13.185)$$

where

$$k_i = (e_i + B_{ij}e^{*j}, -e_i + B_{ij}e^{*j}), \quad \bar{k}^j = \left(\frac{1}{2}e^{*j}, \frac{1}{2}e^{*j}\right). \qquad (13.186)$$

Exercise 13.7.2   Verify that the vectors (13.186) are a basis of the Narain lattice $\Gamma_{d,d}$.

On the Narain lattice there are two natural scalar products

$$(\mathbf{k}_L, \mathbf{k}_R) \bullet (\mathbf{k}'_L, \mathbf{k}'_R) = \mathbf{k}_L^2 + \mathbf{k}_R^2, \quad (\mathbf{k}_L, \mathbf{k}_R) \cdot (\mathbf{k}'_L, \mathbf{k}'_R) = \mathbf{k}_L^2 - \mathbf{k}_R^2. \qquad (13.187)$$

The positive definite scalar product $\bullet$ depends on the moduli G_{ij}, B_{ij} and enters into the mass formula:

$$(\mathbf{k}_L, \mathbf{k}_R) \bullet (\mathbf{k}'_L, \mathbf{k}'_R) = (m^i, n_i) \begin{pmatrix} 2(G_{ij} - B_{ik}G^{kl}B_{lj}) & B_{ik}G^{kj} \\ -G^{il}B_{lj} & \frac{1}{2}G^{ij} \end{pmatrix} \begin{pmatrix} m'^j \\ n'_j \end{pmatrix}$$

$$=: M_a \mathcal{H}^{ab} M_b, \qquad (13.188)$$

where the symmetric matrix $\mathcal{H}^{ab}$ combines the metric G_{ij} and the B-field B_{ij} into a symmetric $2d \times 2d$ matrix. The inverse matrix $\mathcal{H}_{ab}$ is called the *generalised metric*.

---

[24] The sign in front of the B-field has been chosen in order that the following formulae take the form found in the majority of the literature.

The indefinite scalar product $\cdot$ is independent of the moduli, and the Narain lattice $\Gamma_{d,d}$ is an even and self-dual with respect to this scalar product.

Exercise 13.7.3   Show that the Narain lattice $\Gamma_{d,d}$ is even self-dual with respect to the indefinite scalar product (13.187). In particular, verify that

$$k_i \cdot k_j = 0, \quad \bar{k}^i \cdot \bar{k}^j = 0, \quad k^i \cdot \bar{k}_j = \delta^i_j. \tag{13.189}$$

Even self-dual lattices with a scalar product of signature (d, d) are unique up to $O(d, d)$ transformations. Since the mass formula (and as can be shown any other physical quantity) is $O(d) \times O(d)$ invariant, the physically inequivalent Narain lattices are locally parametrised by the symmetric space

$$O(d, d)/(O(d) \times O(d)). \tag{13.190}$$

The dimension d^2 of this space matches with the number of moduli G_{ij}, B_{ij}, which therefore can be used as coordinates. As discussed in Section 13.2.8, even self-dual Narain lattices $\Gamma_{d,d}$ define modular invariant theories. In the dimensional reduction of the string effective action on T^d the d^2 scalars G_{ij}, B_{ij} enter through a non-linear sigma model with target space $O(d, d)/(O(d) \times O(d))$ (see Section 13.7.4).

13.7.2 T-duality for Toroidal Compactifications

There are further identifications between compactifications because $O(d, d)$ contains a discrete subgroup $O(d, d, \mathbb{Z})$ which maps a given lattice $\Gamma_{d,d}$ onto itself. The group $O(d, d)$ acts linearly on the vectors $(M_a) = (m^i, n_i)$

$$M_a \rightarrow A_a{}^b M_b, \tag{13.191}$$

where

$$A = (A_a{}^b) \in O(d, d) = \{A \in GL(2d, \mathbb{R}|A^T \eta A = \eta\}, \quad \eta = \begin{pmatrix} 0 & \mathbb{1}_d \\ \mathbb{1}_d & 0 \end{pmatrix}. \tag{13.192}$$

Note that we are working in a basis where the components of the vectors M_a are the winding and momentum quantum numbers. In this 'winding-momentum basis' the group $O(d, d)$ takes a non-standard form, where the invariant metric is represented by an off-diagonal matrix. We could make a basis transformation which diagonalises η, and thus recover the standard definition of $O(d, d)$ as a matrix group, but since winding and momentum quantum numbers are discrete, the description we have chosen is adapted to T-duality. $O(d, d)$ indices are raised and lowered using η_{ab}, η^{ab}, and objects with upper $O(d, d)$ indices transform in the contragradient representation, where A is replaced by $A^{T,-1}$. The generalised metric $\mathcal{H}_{ab}$ and its inverse $\mathcal{H}^{ab}$ are $O(d, d)$ tensors.

Note that the action of $O(d, d)$ can be interpreted either passively or actively. Viewed passively we transform both M_a and $\mathcal{H}^{ab}$ and the value of the scalar product is invariant. In an active interpretation we apply an $O(d, d)$ transformation to the background fields to deform the theory. As (13.178) shows, transformations in the subgroup $O(d) \times O(d)$ do not change the spectrum and are symmetries. T-duality arises as a further discrete symmetry: if we deform the background by an $O(d, d, \mathbb{Z})$ transformation, then we can compensate this deformation by relabelling windings and momenta while respecting their discreteness. This is the subgroup of active $O(d, d)$ transformations which acts cristallographically on the lattice $\Gamma_{d,d}$.

The T-duality group contains the following types of transformations, which already generate the full T-duality group:

- Automorphisms of the lattice Λ, which defines the space-time torus, form a discrete subgroup $GL(d, \mathbb{Z}) \subset GL(d, \mathbb{R})$. In the winding/momentum basis this acts by

$$A = \begin{pmatrix} a & 0 \\ 0 & a^{T,-1} \end{pmatrix}, \quad a \in GL(d, \mathbb{Z}). \tag{13.193}$$

Note that windings and momenta transform contragradiently. The background fields G_{ij}, B_{ij} transform as $GL(d, \mathbb{R}) \supset GL(d, \mathbb{Z})$ tensors.

- Discrete shifts of the momenta by terms proportional to the windings

$$m^i \rightarrow m^i, \quad n_i \rightarrow n_i + \Theta_{ij} m^j, \quad \Theta_{ij} = -\Theta_{ji} \tag{13.194}$$

with corresponding $O(d, d, \mathbb{Z})$ matrices

$$A = \begin{pmatrix} 1_d & 0 \\ \Theta & 1_d \end{pmatrix}, \quad \Theta = -\Theta^T \in \text{Mat}(d, \mathbb{Z}). \tag{13.195}$$

Since windings are the charges of the B-field, the corresponding action on the background fields are discrete shifts

$$B_{ij} \rightarrow B_{ij} + \frac{1}{2}\Theta_{ij}. \tag{13.196}$$

This is analogous to discrete shifts of axionic scalars in field theory. In the functional integral B-field Wilson 'lines' enter as phases, which have a natural periodicity. By inspection of (13.189) we can see that the effect of (13.196) on the standard basis of the Narain lattice can be undone by adding a suitable integer linear combination of the basis vector $\bar{k}^j$.

- Factorised T-dualities. These are the generalisations of the the the T-duality $R \rightarrow \alpha'/R$ of S^1. On windings and momenta a factorised T-duality acts by exchange in one particular (lattice) direction,

$$m^i \leftrightarrow n_i, \quad m^j \rightarrow m^j, \quad n_j \rightarrow n_j, \quad j \neq i, \tag{13.197}$$

with $O(d, d, \mathbb{Z})$ matrix

$$A = \begin{pmatrix} 0 & e_{ii} \\ e_{ii} & 0 \end{pmatrix}, \quad e_{ii} = (\delta_{ik}\delta_{il})_{k,l=1,\dots,d}. \tag{13.198}$$

For rectangular tori in the absence of a B-field, the corresponding action on background fields is the inversion of the radius of i-th direction, $1/R_i \rightarrow \alpha'/R_i$. For generic backgrounds the transformation formula becomes more complicated.[25]

How then does T-duality act on the background fields in general? The continuous group $O(d, d)$ has a natural action on the symmetric space $O(d, d)/(O(d) \times O(d))$, on which G_{ij}, B_{ij} are coordinates. It is useful to introduce the combination

---

[25] See Section 13.7.6 for explicit expressions.

$E_{ij} = 2(G_{ij} + B_{ij})$, which is a 'projective coordinate' on $O(d,d)/(O(d) \times O(d))$, in the sense that $O(d,d)$ acts on $E = (E_{ij})$ by projective linear transformations:

$$E \to (aE + b)(cE + d)^{-1}, \quad \begin{pmatrix} a & b \\ c & d \end{pmatrix} \in O(d,d). \qquad (13.199)$$

This is analogous to the transformation behaviour of the world-sheet modular parameter τ under $SL(2,\mathbb{Z}) \subset SL(2,\mathbb{R})$.

One interesting example is the product of all factorised T-dualities, which exchanges the momenta and windings in all directions. On the background this acts by mapping the generalised metric to its inverse, and thus for vanishing B-field mapping the metric to its inverse. This inverts the volume of the torus, which is the natural generalisation of inverting the radius of a circle.

Exercise 13.7.4   Show that

$$M_a \mathcal{H}^{ab} M_b = M^a \mathcal{H}_{ab} M^b, \qquad (13.200)$$

where $\mathcal{H}_{ab}$ is the inverse of $\mathcal{H}^{ab}$ while M^a is related to M_a by exchanging momentum and winding quantum numbers.

Exercise 13.7.5   As an example for (13.199), consider the $O(d,d,\mathbb{Z})$ element where $a = d = 0_d$, $b = c = \mathbb{1}_d$, which exchanges all momenta and windings, and thus acts as $E \to E^{-1}$. Show that this inverts the generalised metric $\mathcal{H}_{ab} \to \mathcal{H}^{ab}$. *Hint*: use the matrix version of the geometric series.

13.7.3 Symmetry Enhancement

The generic gauge group of a torus compactification is $U(1)^d \times U(1)^d$, but as we have seen for circle compactifications non-abelian extensions can appear for special values of the moduli. Massless charged vector bosons exist whenever the Narain lattice contains vectors of the form $(\mathbf{k}_L, 0)$ or $(0, \mathbf{k}_R)$ with $\mathbf{k}_{L/R}^2 = 2$. Since we cannot obtain additional massless neutral gauge bosons, the rank of the gauge group is fixed to $2d$. By integrality of the Narain lattice the mutual scalar products between the vectors $\mathbf{k}_L$ associated to massless charged vector bosons can only take the values $\mathbf{k}_L \cdot \mathbf{k}_L' = 0, \pm 1, \pm 2$. This implies that these vectors are the *root vectors* of a *semi-simple* and *simply laced* Lie algebra.[26] Simply laced means that all root vectors have the same length. The real, compact, simple, simply laced Lie algebras are $\mathfrak{su}_n$, $\mathfrak{so}_{2n}$, $\mathfrak{e}_6$, $\mathfrak{e}_7$, and $\mathfrak{e}_8$, with ranks $n-1, n, 6, 7, 8$, respectively. To show that all such Lie algebras with rank $l \leq d$ can be realised, we determine the corresponding values of the moduli G_{ij}, B_{ij}. For simplicity, let us consider first the case where the non-abelian Lie algebra $\mathfrak{g}$ has maximal rank $l = d$. Among the root vectors, we can pick an integral basis of l linearly independent *simple roots* α_i, $i = 1, \ldots, l$, such that $\alpha_i \cdot \alpha_j = 0, -1$ for $i \neq j$. Then the matrix $C_{ij} = \alpha_i \cdot \alpha_j$ is the *Cartan matrix* of $\mathfrak{g}$.[27] Next, we rewrite the lattice basis (13.186) in the form

$$k^i = \left(\frac{1}{2} e^{*i}, \frac{1}{2} e^{*i} \right), \quad k_i = \left(2e_i + \frac{1}{2} D_{ij} e^{*j}, \frac{1}{2} D_{ij} e^{*j} \right), \qquad (13.201)$$

---

[26] See Appendix G for a short review of Lie algebras.
[27] This formula only applies to simply laced Lie algebras, the general formula is $C_{ij} = 2(\alpha_i \cdot \alpha_j)/(\alpha_i \cdot \alpha_i)$.

where

$$D_{ij} = 2(-G_{ij} + B_{ij}).\tag{13.202}$$

Vectors of the form $(\mathbf{k}_L, 0)$ can be obtained by taking linear combinations

$$k_i - D_{ij}k^j = (2e_i, 0) = (\alpha_i, 0),\tag{13.203}$$

and these vectors are in $\Gamma_{d,d}$ if and only if $D_{ij} \in \mathbb{Z}$. The vectors α_i are the simple roots of a simply laced Lie algebra $\mathfrak{g}$ if

$$\alpha_i \cdot \alpha_j = C_{ij} \Rightarrow e_i \cdot e_j = G_{ij} = \frac{1}{4}C_{ij}.\tag{13.204}$$

Thus, up to a normalisation factor the lattice Λ which defines the torus T^d must be the *root lattice* of $\mathfrak{g}$. The integrality of D_{ij} also fixes the B-field up to discrete choices:

$$D_{ij} = -\frac{1}{2}C_{ij} + 2B_{ij} \in \mathbb{Z} \Rightarrow B_{ij} \equiv \frac{1}{4}C_{ij} \mod \frac{1}{2}.\tag{13.205}$$

Note that the condition on B_{ij} can be satisfied, despite that B_{ij} is antisymmetric while C_{ij} is symmetric, since C_{ij} is antisymmetric modulo 2: $C_{ii} = 2 \equiv 0 \mod 2$, $C_{ij} = 0$, or $C_{ij} = 1 \equiv -1 \mod 2$.

Exercise 13.7.6　Work out the condition on root vectors of the form $(0, \alpha_i)$ to be lattice vectors and show that enhanced gauge groups of toroidal compactifications are of the form $G \times G$.

So far we have shown that $U(1)^d \times U(1)^d$ can get enhanced to a product $G^{(d)} \times G^{(d)}$ of simple, simply laced groups of rank d at special points in the moduli space. This can be generalised in two ways. First, $G^{(d)}$ can be a product of simple Lie groups, that is a semi-simple Lie group. In this case C_{ij} has a block decomposition into the Cartan matrices of the simple groups. Second, the root system contained in $\Gamma_{d,d}$ could have a non-maximal rank, $l < d$. In this case G_{ij} and B_{ij} are not fixed completely, and we obtain subspaces of various dimensions where the gauge group takes the form $(G^l \times U(1)^{d-l}) \times (G^l \times U(1)^{d-l})$. Thus, the moduli space decomposes into generic points with an abelian gauge group, special points with a semi-simple gauge group and subspaces where the gauge group is a reductive group. It can be shown that all non-generic points and subspaces can be characterised as fixed points of T-duality transformations. Moreover, the full T-duality group can be generated by the Weyl twists of SU(2) groups which are smallest types of enhanced gauge groups. Like for the circle, T-duality can be interpreted the discrete remnant of a local gauge symmetry.

The world-sheet realisation of non-abelian gauge symmetries uses conserved chiral currents $\partial_z X^i(z)$, $: \exp(i\mathbf{k}_L\mathbf{X})(z) :$, where the vectors $\mathbf{k}_L$ are the root vectors of a semi-simple Lie algebra $\mathfrak{g}$. Similar to the SU(2) case one can build a Kac–Moody algebra out of these currents, but with one interesting complication, which is due to the fact that the scalar product between two roots is an integer, but not necessarily an even integer. The radially ordered OPE between chiral exponential operators takes the form

$$e^{i\mathbf{k}_L\mathbf{X}(z)}e^{i\mathbf{k}'_L\mathbf{X}(w)} = (z-w)^{\mathbf{k}_L\cdot\mathbf{k}'_L}e^{i\mathbf{k}_L\mathbf{X}(z)+i\mathbf{k}'_L\mathbf{X}(w)}, \quad \text{for } |z| > |w|,\tag{13.206}$$

and

$$e^{ik_L X(z)} e^{ik'_L X(w)} = (w-z)^{k_L \cdot k'_L} e^{ik_L X(z) + ik'_L X(w)} \tag{13.207}$$
$$= (-1)^{k_L \cdot k'_L}(z-w)^{k_L \cdot k'_L} e^{ik_L X(z) + ik'_L X(w)}, \quad \text{for } |z| < |w|,$$

We observe that, when $\mathbf{k}_L \cdot \mathbf{k}'_L$ is an odd integer, the resulting correlation function is antisymmetric rather than symmetric. When converting these OPEs into relations between their Laurent modes, one obtains anticommutation rather than commutation relations. One class of algebras which has both commutators and anticommutators are Lie superalgebras, which we will encounter in Chapter 14. But the algebra we have obtained here is not a Lie superalgebra because it does not have a $\mathbb{Z}_2$ grading which classifies individual operators as even or odd. Instead, we have a non-standard 'representation up to signs' or projective representation of a Lie algebra, which however can be shown to be equivalent to a proper representation. One way of doing this is to find a $\mathbb{Z}_2$-valued function on the root lattice, $\mathbf{k}_L \mapsto c(\mathbf{k}_L) = \pm 1$, such that $c(\mathbf{k}_L)c(\mathbf{k}'_L) = (-1)^{k_L \cdot k'_L}$. One can then absorb the unwanted signs by multiplying each vertex operator : $\exp(ik_L X(z))$: by a factor $c(\mathbf{k}_L)$. Group theoretically, finding the function $c(\mathbf{k}_L)$ corresponds to solving a group extension problem where on replaces the root lattice Λ_R, considered as an abelian group, by a central extension with the group $\mathbb{Z}_2$.

Of course, this cannot be done for the root lattice in isolation, but has to be extended to the whole Narain lattice. In fact, the problem with antisymmetric correlators needs to be addressed irrespective of any symmetry enhancement, since for any values of the moduli the operators : $\exp(ik_L X(z) + ik_R \bar{X}(\bar{z}))$: are part of vertex operators of bosonic states and, therefore, should have symmetric Euclidean correlation functions. The spin-statistics theorem is often stated in the way that bosonic (fermionic) fields must be quantised using commutators (anti-commutators) to avoid violations of causality. A more complete statement is that non-standard relations are possible, but always can be related to the standard relations by a field redefinition, known as a *Klein transformation*.[28]

The standard approach for exponential operators is to explicitly construct functions $c(\mathbf{k}_L, \mathbf{k}_R)$ on the Narain lattice such that Laurent modes the modified exponentials : $\exp(ik_L X(z) + ik_R \bar{X}(\bar{z})$: $c(\mathbf{k}_L, \mathbf{k}_R)$ satisfy commutation relations. We will follow another approach by modifying the commutation relations for the zero mode parts of the vertex operators. This can be physically motivated and is very much in the spirit of the idea that string theory forces us to modify geometry. The hidden assumption of the standard approach is that left- and right-moving momenta satisfy the standard commutation relations, $[x_L^i, p_L^i] = iG^{ij} = [x_R^i, p_R^j]$. However these commutators are under-determined, since the mode expansion of $X^i(z, \bar{z})$ only contains three operators, x^i, p^i, w^i, but no operator conjugate to the winding operator w^i. We can introduce a fourth operator v^i by imposing

$$\mathbf{k}_L \cdot \mathbf{x}_L + \mathbf{k}_R \cdot \mathbf{x}_R = \mathbf{k} \cdot \mathbf{x} + \mathbf{q} \cdot \mathbf{v}, \tag{13.208}$$

where $\mathbf{k}_L, \mathbf{k}_R, \mathbf{k}, \mathbf{q}$ are the eigenvalues of the operators $\mathbf{p}_L, \mathbf{p}_R, \mathbf{x}, \mathbf{w}$. In other words, momentum/winding eigenstates are generated by the zero mode parts

---

[28] See, e.g., Streater and Wightman (1964).

$$e^{i\mathbf{k}_L \cdot \mathbf{x}_L + i\mathbf{k}_R \cdot \mathbf{x}_R} = e^{i\mathbf{k} \cdot \mathbf{x} + i\mathbf{q} \cdot \mathbf{v}} \tag{13.209}$$

of the exponential operators $: \exp(i\mathbf{k}_L \cdot \mathbf{X}(z) + i\mathbf{k}_R \bar{\mathbf{X}}(\bar{z})) :$.

If we impose that $\mathbf{x} = \frac{1}{2}(\mathbf{x}_L + \mathbf{x}_R)$ in addition to the previous expressions for p_L^i, p_R^i, then a straightforward calculation allows one to determine v^i. We list the full set of relations between the momentum/winding and left-/right-moving bases:

$$\left. \begin{aligned} x^i &= \tfrac{1}{2}(x_L^i + x_R^i), \\ p^i &= p_L^i + p_R^i + G^{ij}B_{jk}(p_L^k - p_R^k), \\ v^i &= x_L^i - x_R^i + G^{ij}B_{jk}(x_L^k + x_R^k), \\ w^i &= \tfrac{1}{2}(p_L^i - p_R^i), \end{aligned} \right\} \Leftrightarrow \left\{ \begin{aligned} x_L^i &= x^i - G^{ij}B_{jk}x^k + \tfrac{1}{2}v^i, \\ x_R^i &= x^i + G^{ij}B_{jk}x^k - \tfrac{1}{2}v^i, \\ p_L^i &= \tfrac{1}{2}p^i - G^{ij}B_{jk}w^k + w^i, \\ p_R^i &= \tfrac{1}{2}p^i - G^{ij}B_{jk}w^k - w, \end{aligned} \right. \tag{13.210}$$

where p_i is the canonical momentum. Since x^i, p^j and v^i, w^j are conjugate variables, they should satisfy canonical commutation relations

$$[x^i, p^j] = iG^{ij}, \quad [v^i, w^j] = iG^{ij}. \tag{13.211}$$

Naively, we expect that all other commutators vanish, which is equivalent to $[x_L^i, p_L^j] = iG^{ij} = [x_R^i, p_R^j]$. But, as pointed out in Sakamoto (1989), one should ask: what is the meaning of x^i? For strings on $\mathbb{R}^d$ the mode expansion tells us that x^i is the centre of mass position at world-sheet time $\sigma^0 = 0$. On a torus this interpretation makes only sense for strings without winding, but not if the string winds around the torus. In this case, we can still compute the centre of mass position on the covering space $\mathbb{R}^d$ of the torus $T^d = \mathbb{R}^d/2\pi\Lambda$:

$$x^i := x_{\text{cm}}^i(\sigma^0 = 0) = \int_0^\pi \frac{d\sigma^1}{\pi} X^i(\sigma^0 = 0, \sigma^1) = \int_0^\pi \frac{d\sigma^1}{\pi} x_0^i + 2w^i\sigma^i + \cdots = x_0^i + \pi w^i. \tag{13.212}$$

Thus, on the covering space the centre of mass position is shifted compared to the constant term x_0^i of the mode expansion by a term proportional to the winding operator. It is natural to assume that x_0^i has the canonical commutation relation with p^j and commutes with v^j. However, then x^i does not commute with v^j and we obtain the following non-standard commutation relations:

$$[x^i, p^j] = iG^{ij}, \quad [v^i, w^j] = iG^{ij}, \quad [x^i, v^j] = -i\pi G^{ij}. \tag{13.213}$$

It is straightforward to convert these into commutation relations between left- and right-moving coordinates and momenta:

$$[x_L^i, p_L^j] = iG^{ij} = [x_R^i, p_R^j], \quad [x_L^i, x_L^j] = i\pi B^{ij} = [x_R^i, x_R^j], \quad [x_L^i, x_R^j] = i\pi(G^{ij} - B^{ij}). \tag{13.214}$$

If these relations are imposed for the zero mode parts of exponential operators, we obtain a modified representation of translations and windings:

$$e^{i\mathbf{k}_L \cdot \mathbf{x}_L + i\mathbf{k}_R \mathbf{x}_R} e^{i\mathbf{k}_L' \mathbf{x}_l + i\mathbf{k}_R' \mathbf{x}_R} = (-1)^{\mathbf{k}_L \cdot \mathbf{k}_L' + \mathbf{k}_R \cdot \mathbf{k}_R'} e^{i(\mathbf{k}_L + \mathbf{k}_L') \cdot \mathbf{x}_L + i(\mathbf{k}_R + \mathbf{k}_R') \cdot \mathbf{x}_R}. \tag{13.215}$$

With this zero mode algebra the OPEs of exponential operators take their standard, symmetric form, without the need to introduce the functions $c(\mathbf{k}_L, \mathbf{k}_R)$ by hand. We also see explicitly that translations on the doubled space parametrised by $\mathbf{x}_L, \mathbf{x}_R$ have a modified non-commutative group law. In this sense, the geometry 'seen' by a closed string on a torus is non-commutative.

For reference, let us note the commutation relations for the Laurent modes of the vertex operators which realise a representation of the (untwisted affine) Kac–Moody algebra $\hat{g}$ associated to the semi-simple Lie algebra g:

$$[H_m^i, E_m^\alpha] = (e_i \cdot \alpha) E_{m+n}^\alpha, \tag{13.216}$$

$$[E_m^\alpha, E_n^\beta] = \begin{cases} \epsilon(\alpha, \beta) E_{m+n}^{\alpha+\beta}, & \alpha \cdot \beta = -1, \\ \alpha \cdot H_{m+n} + m\delta_{m+n,0}, & \alpha \cdot \beta = -2, \\ 0, & \alpha \cdot \beta = 0, 1, 2, \end{cases} \tag{13.217}$$

where $m, n \in \mathbb{Z}$, where e_i are unit vectors, and where α, β are roots of g. The structure constants $\epsilon(\alpha, \beta)$ are $\mathbb{Z}_2$-valued. Note that for simply-laced Lie algebras the scalar products of two root vectors takes values $0, \pm 1, \pm 2$, with value -1 if and only if $\alpha + \beta$ is a root and with value -2 if and only if $\alpha + \beta = 0$. We remark that the Kac–Moody algebra generators H_m^i, E_m^α can be presented in terms of generators H^i, E^α of g and a formal complex variable z as

$$H_m^i = z^m \otimes H^i, \quad E_m^\alpha = z^m \otimes E^\alpha. \tag{13.218}$$

In a general orthonormal basis T_m^a the relations for an affine Kac–Moody algebra $\hat{g}$ are

$$[T_m^a, T_n^b] = f_c^{ab} T_{m+n}^c + km\delta_{m+n,0}\delta^{ab}, \tag{13.219}$$

where f_c^{ab} are the structure constants of g, and where k generates the centre of $\hat{g}$. In toroidal compactifications we obtain a level 1 representation, where k is represented by the unit operator. Higher level representations with $k > 1$ appear for example in orbifold compactifications. The Kac–Moody algebra has a further generator, the derivation d, which is related to the Virasoro generator L_0 by $d = -L_0$. Moreover, the so-called *Sugawara construction* allows one to construct the Virasoro generators out of the Kac–Moody generators:

$$L_m = \frac{1}{2k + 2\tilde{h}} \sum_{n \in \mathbb{Z}} \sum_a : T_{m+n}^a T_{-n}^a :, \tag{13.220}$$

where $\tilde{h}$ is the so-called dual Coxeter number of the Lie algebra g. This amounts to constructing the ('Sugawara') energy momentum tensor out of the Kac–Moody currents. The Kac–Moody and Virasoro algebra form a semi-direct sum, with the remaining relations given by

$$[L_m, T_n^a] = -n T_{m+n}^a, \tag{13.221}$$

$$[L_m, L_n] = (m - n)L_{m+n} + \frac{c}{12}m(m+1)(m-1)\delta_{m+n,0}, \quad \text{with} \quad c = l,$$

where l is the rank of the Lie algebra g.

13.7.4 Effective Actions for Toroidal Compactifications

It is straightforward to obtain the leading order effective action for the generic massless modes ($N = \tilde{N} = 1, \mathbf{k}_L = \mathbf{k}_R = 0$) of a toroidal compactification. This requires one to perform a dimensional reduction of the effective action (12.23) for the graviton, B-field, and dilaton on a d-dimensional torus from D to $n = D - d$ dimensions. We take the lattice defining the torus to be the standard lattice $\Lambda = \mathbb{Z}^d$, but allow for an overall scale by imposing the identification $X^i \cong X^i + 2\pi R$. It can be shown that in the n-dimensional string frame the effective action takes the form[29]

$$S = \frac{(2\pi R)^d}{2\kappa^2} \int d^n x \sqrt{G_n} e^{-2\phi_n} \left(R_G + 4\partial_\mu \phi \partial^\mu \phi \right. \tag{13.222}$$

$$- \frac{1}{4} G^{ij} G^{kl} \left(\partial_\mu G_{ik} \partial^\mu G_{jl} + \partial_\mu B_{ik} \partial^\mu B_{jl} \right) \tag{13.223}$$

$$- \frac{1}{4} G_{ij} F^i_{\mu\nu} F^{j\mu\nu} - \frac{1}{4} G^{ij} (F'_{i\mu\nu} - B_{ik} F^k_{\mu\nu})(F'^{\mu\nu}_j - B_{jl} F^{l\mu\nu}) \tag{13.224}$$

$$\left. - \frac{1}{12} H_{\mu\nu\rho} H^{\mu\nu\rho} \right). \tag{13.225}$$

The n-dimensional dilaton is defined as

$$\phi_n = \phi - \frac{1}{4} \log \det G_{ij}, \tag{13.226}$$

and the lower dimensional gravitational coupling κ_d is defined by $\kappa_d^2 = \kappa^2/(2\pi R)^d$, which shows that upon reduction couplings are rescaled by the volume of the internal space.

The effective field theory is invariant under the action of the group $O(d, d)$ introduced in Section 13.7.2. The d-dimensional metric $G_{\mu\nu}$ and B-field $B_{\mu\nu}$ are $O(d, d)$ invariant since they do not carry indices in the torus direction. The same is true for the n-dimensional dilaton ϕ_n, while the D-dimensional dilaton ϕ transforms non-trivially. The D-dimensional dilaton enters into the world-sheet action at one-loop level in α', and its non-trivial transformation under T-duality can be seen when looking at T-duality from the world-sheet point of view. We will come back to this in Section 13.7.6, when we discuss the so-called *Buscher rules*. Here we note that it is clear that ϕ_n and ϕ must transform differently since G_{ij} transforms non-trivially. The combination which defines the n-dimensional dilaton is precisely chosen such that it is $O(d, d)$ invariant. It remains to deal with the scalars G_{ij}, B_{ij} and gauge fields $F^i_{\mu\nu}, F'_{i\mu\nu}$ which carry internal indices $i, j = 1, \ldots, d$. The group $O(d, d)$ acts linearly on vectors with $2d$ components, which suggest that we need to repackage these fields if we want to make $O(d, d)$ covariance manifest.[30] For the scalars we introduce matrix notation, $G = (G_{ij}), B = (B_{ij})$, so that the scalar term in the Lagrangian is

$$\mathcal{L}_{scalar} \cong \frac{1}{4} \mathrm{Trace} \left(\partial_\mu G^{-1} \partial^\mu G + G^{-1} \partial_\mu B G^{-1} \partial^\mu B \right), \tag{13.227}$$

where we dropped a multiplicative pre-factor which is not relevant for the argument. Rewriting this using the generalised metric $\mathcal{H} = (\mathcal{H}_{ab}), a, b = 1, \ldots 2d$, introduced (13.188), we find

$$\mathcal{L}_{scalar} \cong \frac{1}{8} \mathrm{Trace} \left(\partial_\mu \mathcal{H} \partial^\mu \mathcal{H}^{-1} \right) = -\frac{1}{8} \mathrm{Trace} \left(\mathcal{H}^{-1} \partial_\mu \mathcal{H} \mathcal{H}^{-1} \partial^\mu \mathcal{H} \right). \tag{13.228}$$

This expression is manifestly $O(d, d)$ invariant since $\mathcal{H}$ is an $O(d, d)$ tensor. To complete showing that the action is $O(d, d)$ invariant, one combines the gauge fields

---

[29] This has been adapted from Maharana and Schwarz (1993).

[30] For the scalars we could use the combination $E_{ij} = 2(G_{ij} + B_{ij})$, which transforms fractionally linearly, see (13.199). In the following, we will identify quantities which have even simpler, linear transformation properties.

$F^i_{\mu\nu}$ and $F'_{i\mu\nu}$ into an $O(d,d)$ vector $(F^a_{\mu\nu}) = (F^i_{\mu\nu}, F'_{j\mu\nu})$ and expresses their couplings using the generalised metric:

$$\mathcal{L}_{\text{vect}} = -\frac{1}{4}\mathcal{H}_{ab}F^a_{\mu\nu}F^{a\mu\nu}.\tag{13.229}$$

The continuous $O(d,d)$ symmetry of the effective field theory is clearly not an exact symmetry of the full string theory. We know that there are special subspaces in the moduli space where the number of massless states is enhanced. If we include the momentum and winding modes then the $O(d,d)$ action deforms the massive spectrum, and only the discrete T-duality subgroup $O(d,d,\mathbb{Z})$ is a symmetry. If we study the effective field theory beyond leading order, then the couplings of massless fields are modified by loop diagrams which include massive modes as intermediate states. Such corrections break the $O(d,d)$ symmetry of couplings but preserve the discrete $O(d,d,\mathbb{Z})$ symmetry. At points in the moduli space where some of the generically massive states that we have 'integrated out' become massless, the loop corrected couplings show characteristic singularities which signal that the effective field theory becomes inconsistent close to these points because it does not include all the light states.

We remark that the matrix $\mathcal{H}$ does not only transform as a tensor under $O(d,d)$, but can also be interpreted as an $O(d,d)$ group element, because it satisfies the group defining relation

$$\mathcal{H}^T\eta\mathcal{H} = \eta.\tag{13.230}$$

Group elements of the form $\mathcal{H}$ form a d^2-dimensional subspace of $O(d,d)$. Since every $O(d,d)$ group element can be brought to the form $\mathcal{H}$ by the right action of the subgroup $O(d) \times O(d)$ on $O(d,d)$, we can view the matrices $\mathcal{H}_{ab}$ as representatives of elements of the homogeneous space $O(d,d)/(O(d) \times O(d))$. In the Lagrangian (13.228) $\mathcal{H}_{ab} = \mathcal{H}_{ab}(x)$ is a group valued function on space-time, which encodes the d^2 moduli scalar fields. In the field theoretic setting, the action of $O(d,d)$ on $\mathcal{H}_{ab}(x)$ is a global symmetry, while the action of $O(d)\times O(d)$ from the right is a local symmetry, which makes sure that we only have d^2 propagating degrees of freedom. One can rewrite the action to make explicit the presence of a corresponding gauge field, which is related to the so-called *Maurer-Cartan form*. We will not describe this here but indicate schematically how (13.228) can be rewritten into the standard form of a sigma model. To do this one expresses the matrix $\mathcal{H}_{ab}$ as a function of coordinates $\varphi^a, a = 1, \ldots d^2$ on G/H, and promotes them to scalar fields $\varphi^a(x)$ which parametrise a map from space-time into G/H:

$$\mathcal{L}_{\text{scalar}} = -\frac{1}{8}\text{Trace}\left(\mathcal{H}^{-1}\partial_a\mathcal{H}\mathcal{H}^{-1}\partial_b\mathcal{H}\right)\partial_\mu\varphi^a\partial^\mu\varphi^b = -\frac{1}{2}g_{ab}(\varphi)\partial_\mu\varphi^a\partial^\mu\varphi^b.\tag{13.231}$$

13.7.5 Compactification on a Two-Torus

In this section, we consider the two-torus as the simplest example of a higher-dimensional compactification. One new feature compared to the circle is that a two-torus can be made into a complex manifold. Complex manifolds, in particular so-called Calabi–Yau manifolds play an important role in the compactifications of superstring theories, and the two-torus can serve as a toy example which already demonstrates many of their characteristic features. In the following we define certain concepts for a general $2n$-dimensional real manifold with coordinates x^i, $i = 1, \ldots 2n$, and illustrate them for a two-torus with coordinates $x^1 = x, x^2 = y$.

The compactification on a two-torus has four real moduli, the independent components of $G_{ij}, B_{ij}, i, j = 1, 2$, which encode the shape and size of the torus as well as a constant B-field. The moduli space is

$$\mathcal{M} = O(2, 2, \mathbb{Z}) \backslash O(2, 2) / (O(2) \times O(2)), \tag{13.232}$$

where we have taken into account that we need to identify points which are related by the action of the T-duality group $O(2, 2, \mathbb{Z}) \subset O(2, 2)$. Like the two-torus itself, this moduli space is a complex manifold, more precisely a Kähler manifold.[31]

A *complex manifold* is a manifold which can be described using local complex coordinate systems related by holomorphic coordinate transformations. Complex one-dimensional examples are the complex plane $\mathbb{C}$ and two-tori, which can be defined by dividing $\mathbb{C}$ by a rank-2 lattice. In terms of a complex coordinate z on $\mathbb{C}$ this amounts to the identifications

$$z \cong z + \omega_1 \cong z + \omega_2, \tag{13.233}$$

where the *periods* $\omega_1, \omega_2 \in \mathbb{C}$ do not lie on the same real line, that is, their ratio is not a real number. The corresponding lattice is

$$\Lambda = \{z \in \mathbb{C} | z = \alpha \omega_1 + \beta \omega_2, \quad \alpha, \beta \in \mathbb{Z}\}. \tag{13.234}$$

Two complex manifolds are isomorphic if they are related by a bijective holomorphic map. In complex dimension one, holomorphic maps are conformal, and vice versa. As we have seen in Section 6.1 conformal structures (conformal equivalence classes) of two-tori determine their shape (angles), while not being sensitive to their size (volume). The shape of a two-torus is encoded by the ratio of its periods,

$$\tau = \frac{\omega_2}{\omega_1}, \tag{13.235}$$

where τ is the modular parameter that we already know from world-sheet tori. Thus τ labels distinct complex structures on a two-torus. We fix the orientation of the two-torus by choosing $\mathrm{Im}\,\tau > 0$. On the upper halfplane $\mathcal{H} = \{\tau \in \mathbb{C} | \mathrm{Im}\,\tau > 0\}$ we have an action of the group $SL(2, \mathbb{R})$ by

$$\begin{pmatrix} \omega_2 \\ \omega_1 \end{pmatrix} \to \begin{pmatrix} a & b \\ c & d \end{pmatrix} \begin{pmatrix} \omega_2 \\ \omega_1 \end{pmatrix}, \tag{13.236}$$

equivalently,

$$\tau \to \frac{a\tau + b}{c\tau + d}, \quad \begin{pmatrix} a & b \\ c & d \end{pmatrix} \in SL(2, \mathbb{R}). \tag{13.237}$$

The subgroup $SL(2, \mathbb{Z}) \subset SL(2, \mathbb{R})$ preserves the lattice Λ and thus the corresponding two-tori are equivalent as complex manifolds. This is the *modular group* of the two-torus, which now acts on a space-time torus rather than a world-sheet torus. We will see that the space-time modular group is part of the T-duality group, which explains why T-duality is also called *target space modular symmetry*.

---

[31] To obtain a smooth manifold, one needs to exclude the fixed points of the action of the T-duality group.

The relation between the modular parameter and the metric $G = G_{ij}dx^i dx^j$ of the torus is[32]

$$\tau = \tau_1 + i\tau_2 = \frac{G_{12}}{G_{22}} + i\frac{\sqrt{\det G}}{G_{22}} \Leftrightarrow G = \frac{\rho_2}{\tau_2}\begin{pmatrix} |\tau|^2 & \tau_1 \\ \tau_1 & 1 \end{pmatrix}, \quad \text{where} \quad \rho_2 = \sqrt{\det G}.$$

$$(13.238)$$

To understand this relation it is useful to define the two-torus by the standard lattice $\Lambda = \mathbb{Z}^2$, and to choose the metric

$$ds^2 = \frac{\rho_2}{\tau_2}|\tau dx + dy|^2 = G_{ij}dx^i dx^j, \quad x^1 = x, x^2 = y \tag{13.239}$$

on $\mathbb{C}$, which then induces a metric on the two-torus. We will now introduce some more concepts in order to explore the relation between metric and complex structure. An *almost complex manifold* is an even-dimensional real manifold equipped with a $(1, 1)$ tensor field J with the property $J^2 = -\mathrm{Id}$. This tensor is called an *almost complex structure* and its action on tensors corresponds to the multiplication with the imaginary unit i. If J satisfies an additional integrability condition, then one can construct holomorphic coordinate systems and the almost complex manifold is a complex manifold. Imposing compatibility between a complex structure and a metric leads to the concept of a Hermitian manifold. A complex manifold is called a *Hermitian manifold* if it carries a Riemannian metric $G = G_{ij}dx^i dx^j$ which is invariant under the action of the complex structure, $G_{ij} = G_{kl}J^k_i J^l_j$. Combined with the defining property $J^i_k J^k_j = -\delta^i_j$ of a complex structure this implies that $G_{jk}J^k_i = -G_{ik}J^k_j$, and therefore the tensor $\omega_{ij} := G_{ik}J^k_j$ is antisymmetric. Since ω_{ij} is also invertible, the *fundamental two-form* $\omega = \frac{1}{2}\omega_{ij}dx^i \wedge dx^j$ has maximal rank and defines an *almost symplectic* structure. On a Hermitian manifold any two of the three data G, J, ω determine the third.

Exercise 13.7.7 The metric (13.239) on $\mathbb{C}$ defines a metric on the two-torus $\mathbb{R}^2/\mathbb{Z}^2$. Similarly $\omega = \frac{1}{2}\omega_{ij}dx^i \wedge dx^j = \rho_2 dx \wedge dy$ defines an almost symplectic structure. Find the complex structure $J(\tau)$ which makes the two-torus a Hermitian manifold, and determine the value of τ for which this is the standard complex structure on $\mathbb{C}$ (which acts by multiplication with the imaginary unit i).

So far, we have expressed the data (G, ω, J) in terms of real coordinates x^i (for the specific case of a two torus the real coordinates are $x^1 = x, x^2 = y$). On a Hermitian manifold of dimension $2n$ one can find complex coordinates z^i, $i = 1, \ldots, n$ which are adapted to the complex structure. In such a coordinate system

$$J = \begin{pmatrix} i\,\mathbb{1}_n & 0 \\ 0 & -i\,\mathbb{1}_n \end{pmatrix}, \tag{13.240}$$

and the metric and fundamental form take the form

$$ds^2 = G_{i\bar{j}}dz^i d\bar{z}^j, \quad \omega = \frac{i}{2}G_{i\bar{j}}dz^i \wedge d\bar{z}^j. \tag{13.241}$$

On the two-torus with modular parameter τ the adapted complex coordinate is $z := x + \bar{\tau}^{-1}y$, and the metric takes the form $ds^2 = \frac{\rho_2}{\tau_2}|\tau|^2 dz d\bar{z}$. For $\tau = i$ we recover

---

[32] In the following, we use the same symbol for rank 2 tensors and the matrices which represent them in the chosen coordinate system.

the standard complex structure where $z = x + iy$ and the corresponding metric is proportional to the standard metric, $ds^2 = \rho_2 dz d\bar{z}$.

The two-torus and its moduli space have further properties. We now define the relevant concepts. An almost symplectic structure is called a *symplectic structure* if the fundamental two-form is closed:

$$d\omega = 0 \Leftrightarrow \partial_{[i}\omega_{jk]} = 0. \tag{13.242}$$

A *Kähler manifold* is a Hermitian manifolds which is symplectic, that is, the fundamental form is closed, and then is called the *Kähler form*. The closedness of the Kähler form implies a local integrability condition which allows one to write the metric coefficients as the mixed derivatives of a real analytic function, called the *Kähler pontential* $K(z^i, \bar{z}^{\bar{j}})$:[33]

$$G_{i\bar{j}} = \frac{\partial^2}{\partial z^i \partial \bar{z}^{\bar{j}}} K. \tag{13.243}$$

If the metric is flat (as is the case for the two-torus), we can choose coordinates such that the metric coefficients $G_{i\bar{j}}$ are constant. In this case, the Kähler potential simply is

$$K = G_{i\bar{j}} z^i \bar{z}^{\bar{j}}. \tag{13.244}$$

For a real two-dimensional manifold the closure of ω holds automatically, and in holomorphic coordinates the metric has a single independent component $G_{1\bar{1}}$. The Kähler potential for the flat Kähler metric is $K = \frac{\rho_2}{\tau_2}|\tau|^2 z\bar{z}$.

So far, we have seen that flat metrics on a two-torus are Kähler metrics, and that the three real moduli G_{ij} encode the complex structure and the volume. In string theory we also have a fourth modulus, resulting from the B-field. We can combine the B-field with the volume of the torus into a second complex modulus

$$\rho = B_{12} + i\sqrt{\det G}. \tag{13.245}$$

This can be viewed as a complexification of the Kähler form, $B + i\omega$. Since $\rho_2 = \sqrt{\det G} > 0$, the parameter ρ takes values on the upper half plane. In fact, the situation is completely symmetric between τ and ρ in that the T-duality group $O(2,2,\mathbb{Z})$ contains a subgroup $SL(2,\mathbb{Z})_\rho$ which acts as a modular group on ρ.

This also reflects itself in that the complex moduli τ and ρ appear symmetrically in the effective action. By rewriting the moduli terms in the effective action one can show that

$$\mathcal{L}_{\text{moduli}} \cong -\frac{1}{4} G^{ij} G^{kl} \left(\partial_\mu G_{ik} \partial^\mu G_{jl} + \partial_\mu B_{ik} \partial^\mu B_{jl} \right) = -\frac{1}{2} \frac{\partial_\mu \tau \partial^\mu \bar{\tau}}{(\text{Im}\tau)^2} - \frac{1}{2} \frac{\partial_\mu \rho \partial^\mu \bar{\rho}}{(\text{Im}\rho)^2}. \tag{13.246}$$

Exercise 13.7.8   As a quick check, work out the B-field term in (13.246) and compare to the rhs.

The fields τ, ρ take values in the upper half plane, and the corresponding terms in the action are non-linear sigma models with values in the symmetric space

---

[33] This is a refinement of the Poincaré lemma which uses that on Kähler manifolds the exterior derivative d decomposes into a holomorphic and antiholomorphic part, $d = \partial + \bar{\partial}$, where $\partial^2 = \bar{\partial}^2 = \partial\bar{\partial} + \bar{\partial}\partial = 0$.

$$\mathcal{H} \cong \frac{SL(2, \mathbb{R})}{SO(2)}, \tag{13.247}$$

which is a Kähler manifold.

Exercise 13.7.9 Show that $K = -\log[i(\tau - \bar{\tau})]$ is a Kähler potential for the metric $ds^2 = \frac{d\tau d\bar{\tau}}{4(\text{Im}\tau)^2}$ on $\mathcal{H} = SL(2, \mathbb{R})/SO(2)$.

In applications one often uses other, equivalent realisations of this symmetric Kähler space. For example

$$\mathcal{H} \cong \frac{SL(2\mathbb{R})}{SO(2)} \cong \frac{SU(1, 1)}{U(1)} \cong D, \tag{13.248}$$

where D is the unit disk. The map between $\mathcal{H}$ and D is the conformal map which maps the real line to the unit circle. It belongs to the Moebius group $SL(2, \mathbb{C})$ which allows one to map any three points in the complex plane to any other three points. Each configuration of three points uniquely specifies a circle or a line (which can be viewed as a 'generalised circle through ∞').

The equations of motion of the effective field theory are invariant under the action of the continuous group $SL(2, \mathbb{R})_\tau \otimes SL(2, \mathbb{R})_\rho$. In the full string theory this is broken to the discrete subgroup $SL(2, \mathbb{Z})_\tau \otimes SL(2, \mathbb{Z})_\rho$ which leaves the massive spectrum and interactions invariant.

Taking this into account, the moduli space becomes

$$\mathcal{M} \cong (SL(2, \mathbb{Z})\backslash SL(2, \mathbb{R})/SO(2))^2. \tag{13.249}$$

Howeover, the full T-duality group contains further elements. In particular, there is a symmetry under the exchange of the complex structure modulus τ and the (complexified) Kähler modulus ρ

$$M : \tau \leftrightarrow \rho. \tag{13.250}$$

This is a simple example of what is called *mirror symmetry*, which plays an important role for Calabi–Yau manifolds. A *Calabi–Yau manifold* is a compact complex manifold which admits a Ricci flat Kähler metric. In complex dimension one Calabi–Yau manifolds are two-tori. The generic Calabi–Yau twofolds (where we discard the special case of four-tori) are the K3-surfaces. These are complex surfaces which all have the same topology, and which form a single connected moduli space. In contrast, there is a huge number of Calabi–Yau threefolds with different topologies, each with its own moduli space. According to a conjecture known as *Reid's fantasy* all these moduli space intersect at special points and form a single connected web, similar to the circle and orbicircle line for one-dimensional compactifications. For Calabi–Yau threefolds the mirror symmetry conjecture becomes highly non-trivial because it relates manifolds X, Y which have different topology, while the complex structure moduli space of X is identical to the moduli space of complexified Kähler structures on Y, and vice versa. Mirror symmetry is far from obvious from the geometric point of view, but it is a prediction of string theory where mirror symmetry is manifest for a class of internal world-sheet CFTs as the invariance under the sign flip of charges with respect to an extended chiral algebra. The *SYZ conjecture* (formulated by Strominger, Yau, Zaslov) claims that mirror symmetry for Calabi–Yau threefolds quite literally is T-duality, in the sense that it can be formulated as a fibre-wise

Table 13.4. Enhanced symmetry points for two-torus compactifications	
Moduli	Gauge group
τ, ρ generic	$U(1)_L^2 \times U(1)_R^2$
$\tau = \rho$	$U(1)_L \times SU(2)_L \times U(1)_R \times SU(2)_R$
$\tau = \rho = i$	$SU(2)_L^2 \times SU(2)_R^2$
$\tau = \rho = \frac{1}{2} + i\frac{\sqrt{3}}{2}$	$SU(3)_L \times SU(3)_R$

T-duality acting on three-tori which represent three of the six directions. For a two-torus mirror symmetry simply is one of the factorised T-dualities. This is seen most easily by setting $B_{12} = G_{12} = 0$ where it acts as $G_{22} \to \frac{1}{G_{22}}$.

The full T-duality group of the two-torus contains one further generator, the reflection $X^1 \to -X^1$, which acts by

$$R : (\tau, \rho) \to (-\bar{\rho}, -\bar{\tau}). \tag{13.251}$$

The full T-duality group is

$$O(2, 2, \mathbb{Z}) = (SL(2, \mathbb{Z})_\tau \otimes SL(2, \mathbb{Z})_\rho) \otimes_{s.d.} (\mathbb{Z}_2^M \otimes \mathbb{Z}_2^R), \tag{13.252}$$

where $\otimes_{s.d.}$ indicates that the product is semi-direct, and where $\mathbb{Z}_2^M, \mathbb{Z}_2^R$ are the groups generated by the transformations M and R. The full moduli space can be written in the form

$$\mathcal{M} = (SL(2, \mathbb{Z}) \backslash SL(2, \mathbb{R}) / SO(2) \otimes SL(2, \mathbb{Z}) \backslash SL(2, \mathbb{R}) / SO(2)) / (\mathbb{Z}_2^M \times \mathbb{Z}_2^R). \tag{13.253}$$

Finally, compactifications on two-tori have one further $\mathbb{Z}_2$-symmetry which is not contained in the T-duality group, namely the world-sheet parity transformaton $\sigma^1 \to -\sigma^1$ which acts on the background as $B \to -B$, or

$$P : (\tau, \rho) \to (\tau, -\bar{\rho}). \tag{13.254}$$

This transformation changes the sign of $\mathbf{k}_L^2 - \mathbf{k}_R^2$, which makes it an *anti-isometry*, which therefore is not contained in $O(2, 2, \mathbb{Z})$. The full discrete symmetry group can already be generated by four elements, for examples the generators S, T of one of the $SL(2, \mathbb{Z})$ groups together with mirror symmetry M and world-sheet parity P.

Let us briefly review at which points of the moduli space the gauge symmetry is enhanced. Non-abelian symmetry enhancement occurs at fixed points of the action of the T-duality group. The points i and $\frac{1}{2} + i\frac{\sqrt{3}}{2}$ are the inequivalent finite order fixed points of the action of $SL(2, \mathbb{Z})$ on the modular domain. Mirror symmetry has the additional fixed line $\tau = \rho$. This resulting pattern of symmetry enhancement is summarised in Table 13.4.

The modular groups $SL(2, \mathbb{Z})_{\tau, \rho}$ also have an infinite order fixed point $i\infty$ under the generator T (that is $\tau \to \tau + 1$ or $\rho \to \rho + 1$). These fixed points correspond to decompactification limits where infinitely many states become massless.

Exercise 13.7.10　Find the T-duality group elements for which enhancement points are fixed points in terms of $SL(2, \mathbb{Z})$ group generators S, T. Compare the order of these group elements (the power which gives the unit element) to the number of extra massless states.

Exercise 13.7.11   Show that for $\rho = \tau \to i\infty$ the decompactification limit is the self-dual circle with gauge group $SU(2)_L \times SU(2)_R$.

13.7.6 The Buscher Rules

T-duality can be also be derived from the world-sheet point of view, where it works similarly to the electric-magnetic duality, or Hodge-duality discussed in Section 9.2.2. Consider the general string sigma model with background fields $G_{\mu\nu}, B_{\mu\nu}, \phi$. It is not necessary to assume that the background contains a factor which is a torus, just that it is invariant under a one-parameter group of transformations generated by a vector field ξ^μ. Then we can choose coordinates such that the background is independent of one of the coordinates, say X^1. For simplicity, we will consider the most basic example, where the metric and B-field do not have cross terms: $G_{1j} = 0, B_{1j} = 0$ for $j \neq 1$, and where the dilaton is trivial, $\phi = 0$. Then we can split off the terms in the world-sheet action which involve X^1:

$$S[X^1] = -\frac{1}{4\pi\alpha'} \int d^2\sigma \, G_{11}(X^2, \ldots) \partial_+ X^1 \partial_- X^1, \qquad (13.255)$$

where we have imposed the conformal gauge and use light-cone coordinates on the world-sheet. From the world-sheet point of view, the target-space isometry $X^1 \to X^1 + \xi^1$ is a global symmetry. We now *gauge* this global symmetry, that is we promote it to a local symmetry $X^1 \to X^1 + \xi^1(\sigma)$. The action (13.255) is only invariant under global shifts of X^1, but we can obtain a locally invariant action through replacing the partial derivatives $\partial_\pm$ by covariant derivatives $D_\pm$, where

$$D_\pm X^1 = \partial_\pm X^1 + A_\pm, \qquad (13.256)$$

and where the gauge field $A_\pm$ transforms as

$$A_\pm \to A_\pm - \partial_\pm \xi, \qquad (13.257)$$

so that $D_\pm X^1 \to D_\pm X^1$. The new action is locally equivalent to (13.255) if we impose that the connection is flat, $F_\pm = \partial_+ A_- - \partial_- A_+ = 0$, in which case we can at least locally impose the gauge $A_\pm = 0$.[34] We can impose this condition as an Euler–Lagrange equation by adding it to the Lagrangian using a Lagrange multiplier $\tilde{X}^1$:[35]

$$S = -\frac{1}{2\pi} \int d^2\sigma \left(\frac{1}{2\alpha'} G_{11} D_+ X^1 D_- X^1 + \tilde{X}^1 F_{+-} \right). \qquad (13.258)$$

Instead of going back to (13.255), we can eliminate the gauge field $A_\pm$ by its algebraic equation of motion,

$$A_\pm = -\partial_\pm X^1 \pm \frac{2\alpha'}{G_{11}} \partial_\pm \tilde{X}^1. \qquad (13.259)$$

Substituting this back we obtain the dual action

$$\tilde{S} = -\frac{1}{2\pi} \int d^2\sigma \frac{2\alpha'}{G_{11}} \partial_+ \tilde{X}^1 \partial_- \tilde{X}^1 = -\frac{1}{2\pi} \int d^2\sigma \frac{\tilde{G}_{11}}{2\alpha'} \partial_+ \tilde{X}^1 \partial_- \tilde{X}^1. \qquad (13.260)$$

[34] Our analysis is only local. See, e.g., Rocek and Verlinde (1992) for a discussion of global aspects.
[35] The relative normalisation of the two terms has been chosen for convenience.

Thus, for the dual string coordinate $\tilde{X}^1$ the metric is inverted in string units. This is precisely T-duality for the circle, and if we take a decompactification limit we can T-dualise along a non-compact isometric direction as well.

We have carried along the constant α' to make explicit that the world-sheet coupling is inverted. Thus, T-duality is perturbative with respect to space-time (the string coupling is invariant, see (13.264)), but a strong-weak coupling duality (an 'S-duality') from the world-sheet point of view. Technically, the process of dualisation is similar to the one for Maxwell-like actions for antisymmetric tensor gauge fields of any rank, with the difference that the Lagrange multiplier is used to impose the flatness of the connection and not the Bianchi identity. Compare Exercise 9.2.2.

Our example admits the following generalisations. First, one can relax the condition that the metric and B-field factorise. This leads to the so-called *Buscher rules*, where we now set $\alpha' = \frac{1}{2}$:

$$\tilde{G}_{11} = \frac{1}{G_{11}}, \quad \tilde{G}_{1j} = \frac{B_{1j}}{G_{11}}, \quad \tilde{B}_{1j} = \frac{G_{1j}}{G_{11}}, \tag{13.261}$$

$$\tilde{G}_{ij} = G_{ij} - \frac{G_{1i}G_{j1} + B_{i1}B_{j1}}{G_{11}}, \quad \tilde{B}_{ij} = B_{ij} - \frac{G_{i1}B_{j1} + B_{i1}G_{j1}}{G_{11}}. \tag{13.262}$$

Observe the characteristic mixing of metric and B-field. Moreover, the dilaton transforms as

$$\tilde{\phi} = \phi - \frac{1}{2}\log G_{11}. \tag{13.263}$$

In the world-sheet theory the transformation behaviour of the dilaton is a one-loop effect. From the target space perspective we know that upon dimensional reduction we need to redefine the higher-dimensional dilaton by shifting it by a term proportional to the Kaluza–Klein scalar in order to bring the action to the standard form,

$$e^{-2\bar{\phi}} = e^{-2\phi+\sigma} = e^{-2\phi}\sqrt{G_{11}}. \tag{13.264}$$

The transformation rule (13.263) is equivalent to the statement that the lower-dimensional dilaton $\bar{\phi}$ is T-duality invariant.

13.8 The Vacuum Selection Problem

13.8.1 Beyond Toroidal Compactifications

Toroidal compactifications allow us to obtain four-dimensional string theories. However, the states of the *bosonic string theory* we have considered so far fall into tensor representations of the Poincaré group and thus are bosonic states, which account for the graviton, gauge bosons, and scalars, including Higgs bosons. To describe matter we need fermionic states, which transform as space-time spinors. In Chapter 14, we will see how this is achieved in *superstring theories*. Apart from this, toroidal compactifications have a highly degenerate ground state, parametrised by scalar vacuum expectation values, a moduli space of vacua. At generic points in this moduli space gauge symmetry is abelian and massless matter is neutral, though we have special subspaces with non-abelian gauge symmetry and charged matter.

One way to obtain spectra with a lower number of massless states is to apply the orbifold construction we have studied for one-dimensional compactifications. In toroidal compactifications one divides out by discrete rotations $\theta \in O(d)$, $\theta^N = 1$, or, more generally, by discrete, possibly non-abelian subgroups of the Euclidean group. Transformations which act with fixed points will introduce singularities in the target space. For example, if we act with a finite order rotation $\theta \in O(2)$, $\theta^N = 1$ on $\mathbb{R}^2$, space around the origin will look like a cone with its tip at the origin. While this would make a theory of point particles ill-defined, string theory on toroidal orbifolds is consistent, because modular invariance implies the existence of twisted sectors. The twisted sector states are located at the singularities and screen them in the sense that if we probe the geometry with strings all amplitudes are finite.

Since the background must be consistent with the group action we mod out, orbifolds also have a reduced number of moduli, or to be precise, of untwisted moduli. The twisted sectors can contribute new integrable marginal operators, and the corresponding deformation parameters are called *twisted moduli*. Some of these deformations correspond to a smoothing of the orbifold, that is, to deformations into smooth Ricci-flat manifolds. Conversely orbifolds can be seen as singular limits of smooth manifolds where all the curvature is concentrated at the fixed points.

This brings us to compactifications of the form $M_4 \times X$, where M_4 is four-dimensional Minkowski space and where X is a generic compact Ricci-flat space. Here the partition function, amplitude, and spectrum can, in general, not be computed exactly any more. The generic massless spectrum can be obtained by reduction of the effective field theory and is determined by topological data of X. The decomposition of the higher-dimensional metric can give rise to massless vector and scalar fields. By analyzing the decomposition of the field equations one finds that massless vector fields are in one to one correspondence with Killing vector fields on X, that is with vector fields which generate symmetries of metric (isometries). The resulting gauge group is isomorphic to the isometry group of X. Massless scalars are in one-to-one correspondence with solutions of a non-linear Laplace-like equation, and parametrise the moduli space of Ricci flat metrics. The B-field can also give rise to vector and scalar fields, which are in one-to-one correspondence to solutions of Laplace equations for antisymmetric tensors (differential forms). While not much more can be said for generic X, one characteristic feature is that generically there is no unique vacuum, but a moduli space of vacua.

13.8.2 From Moduli Spaces to the Landscape

While moduli spaces of vacua allow one to investigate many interesting physical and mathematical problems analytically, they are ultimately unrealistic. In fact, without supersymmetry, quantum corrections will in general lift such degeneracies, but then one faces the problem of *moduli stabilisation*, because the resulting scalar potential generally has run-away directions rather than stable and physically interesting stable or meta-stable vacua. The problem of moduli stabilisation can be addressed by moving to a wider class of compactifications, where a potential is already generated at the classical level by switching on non-trivial matter fields along the internal manifold X. Through the Einstein equations $R_{\mu\nu} - \frac{1}{2}Rg_{\mu\nu} = \kappa^2 T_{\mu\nu}$ this implies that the metric on X is no longer Ricci-flat, and one generically generates a potential for some of the moduli. One class of constructions are *flux compactifications*, where the background contains a gauge field with a non-vanishing covariantly constant field strength. For example, one could switch on a non-vanishing, but covariantly constant field strength H for the B-field. Through the field equations the flux will change the admissible choices for other background fields, in particular for the metric. One can also introduce D-branes into the background, which act as sources for closed string background fields and contribute an open string sector to the spectrum. There are various further generalisations, of which we only list a few: one can relax the assumption that the non-compact and compact parts of space-time form a metric product and consider *warped products*, where the metric of the internal space varies over space-time. And one can allow field configurations which along non-contractible closed curves X only close up to symmetry transformations. *Scherk–Schwarz compactifications* are an example where the symmetries one uses are the shift symmetries of axion-like scalars. This can be regarded as a special case of a *T-fold*, where space-times are glued using T-duality transformations.

In the context of superstring theories there are several quite involved constructions, which manage to fix the moduli and have stable or semi-stable vacua with 'semi-realistic' particle spectra. While lack of space prevents us from going into the details, let us talk about the main features. What has essentially been achieved is to replace a moduli space of vacua by a landscape of hills and valleys, with the minima being stable or at least sufficiently long-lived meta-stable vacua. This raises the question: is there one vacuum which corresponds to the observed universe, and if yes, how does it get selected? For the first question one needs to satisfy more conditions than reproducing the known particle spectrum of the standard model plus making predictions which can be tested in future particle physics experiment. In particular, one needs to also have a model for cosmology which includes inflation (or otherwise accounts for the physics of the early universe), and which contains a small positive cosmological constant (or otherwise accounts for the late time behaviour of the universe). The number of string vacua, for which only estimates exist, is very large, which does not only make it difficult to find the 'right one' but also raises the question to which extent string theory has the predictive power expected of a good theory. This depends on the expectations one has regarding the explanatory power of theories. A conservative point of view is that which solution of a theory is realised is determined by initial conditions, so that the only relevant problem is to find a solution which fits the data. The other extreme is to hope for a theory which predicts its own initial condition and has a unique solution corresponding to our universe. If one accepts the existence of a huge landscape of string vacua, one has to abandon uniqueness. As a consequence, various versions of *anthropic reasoning* have been advocated, which draw on the fact that a universe admitting complex systems and thus life seems to need rather special values of the fundamental physical constants and parameters. This is then combined with the idea of a *multiverse*, consisting of 'pocket universes' which populate the string landscape. One may then envisage that in an overall inflating universe, there are regions where inflation stops, with moduli fixed to select a particular string vacuum. It remains an open question whether there is a dynamical selection effect with a preference for 'universes like ours'. One approach is to apply statistical reasoning, which assumes that it is meaningful to talk about ensembles of universes and to assign probabilities to particular types of universes being realised. Not surprisingly, this has lead to controversial, sometimes heated debates about the boundaries of scientific methodology.

Without entering into the debate I would like to remark that the problem is not a problem of string theory as such, though string theory brings it into focus. If one remains within the framework of quantum field theories (considered either as fundamental or effective), one either simply does not ask the question, or phrases it as the question of selecting the 'right theory'. Note that the term 'theory' can refer to both a general framework (such as quantum field theory) and to a specific model (such as the standard model), and what we have to select is the right 'specific theory' or 'model'. In our current understanding string theory is a single theory with a vast landscape of solutions, and the problem is to choose the 'right solution' of this unique theory. While, in both cases, the question is how our to select our universe among other 'possible worlds', in string theory this becomes a question of dynamics, which makes it harder to avoid the question, but also raises the perspective that there might be an answer.

13.8.3 The Landscape and the Swampland

In the past years, a lot of effort has been put into understanding how generic string vacua look like, and how they can be distinguished from effective field theories that do not have an ultraviolet completion. This is often phrased as separating the landscape of effective field theories which admit a consistent UV completion from the swampland of theories which are consistent as effective field theories, but cannot be consistently coupled to quantum gravity. Within this swampland programme, string theory is used as a concrete model for quantum gravity, and the landscape of string vacua as a concrete proposal for effective field theories that admit an UV completion. This programme has lead to a variety of interesting conjectures, at least part of which are mathematically precise and can be tested against our knowledge of the string landscape. Among other things, this leads to interesting insights into the qualitative properties of effective field theories resulting from string theory. For example, one persistent feature is that it seems to be extremely hard if not impossible to have stable de Sitter vacua in string theory, which forces one to either look for other ways to realise inflation or to find alternative explanations for cosmological data.

Example: The Swampland Distance Conjecture

As a taster, we give a somewhat simplified account of the so-called *swampland distance conjecture* and ask the reader to verify that it holds in the simple case of circle compactifications.[36] The conjecture concerns effective theories with a moduli space M parametrised by scalar vacuum expectation values. It claims that for any given point $p_0 \in M$ and any given distance $T > 0$ there is a point $p \in M$ such that

---

[36] This example is contained, together with other instructive examples, in Ooguri and Vafa (2007).

the distance beween p, p_0 is larger than T. Here, distance is defined by the kinetic energy term of the moduli fields. Moreover, for sufficiently large T the effective field theory breaks down in a particular way, namely through the presence of infinitely many exponentially light states with masses $M \propto e^{-\alpha T}$, where $\alpha > 0$ is a (non-universal) constant. Thus when moving towards the boundary of moduli space, one encounters either a decompactification limit (as in Exercises 13.8.1 to 13.8.3) or another modification of low energy physics which involves infinitely many states becoming light (for example, a tensionless string).

In the context of circle compactifications, the points at infinite distance are the decompactification limits $R \to \infty$ and $R \to 0$, and the associated light states are the momentum and winding states around the circle. To make this precise, we need to determine the metric on the moduli space so that we can compute the value of the parameter α.

Exercise 13.8.1 Show that the metric on the moduli space $\{R \in \mathbb{R} | 0 < R < \infty\}$ of circle compactifications is[37]

$$ds^2 = \frac{dR^2}{R^2}. \tag{13.265}$$

Compute the distance between a fixed, but arbitrary point R_0 and a point corresponding to radius R.

Exercise 13.8.2 Show that if we reduce on a circle, the higher-dimensional and the lower-dimensional mass differ by a specific power of the volume of the circle, which can be expressed in terms of R. *Instruction:* When we dimensionally reduce an effective action from $d + 1$ to d dimension, this generates a factor of e^σ in front of the gravitational term:

$$S_d \propto \int d^d x \sqrt{\bar{G}} e^\sigma R_{\bar{G}} + \cdots \tag{13.266}$$

For simplification, we have set the higher-dimensional dilaton to zero, thus setting equal the higher-dimensional Planck and string scale, so that we can focus on the effect of the Kaluza–Klein scalar. By transforming the action (13.266) to the standard Einstein–Hilbert form, you can work out the factors by which the metric, lengths and masses get rescaled. Give your answer in the form $M_{(d)} = R^\gamma M_{(d+1)}$, where $M_{(n)}, n = d, d + 1$ is the mass scale of the n-dimensional theory.

Exercise 13.8.3 When we reduce a $(d + 1)$-dimensional string theory on a circle to d dimensions, then the $(d + 1)$-dimensional masses of momentum and winding states depend on the radius R like $M_{(d+1)} \propto 1/R$ and $M_{(d+1)} \propto R$, respectively. Use your results from Exercise 13.8.2 to find the R-dependence of the d-dimensional masses. Then show that in the limits $R \to \infty$ and $R \to 0$ the d-dimensional masses behave like $e^{-\alpha T}$, where $T \to \infty$ is the distance measured from a fixed, but arbitrary interior point, and where α is a (non-universal) constant. Hence, show that there is an infinite tower of states that become massless at exponential rate measured in terms of the distance in the two decompactification limits.

---

[37] This is actually a double cover of the moduli space, which extends to zero radius. It is often more convenient to work on such a larger space.

13.9 Literature

More on $c = 1$ conformal field theories and their relation to circle and orbicircle compactifications can be found in Dijkgraaf et al. (1988), Ginsparg (1988), and the relevant chapters in Polchinski (1998a) and Becker et al. (2007) which provide complementary reading for this chapter. A good reference for the vertex operator construction of representations of Kac–Moody algebras is Goddard and Olive (1986). More on Kac–Moody algebras and their representation theory can be found in Fuchs and Schweigert (1997), and, for the mathematically inclined, in Kac (1990) and Frenkel and Ben-Zvi (2001). The proper treatment of the zero modes of vertex operators in toroidal compactifications is related to a group extension problem. For general background on Lie group and Lie algebra cohomology (which besides extension problems is useful for many other things, including anomalies), see de Azcarraga and Izquierdo (1995). Our non-standard treatment of the zero modes, which avoids the explicit introduction of 'cocycle factors', is based on Sakamoto (1989). For the standard treatment, see Goddard and Olive (1986) (and also Green et al. (1987)). See Narain (1986) and Narain et al. (1987) for toroidal compactifications of string theory. A comprehensive review of T-duality is found in Giveon et al. (1994). T-duality for open strings and D-branes is reviewed in Polchinski (1996). The $O(d, d)$-covariant actions for toroidal compactifications were adapted from Maharana and Schwarz (1993). See also Duff et al. (1986); Scherk and Schwarz (1979), and the textbooks Ortin (2004) and Kiritsis (2007) for details about the compactification of field theories. The role of self-dual lattices in string theory is reviewed in Lerche et al. (1989). Loop corrections to string effective actions were studied in Kaplunovsky and Louis (1994), Kaplunovsky and Louis (1995). String orbifold compactifications were introduced in Dixon et al. (1985), Dixon et al. (1986), and the CFT aspects were developed in Dixon et al. (1987), Hamidi and Vafa (1987). Instead of compactifying ten-dimensional string theories, one can construct modular invariant string theories directly in four dimensions. In the so-called free fermionic construction all internal degrees of freedom are realised by world-sheet fermions, see, for example, Antoniadis et al. (1987), Kawai et al. (1987). See also Athanasopoulos et al. (2016) for the relation between the free fermionic construction and orbifolds. For Calabi–Yau compactifications, see for example the lecture notes Candelas (1987), Greene (1996) and the book Hubsch (1991). Mirror symmetry is a huge subject at the interface of theoretical physics and mathematics, see, for example, the volumes Hori et al. (2003) and Aspinwall et al. (2009). The SYZ-conjecture, which regards mirror symmetry as a 'fibre-wise' version of T-duality, was formulated in Strominger et al. (1996). There are various approaches aiming at making T-duality and other string dualities manifest using new types of geometries and field theories. See, for example, Aldazabal et al. (2013), Plauschinn (2019), Hohm and Samtleben (2019). See van Beest et al. (2021) for a review of the Swampland programme. Our example for the Swampland distance conjecture is based on Ooguri and Vafa (2007).

14 Fermions and Supersymmetry

The distinction between bosons ('forces') and fermions ('matter') is fundamental to quantum field theory. The *spin-statistics theorem* states that the absence of a-causal, faster-than-light effects requires that fields with integer spin must be quantised using commutation relations, while fields with half-integer spin must be quantised using anticommutation relations.[1] The resulting *Bose–Einstein* and *Fermi–Dirac* statistics explain why multiparticle systems of bosons and fermions behave qualitatively differently. In particular, the fact that in fermionic systems each state can be occupied at most once is crucial for the *stability of matter* and the existence of stable complex structures in nature, such as ourselves.

The bosonic string theories we have studied so far only have states transforming in tensor representations of the Poincaré group, which are the higher-dimensional analogues of integer spin representations. In this chapter, we will explain how more general string theories can be constructed which contain matter, that is states transforming in spinor representations. We will use the *RNS (Ramond–Neveu–Schwarz) approach*, which allows one to maintain manifest Lorentz covariance.[2] This construction addresses the issue somewhat indirectly, since one adds degrees of freedom to the world-sheet action that are spinors with respect to the world-sheet, but Lorentz vectors with respect to space-time. The resulting world-sheet action is invariant under two-dimensional supersymmetry transformations. This is referred to as *world-sheet supersymmetry*. Space-time fermions emerge indirectly, through a choice of boundary conditions for the world-sheet fermions. Similar to orbifold constructions, boundary conditions cannot be chosen arbitrarily, but are highly restricted by modular invariance. This leaves one with a small number of consistent theories, including the five ten-dimensional *superstring theories*, which realise *supersymmetry* in space-time. The five ten-dimensional superstring theories are the starting points for *string phenomenology*, that is for the attempt to construct realistic models of particle physics and cosmology based on string theory. Since nature is not manifestly supersymmetric at the currently accessible energy scales, understanding supersymmetry breaking is one of its central problems.

To familiarise ourselves with supersymmetry, we will start with the supersymmetric generalisation of the harmonic oscillator. Since the states of the bosonic string are built out of harmonic oscillators, this can be viewed as asking how a single vibration mode of a string can be supersymmetrised.

---

[1] More precisely this holds up to to field redefinitions, as we have previously discussed in Chapter 13.
[2] The complementary 'classical' approach is the *GS (Green–Schwarz) superstring* where space-time supersymmetry is manifest but Lorentz covariance is not.

14.1 The Supersymmetric Harmonic Oscillator

14.1.1 The Simple Supersymmetric Harmonic Oscillator

Let us start by reviewing the simple harmonic oscillator. We set $\omega\hbar = 1$, and express the Hamiltonian in terms of the ladder operators, $a, a^\dagger$, which satisfy

$$[a, a^\dagger] = 1. \tag{14.1}$$

Then the Hamiltonian is

$$H_B = \frac{1}{2}\left(aa^\dagger + a^\dagger a\right) = \frac{1}{2}a^\dagger a + \frac{1}{2}, \tag{14.2}$$

where the constant shift arises from normal ordering. The ground state $|0\rangle$ is defined by $a|0\rangle = 0$, and an orthonormal basis of the Hilbert space is provided by the energy eigenstates

$$|m\rangle = \frac{(a^\dagger)^m}{\sqrt{m!}}|0\rangle, \quad m = 0, 1, 2, \dots \tag{14.3}$$

with energies

$$H_B|m\rangle = \left(m + \frac{1}{2}\right)|m\rangle. \tag{14.4}$$

The ground state energy is $E_0 = \frac{1}{2}$ (or $E_0 = \frac{1}{2}\hbar\omega$ if we reinstate the frequency).

We now introduce an extension by adding operators $d, d^\dagger$, which satisfy the anti-commutation relations

$$\{d, d^\dagger\} = 1, \quad \{d, d\} = \{d^\dagger, d^\dagger\} = 0. \tag{14.5}$$

If we define the ground state $|-\rangle$ by $d|-\rangle = 0$, then

$$d^\dagger|-\rangle =: |+\rangle, \quad d^\dagger|+\rangle = (d^\dagger)^2|-\rangle = 0. \tag{14.6}$$

Since the creation operators anticommute, we can only apply a creation operator once, and the Hilbert space is two-dimensional. Generalised to quantum field theory this has the effect that the occupation number of any single particle state is either zero or one: this is the Fermi–Dirac statistics. We define the Hamiltonian for the operators $d, d^\dagger$ by

$$H_F = \frac{1}{2}\left(dd^\dagger + d^\dagger d\right) = d^\dagger d - \frac{1}{2}. \tag{14.7}$$

The operators $d, d^\dagger$ can be realised as matrices,

$$d = \begin{pmatrix} 0 & 0 \\ 1 & 0 \end{pmatrix}, \quad d^\dagger = \begin{pmatrix} 0 & 1 \\ 0 & 0 \end{pmatrix} \Rightarrow d^\dagger d - \frac{1}{2} = \frac{1}{2}\begin{pmatrix} 1 & 0 \\ 0 & -1 \end{pmatrix} = \frac{1}{2}\sigma_3. \tag{14.8}$$

H_F can be interpreted as the Hamiltonian $\frac{1}{2}\mu B\sigma_3$ of a particle with spin $\frac{1}{2}$ and magnetic moment μ in a magnetic field B, where we have set $\mu B = 1$. In this context, the states $|\pm\rangle$ can be interpreted as spin up and spin down, respectively.

To combine the two systems, we postulate that the operators $a, a^\dagger$ commute with that the operators $d, d^\dagger$ and that the total Hamiltonian is

$$H = H_B + H_F = a^\dagger a + d^\dagger d. \tag{14.9}$$

We observe that the ground states energy is now zero due to mutual cancellation of the normal ordering constants. The Hamiltonian is invariant under exchanging $a, a^\dagger$ and $d, d^\dagger$, and to formulate the underlying symmetry transformation, we introduce two new quantities.

1. The fermion number operator $F := N_F = d^\dagger d$ counts the excitations created by $d^\dagger$:

$$[F, d^\dagger] = d^\dagger, \quad [F, d] = -d. \tag{14.10}$$

Its (± 1)-valued version $(-1)^F$ satisfies

$$\{(-1)^F, d^\dagger\} = 0 = \{(-1)^F, d\} \tag{14.11}$$

and allows one to introduce a $\mathbb{Z}_2$-grading on the combined Hilbert space $\mathcal{H}$ with basis $\mathcal{B} = \{|m\rangle \otimes |-\rangle, |m\rangle \otimes |+\rangle | m = 0, 1, 2, \ldots\}$. States $|B\rangle$ with eigenvalue $+1$ are called *bosonic states*, states $|F\rangle$ with eigenvalues -1 are called *fermionic states*.

$$(-1)^F |B\rangle = |B\rangle, \quad (-1)^F |F\rangle = -|F\rangle. \tag{14.12}$$

2. The *supercharge* or *supersymmetry generator* Q and its Hermitian conjugate $Q^\dagger$:

$$Q = \sqrt{2} a^\dagger d, \quad Q^\dagger = \sqrt{2} a d^\dagger. \tag{14.13}$$

The supercharges anticommute with $(-1)^F$,

$$\{Q, (-1)^F\} = 0 = \{Q^\dagger, (-1)^F\}, \tag{14.14}$$

and therefore map bosonic states to fermionic states and vice versa. Such operators are called *fermionic operators*, while operators which commute with $(-1)^F$ and therefore map bosons (fermions) to bosons (fermions) are called *bosonic operators*. The supercharges satisfy the *supersymmetry algebra*

$$\{Q, Q^\dagger\} = 2H, \quad \{Q, Q\} = 0 = \{Q^\dagger, Q^\dagger\}. \tag{14.15}$$

The supercharges commute with the the Hamiltonian:

$$[H, Q] = 0 = [H, Q^\dagger], \tag{14.16}$$

and the symmetry generated by them is called *supersymmetry*.

The Hilbert space $\mathcal{H}$ has a unique, bosonic, supersymmetry invariant ground state $|\Omega\rangle$ of energy $E_0 = 0$,

$$a|\Omega\rangle = d|\Omega\rangle = 0, \quad Q|\Omega\rangle = Q^\dagger |\Omega\rangle = H|\Omega\rangle = 0. \tag{14.17}$$

Excited states have positive energies $E > 0$ and organise into pairs of bosonic and fermionic states, which are related to one another by supersymmetry transformations.

Exercise 14.1.1   Verify the relations and statements made in this section. Where possible use the supersymmetry algebra (14.15), instead of the concrete realisation of the Hamiltonian and supercharges in terms of $a, a^\dagger, d, d^\dagger$.

It is useful to note that the degeneracy between bosonic and fermionic non-ground states implies that the trace of the operator $(-1)^F$ taken over non-ground states is zero. This motivates the definition of the modified partition function

$$Z = \text{Trace}\left((-1)^F e^{-\beta H}\right), \tag{14.18}$$

called the *Witten index*, which is an important tool for analysing supersymmetric theories.

14.1.2 Supersymmetrisation of the Light-Cone Hamiltonian

The light-cone Hamiltonian of the bosonic string (with the centre of mass momentum neglected), is the Hamiltonian for an infinite set of harmonic oscillators with frequencies $\omega_n = n, n = 1, 2, \ldots,$

$$H_B = \sum_{i=1}^{d} \sum_{n=1}^{\infty} \left(\alpha_{-n}^i \alpha_n^i + \frac{n}{2} \right)_\zeta = \sum_{i=1}^{d} \sum_{n=1}^{\infty} \left(n(a_n^i)^\dagger a_n^i + \frac{n}{2} \right)_\zeta. \tag{14.19}$$

The additional index $i = 1, \ldots, d$, where $d = D - 2$, reminds us that we have one independent oscillator for each transverse direction. The label ζ indicates that the divergent sum over the ground state energies has been regularised using the ζ-function method.

To generalise the construction of the previous section, we introduce fermionic creation and annihilation opertors d_n^i, $n \in \mathbb{Z}$, where $(d_n^i)^\dagger = d_{-n}^i$, $n \in \mathbb{Z}$ and i is an SO(d) index. We postulate the anticommutation relations

$$\{d_m^i, d_n^j\} = \delta^{ij} \delta_{m+n,0}, \tag{14.20}$$

and the Hamiltonians

$$H_F = \sum_{i=1}^{d} \sum_{n=1}^{\infty} \left(n d_{-n}^i d_n^i - \frac{n}{2} \right)_\zeta, \tag{14.21}$$

$$H = H_B + H_F = \sum_{i=1}^{d} \sum_{n=1}^{\infty} \left(n(a_n^i)^\dagger a_n^i + n(d_n^i)^\dagger d_n^i \right), \tag{14.22}$$

where we observe that the ground state energies cancel frequency by frequency, leading to a zero ground state energy.

The fermion number operator and supercharges generalise as

$$F = \sum_{i=1}^{d} (d_n^i)^\dagger d_n^i, \quad Q = \sqrt{2} \sum_{i=1}^{d} \sum_{n=1}^{\infty} (a_n^i)^\dagger d_n^i, \quad Q^\dagger = \sqrt{2} \sum_{i=1}^{d} \sum_{i=1}^{n} a_n^i (d_n^i)^\dagger. \tag{14.23}$$

The algebra (14.15) remains the same, but the properties of the ground state change. The operators d_0^i are Hermitian, and commute with the Hamiltonian:

$$[H, d_0^i] = 0. \tag{14.24}$$

If a state $|\Omega\rangle$ satisfies the defining properties of a ground state,

$$a_m^i |\Omega\rangle = d_m^i |\Omega\rangle = 0, \quad m = 1, 2, \ldots, i = 1, \ldots d, \tag{14.25}$$

then all the 2^d linearly independent states

$$|\Omega\rangle, d_0^i d_0^j |\Omega\rangle, \ i < j, \quad d_0^i d_0^j d_0^k |\Omega\rangle, \ i < j < k, \ldots d_0^1 \cdots d_0^d |\Omega\rangle \tag{14.26}$$

satisfy these conditions, that is, the ground state is degenerate. The anticommutation relations

$$\{d_0^i, d_0^j\} = \delta^{ij} \tag{14.27}$$

show that the operators d_0^i are proportional to the standard generators γ^i of the Clifford algebra Cl_d,

$$\{\gamma^i, \gamma^j\} = 2\delta^{ij}. \tag{14.28}$$

Clifford algebras have one (two) inequivalent irreducible representations for even (odd) d, which can be generated by Dirac γ-matrices, as indicated by our notation.[3]

A linear basis for the Clifford algebra is provided by the antisymmetrised products of the generators:

$$\mathbb{1}, \gamma^i, \gamma^{ij} := \gamma^{[i}\gamma^{j]} = \frac{1}{2}[\gamma^i, \gamma^j], \gamma^{ijk} := \gamma^{[i}\gamma^j\gamma^{k]}, \gamma^{[1...d]} := \gamma^1 \cdots \gamma^d. \tag{14.29}$$

The subset $M^{ij} = \frac{1}{2}\gamma^{ij}$ of generators is a basis for the Lie algebra $\mathfrak{so}(d)$.

Exercise 14.1.2   Show that the generators $M^{ij} = \frac{1}{2}\gamma^{ij} = -M^{ji}$ satisfy the Lie algebra relations for $\mathfrak{so}(d)$,

$$[M^{ij}, M^{kl}] = \delta^{kj}M^{il} - \delta^{ki}M^{jl} - \delta^{lj}M^{ik} + \delta^{li}M^{jk} = 4\delta^{[k[j}M^{i]l]}.$$

The Lie algebra $\mathfrak{so}(d)$ and the double cover Spin(d) of the Lie group SO(d) are naturally embedded into the Clifford algebra Cl_d. Representations of the Clifford algebra Cl_d become *spinor representations* of $\mathfrak{so}(d)$ and of SO(d) upon restriction. The degenerate ground state of the supersymmetric harmonic oscillator transforms as a spinor under the transverse rotation group SO(d). Moreover, since the normal ordering effects in the light-cone Hamiltonian cancel between bosons and fermions, the supersymmetrisation of the world-sheet action seems to allow us to construct a tachyon-free string theory with space-time fermions. In addition, we would like have massless bosonic states, including a photon and a graviton, and we would like to maintain manifest Lorentz covariance. Before we introduce world-sheet theories which achieve this, we need to discuss supersymmetry in a more general setting.

14.2 General Discussion of Supersymmetry

In supersymmetric field theories supersymmetry appears as an extension of the Poincaré Lie algebra into a Poincaré Lie superalgebra. In this section, we review the relevant background on Lie superalgebras, Clifford algebras, and spinors. We keep the signature (p, q), $p + q = D$ of the space-time metric arbitrary, so that we cover both the cases of Lorentz signature ($p = 1$) and of Euclidean signature ($p = 0$). The Euclidean case is relevant for the little groups.

The Clifford algebra $Cl_{p,q}$ has generators γ^μ, $\mu = 1, \cdots, p + q$, subject to the relations[4]

$$\{\gamma^\mu, \gamma^\nu\} = 2\eta^{\mu\nu}, \quad \eta^{\mu\nu} = \mathrm{diag}(\underbrace{-1, \ldots, -1}_{p}, \underbrace{1, \ldots, 1}_{q}). \tag{14.30}$$

---

[3] Explicit matrix realisations can be found in Van Proeyen (1999).
[4] For Lorentz signature where ($p = 1, q = D - 1$), one usually shifts the range of the indices to $\mu = 0, \ldots p + q - 1$.

The generators γ^μ can be realised as matrices, which for Lorentz signature are Dirac's γ-matrices. Explicit constructions in terms of Pauli matrices can be found, for arbitrary dimension and signature, in Van Proeyen (1999). The antisymmetrised products $\gamma^{[\mu}\gamma^{\nu]}$ of generators γ^μ are generators of the Lie algebra $\mathfrak{so}(p,q)$. Moreover, the group $Cl_{p,q}^*$ of invertible elements of $Cl_{p,q}$ has a subgroup $\mathrm{Spin}(p,q)$, which is a double cover of the orthogonal group $\mathrm{SO}(p,q)$. Therefore representations of Clifford algebras become spinor representations of $\mathrm{SO}(p,q)$ upon restriction.[5] In particular, *Dirac spinors* arise from restricting irreducible complex Clifford representations to $\mathrm{Spin}(p,q)$.[6]

In even dimensions, $Cl_{p,q}$ has a unique irreducible representation, but the associated spinor representation is reducible. The reason is that the product γ_* of all γ-matrices anticommutes with the Clifford generators γ^μ, but commutes with the Lorentz generators $\gamma^{[\mu}\gamma^{\nu]}$. When including a suitable phase factor it satisfies $(\gamma_*)^2 = \mathrm{Id}$, and provides a generalisation of the chirality or γ_5-matrix. One can then define projectors $\frac{1}{2}(\mathbb{1} \pm \gamma_*)$ which decompose Dirac spinors ψ into chiral or *Weyl spinors* $\psi_\pm$, which satify $\gamma_*\psi_\pm = \pm\psi_\pm$. In odd dimensions, γ_* commutes with the Clifford generators γ^μ, and therefore it is proportional to $\mathbb{1}$ on irreducible representations by Schur's lemma. The factor of proportionality distinguishes the two inequivalent irreducible representations of $Cl_{p,q}$, which are equivalent as $\mathrm{Spin}(p,q)$ representations.

There exist matrices A, B, C which relate the γ-matrices to the Hermitian conjugate, complex conjugate, and transposed matrices:

$$(\gamma^\mu)^\dagger = (-1)^t A\gamma^\mu A^{-1}, \quad (\gamma^\mu)^* = (-1)^t \tau B\gamma^\mu B^{-1}, \quad (\gamma^\mu)^T = \tau C\gamma^\mu C^{-1}, \quad (14.31)$$

where t is the number of time-like directions, and where $\tau = \pm1$. The charge conjugation matrix is either symmetric or antisymmetric, $C^T = \sigma C$, $\sigma = \pm1$. The values of σ, τ depend on dimension and signature, and can be found in Van Proeyen (1999), where the notation $\eta = -\tau$, $\epsilon = -\sigma$ is used. In odd space-time dimensions the matrices B, C are unique up to equivalence, while in even space-time dimensions there are two inequivalent choices, which are distinguished by the value of τ. In Lorentz signature $A \propto \gamma^0$ defines Dirac's Hermitian scalar product $\bar{\psi}\phi \propto \psi^\dagger A\phi$ between spinors. The matrix C is the *charge conjugation matrix* which relates fermionic particles to their antiparticles. The matrix B satisfies $BB^* = \pm\mathbb{1}$, and can be used to impose reality conditions on spinors. *Dirac spinors*, which are obtained by restricting irreducible *complex* Clifford representations to $\mathrm{Spin}(p,q)$ exist for all signatures. In signatures where $BB^* = \mathbb{1}$ one can define real or *Majorana spinors* by imposing

$$\psi^* = \alpha B\psi, \tag{14.32}$$

where α is a conventional phase factor, $|\alpha| = 1$. In signatures where $BB^* = -\mathbb{1}$ one can impose the reality condition

$$\psi^i = \alpha B\psi^j \epsilon_{ji}, \quad i,j = 1,2 \tag{14.33}$$

---

[5] Here 'spinor representations' refers to the analogues of spin 1/2 representations. Analogues of higher spin representations are obtained by tensoring these with tensor representations.

[6] Real representations and Majorana spinors will be explained later in this section.

on a pair of Dirac spinors. While this does not reduce the number of degrees of freedom compared to a single Dirac spinor, replacing a Dirac spinor by a pair of *symplectic Majorana spinors* is sometimes useful. Whether reality (Majorana) conditions and chirality (Weyl) conditions are compatible and thus can be imposed simultaneously depends on the signature. *Majorana–Weyl spinors* exist if and only if $p - q = 0$ modulo 8. For four-dimensional Lorentz signature $p = 1, q = 3$, the Majorana and Weyl condition are not compatible, but instead Majorana and Weyl spinors are equivalent as real Spin$(1, 3)$ representations, so that one can rewrite one type of spinor in terms of the other.

Lie superalgebras are generalisations of Lie algebras which involve both commutators and anticommutators. The underlying vector space $\mathfrak{g}$ is $\mathbb{Z}_2$-graded, $\mathfrak{g} = \mathfrak{g}_0 \oplus \mathfrak{g}_1$ and decomposes into an even or bosonic part $\mathfrak{g}_0$ and an odd or fermionic part $\mathfrak{g}_1$. The bracket is defined such that it respects the grading,

$$[\mathfrak{g}_0, \mathfrak{g}_0] \subset \mathfrak{g}_0, \quad [\mathfrak{g}_0, \mathfrak{g}_1] \subset \mathfrak{g}_1, \quad [\mathfrak{g}_1, \mathfrak{g}_1] \subset \mathfrak{g}_0. \tag{14.34}$$

The bracket on $\mathfrak{g}_0$ is antisymmetric, making it a Lie algebra. Mixed brackets are also antisymmetric, so that $\mathfrak{g}_1$ carries a representation of the Lie algebra $\mathfrak{g}_0$. Finally, the bracket on $\mathfrak{g}_1$ is symmetric. The bracket is required to satisfy a graded version of the Jacobi identity.

For *supersymmetry algebras*, or more precisely, *Poincaré Lie superalgebras*, the bosonic part $\mathfrak{g}_0$ is the Poincaré Lie algebra. The extension into a Lie superalgebra works as follows:[7]

1. The odd generators Q_α commute with the generators P_μ of translations and transform in a spinor representation of the Lorentz Lie algebra

$$[M_{\mu\nu}, Q_\alpha] = R(M_{\mu\nu})_\alpha{}^\beta Q_\beta. \tag{14.35}$$

The representation R need not be irreducible, but can be a sum of irreducible spinor representations.

2. The anticommutators of supercharges close into translations. If we are in a dimension where the supercharges are Majorana spinors, which holds in particular in dimensions $D = 2, 4, 10$, then the anticommutators take the form

$$\{Q_\alpha, Q_\beta\} = (\gamma^\mu C^{-1})_{\alpha\beta} P_\mu, \tag{14.36}$$

where C is an invertible matrix such that $(\gamma^\mu C^{-1})_{\alpha\beta}$ is a symmetric matrix. If the Q_α form an irreducible representation, then C is the charge conjugation matrix C. If the spinor representation is reducible, then we can decompose Q_α into $\mathcal{N}$ irreducible representations $Q^i_{\tilde{\alpha}}$, $i = 1, \ldots, \mathcal{N}$, and the supersymmetry algebra[8] takes the form

$$\{Q^i_{\tilde{\alpha}}, Q^j_{\tilde{\beta}}\} = (\gamma^\mu C^{-1})_{\tilde{\alpha}\tilde{\beta}} P_\mu N^{ij}, \tag{14.37}$$

---

[7] See, e.g., Wess and Bagger (1992) for the standard physics argument why this is the largest non-trivial extension of the space-time symmetries of a relativistic S-matrix, except for central extensions which will be introduced later in this section.

[8] The Q-Q anticommutator defines a subalgebra of the Poincaré Lie superalgebra, which is called the supertranslation algebra or simply the supersymmetry algebra. Note that this is the only relation we need to specify when extending a Poincaré Lie algebra into a Poincaré Lie superalgebra.

where C is the charge conjugation matrix and where $(\gamma^\mu C^{-1})_{\tilde{\alpha}\tilde{\beta}}$ and N^{ij} are either both symmetric or both antisymmetric. In dimensions where no Majorana spinors exist, supercharges can be chosen to be Dirac spinors, or, equivalently, symplectic Majorana spinors. When using symplectic Majorana spinors, the minimal supersymmetry algebra takes the form (14.37) with $N^{ij} = \epsilon^{ij}$, $i,j = 1,2$. Supersymmetry algebras involving a single irreducible Spin(p,q) representation (depending on the signature, this is either a complex (Dirac) or a real (Majorana) representation) are called *minimal* or $\mathcal{N} = 1$ supersymmetry algebras, while those involving $\mathcal{N} > 1$ irreducible Spin(p,q) representations are called *extended supersymmetry algebras*.

In some dimensions there are two inequivalent irreducible Spin(p,q) representations, and one can have *chiral supersymmetry algebras*. These are based on chiral supercharges $Q_{\tilde{\alpha}}^{(+)i}, Q_{\tilde{\alpha}}^{(-)m}$, $i = 1,\ldots \mathcal{N}_+$, $m = 1,\ldots \mathcal{N}_-$, which separately satisfy relations of the form (14.37). One can then choose the numbers $\mathcal{N}_+$ and $\mathcal{N}_-$ of left- and right-handed supersymmetries independently. We remark that chiral supersymmetry algebras do not only require the existence of two inequivalent irreducible Spin(p,q) representations, but also the existence of a non-zero superbracket which closes on supercharges of the same chirality. This happens in dimensions $D = 2, 6, 10, \ldots$. In Lorentz signature the irreducible supercharges are Majorana–Weyl spinors for $D = 2, 10, \ldots$ and Weyl spinors for $D = 6, \ldots$.

Supersymmetry algebras can be further generalised by admitting *central extensions*, which amounts to adding generators which are Lorentz scalars and which only appear on the r.h.s of the relations (14.37). As a further generalisation, one can admit *poly-vector extensions*, usually called *BPS extensions* in the physics literature, where the additional generators appearing on the r.h.s of the relations (14.37) are antisymmetric Lorentz tensors, contracted with γ-matrices.

To classify one-particle states in supersymmetric field theories, one needs to extend the representation theory of the Poincaré Lie algebra to the Poincaré Lie superalgebra. Irreducible unitary Poincaré representations are labelled by their mass and a representation of the little group.[9] Irreducible unitary representations of the Poincaré Lie superalgebra combine several representations of the little group with the same mass through the action of the supercharges. For concreteness, let us consider massive representations where $M^2 = -P^\mu P_\mu > 0$, and where the little group is the full rotation subgroup. In this case, we can choose $P_\mu = (M, 0, \ldots 0)$, so that (14.36) reduces to

$$\{Q_\alpha, Q_\beta\} = (\gamma^0 C^{-1})_{\alpha\beta} M. \tag{14.38}$$

These are, up to normalisation, the defining relations of a Clifford algebra. One can therefore construct the *supermultiplets* which combine representations of the Lie algebra of the little group into representations of the Poincaré Lie superalgebra by taking linear combinations of supercharges which act as fermionic creation and annihilation operators. This works in the same way as for the supersymmetric

---

[9] To include spinor representations, we have to admit projective representations, or, equivalently, replace the Poincaré group and the little groups by the corresponding twofold covering groups $\mathbb{R}^D \ltimes \mathrm{Spin}(1, D-1)$, $\mathrm{Spin}(D-1)$, and $\mathrm{Spin}(D-2)$. This is understood in the following.

harmonic oscillator. The dimension of the supermultiplet is the product of the dimension of the Clifford algebra (14.38) with the dimension of the representation of the little group that we assign to the Clifford vacuum. The creation and annihilation operators map bosons to fermions and vice versa, and each representation contains an equal number of bosonic and fermionic states.

Exercise 14.2.1 Show that

$$\text{Trace}\left((-1)^F \{Q_\alpha, Q_\beta\}\right) = 0, \tag{14.39}$$

and use this to prove that massive representations have the an equal number of bosonic and fermionic states.

For massless representations the little group is the helicity or transverse rotation subgroup, and the standard form of the momentum vector is $P_\mu = (E, 0, \ldots, E)$. One can show by taking suitable linear combinations that half of the supercharges act trivially, and therefore one only obtains half the number of fermionic creation and annihilation operators. Thus massless supersymmetry representations are smaller than massive representations.

If the supersymmetry algebra contains central charges Z_i, $i = 1, \ldots, l$, these are represented by multiples of the unit operator on irreducible representations. We can then order the central charges by size, $|Z_1| \geq |Z_2| \cdots$. It can be shown that all states in supersymmetry representations satisfies the mass bound (called *BPS bound*), $M \geq |Z_1| \geq |Z_2| \cdots$. The saturation of one or of several mass bounds ($M = |Z_1| > |Z_2|$, $M = |Z_1| = |Z_2| > |Z_3|$, etc.) reduces the number of supercharges which act non-trivially, and thus the dimension of the representation. Representations saturating BPS bounds are called short or *BPS representations*. The shortest BPS representations where all BPS bounds are saturated, $M = |Z_1| = \cdots |Z_l|$, have as many states as massless representations.

Exercise 14.2.2 A Dirac (Majorana) spinor in D dimensions has $2^{[D/2]}$ complex (real) components, where $[\cdot]$ denotes the integer part. Find the dimensions of the smallest massless representations of the ten-dimensional Poincaré Lie superalgebras with $(\mathcal{N}_+, \mathcal{N}_-) = (1, 0), (1, 1), (2, 0)$, which are also known as the Type-I, Type-IIA, Type-IIB, or $\mathcal{N} = 1, 2A, 2B$ supersymmetry algebras.

14.3 Supersymmetric Field Theories in Two Dimensions

In this section, we start with the Polyakov action in the conformal gauge and extend it to a two-dimensional supersymmetric field theory. This can be viewed as a generalisation of the supersymmetrisation of the harmonic oscillator. We will consider this as a purely field-theoretic problem and discuss the string theory interpretation in the next section.

From the world-sheet point of view, the string coordinates X^μ are a collection of scalar fields which transform as a vector under a global internal $SO(1, D - 1)$ symmetry. The natural guess for the corresponding fermionic fields is a collection

ψ^μ of two-dimensional spinor fields which also transform as an $SO(1, D-1)$ vector. In two-dimensional Minkowski space we can choose the γ-matrices to be real:

$$\gamma^0 = \left((\gamma^0)_a{}^b\right) = \begin{pmatrix} 0 & -1 \\ 1 & 0 \end{pmatrix}, \quad \gamma^1 = \left((\gamma^1)_a{}^b\right) = \begin{pmatrix} 0 & 1 \\ 1 & 0 \end{pmatrix}, \quad \{\gamma^\alpha, \gamma^\beta\} = 2\eta^{\alpha\beta}.$$

(14.40)

Therefore, we can choose the spinor fields $\psi^\mu = (\psi^\mu)_a, a = 1, 2$ to be Majorana spinors. The Majorana condition is defined using the matrix B, introduced in (14.31). Equivalently, one can impose that the Dirac and Majorana conjugate of a spinor agree, up to a constant phase. The Dirac and Majorana conjugate are defined as

$$\overline{\psi}^{(\text{Dirac})} = \psi^\dagger \gamma^0, \quad \overline{\psi}^{(\text{Majorana})} = \psi^T C,$$

(14.41)

where C is the charge conjugation matrix. For our representation we can choose $C = \gamma^0$ and the reality condition simply becomes $\psi^* = \psi$, that is $B = \mathbb{1}$, and the spinor fields are real valued. To be precise, classical fermionic fields have anti-commuting components, therefore the components are real Grassmann numbers.

Exercise 14.3.1   Show that $C = \gamma^0$ satisfies the properties of a charge conjugation matrix.

Since the number of space-time dimensions is even, we can impose a Weyl condition. For our representation the chirality matrix is

$$\gamma_* = \gamma^0\gamma^1 = \begin{pmatrix} -1 & 0 \\ 0 & 1 \end{pmatrix}.$$

(14.42)

Since γ^α, γ_* are real, it is manifest that spinors in two-dimensional Minkowski space can be simultanously chiral and real, that is, there exist Majorana–Weyl spinors.

Exercise 14.3.2   Compute the transformation matrix $\exp\left(\frac{\chi}{2}\gamma^{01}\right)$ for spinors under two-dimensional Lorentz boosts, where $\gamma^{01} = \frac{1}{2}[\gamma^0, \gamma^1]$ is the infinitesimal Lorentz generator. Since space is one-dimensional, the usual concept of spin does not apply. Show that upon continuation to Euclidean signature, this becomes becomes a double-valued representation of the rotation group $SO(2)$. This is relevant if we formulate the world-sheet theory in Euclidean signature. Start by modifying the Clifford generator to obtain the Euclidean version of the Clifford algebra, then find the expression for finite rotations by exponentiating the generator.

The natural candidate for extending the gauge-fixed Polyakov action to an action for fields (X^μ, ψ^μ) is to add the action for massless Majorana spinors. This is the RNS-action

$$S_{RNS} = -\frac{1}{2\pi} \int d^2\sigma \left(\partial_\alpha X^\mu \partial^\alpha X_\mu + i\overline{\psi}^\mu \gamma^\alpha \partial_\alpha \psi_\mu\right).$$

(14.43)

This action is invariant under the two-dimensional supersymmetry transformations

$$\delta X^\mu = i\overline{\epsilon}\psi^\mu, \quad \delta\psi^\mu = \gamma^\alpha \partial_\alpha X^\mu \epsilon,$$

(14.44)

where $\epsilon = (\epsilon_a)$ is the parameter of the superysmmtry transformation. Like ψ^μ, the parameters ϵ form an anticommuting Majorana spinor. Spinor indices have been suppressed. Supersymmetry is realised 'on-shell', that is the action is invariant up to terms which vanish if we impose the equations of motion. It is possible to realise supersymmetry off-shell, that is to find a representation of the supersymmetry algebra on fields which does not involve imposing the equations of motion. Such representations, if available, have the advantage that they disentangle the superymmetry representations from the dynamics, that is the choice of equations of motion (normally through the choice of an action). Off-shell representations involve additional auxiliary fields, which are non-dynamical. Upon eliminating the auxiliary fields by their equations of motion, one obtains a theory where supersymmetry is realised on shell.[10]

The equations of motion obtained from variation of the RNS action are

$$\Box X^\mu = 0, \quad i\gamma^\alpha \partial_\alpha \psi^\mu = 0. \tag{14.45}$$

Besides the wave equation for the X^μ we recognise the massless Dirac equation for the spinors ψ^μ. If we write this out explicitly we obtain

$$i \begin{pmatrix} 0 & -\partial_0 + \partial_1 \\ \partial_0 + \partial_1 & 0 \end{pmatrix} \begin{pmatrix} \psi_1 \\ \psi_2 \end{pmatrix} = \begin{pmatrix} 0 \\ 0 \end{pmatrix}. \tag{14.46}$$

Setting $\psi_- := \psi_1$ and $\psi_+ := \psi_2$ we can write this in the suggestive form

$$\partial_+ \psi_- = 0, \quad \partial_- \psi_+ = 0. \tag{14.47}$$

The Dirac equation decouples into two one-component equations for the fields $\psi_\pm^\mu$, which are the Majorana–Weyl spinors into which the Majorana spinors ψ^μ can be decomposed. After imposing the Dirac equation positive chirality spinors are purely left-moving, $\psi_+ = \psi_+(\sigma^+) = \psi_+(\sigma^0 + \sigma^1)$, while negative chirality spinors are purely right-moving, $\psi_- = \psi_-(\sigma^-) = \psi_-(\sigma^0 - \sigma^1)$. This reflects that supersymmetry is realised chirally. We can think of $\psi_\pm$ as superpartners of the left- and right-moving parts of the coordinates X^μ. The fact that supersymmetry can be realised chirally in two-dimensions is important for the construction of *heterotic string theories*.

Exercise 14.3.3   Show that the RNS action is invariant under supersymmetry transformations (14.44) up to the equations of motion.

Exercise 14.3.4   Show that the supersymmetry transformations (14.44) represent the two-dimensional supersymmetry algebra. Since supertransformations must anticommute into translations, and given that we take supersymmetry parameters to be anticommuting, this means that the commutator of two supersymmetry transformations must take the form

$$[\delta_{\epsilon_1}, \delta_{\epsilon_2}](\cdot) = a^\alpha \partial_\alpha(\cdot), \tag{14.48}$$

where the parameter a^α of the translation depends on ϵ_1, ϵ_2. This relation needs to be verified for both X^μ and ψ^μ. *Remark:* At some point you will need to rewrite terms of the form $\epsilon_1 \bar\epsilon_2$. To do this you need to use (or even better to derive) that any real 2×2 matrix can written in the form

$$M = \frac{1}{2}\mathrm{Tr}(M)\mathbb{1} + \frac{1}{2}\mathrm{Tr}(M\gamma_\alpha)\gamma^\alpha + \frac{1}{2}\mathrm{Tr}(M\gamma_*)\gamma_*. \tag{14.49}$$

---

[10]   If theories are sufficiently complicated, it might not be possible to eliminate the auxiliary fields in closed form, but only iteratively.

For the derivation of (14.49) note that $\mathbb{1}, \gamma^\alpha, \gamma_*$ are a basis for the real 2×2 matrices, and that only $\mathbb{1}$ has a non-vanishing trace.

The conserved current associated with supersymmetry transformations can be derived using the Noether method (see Section 2.2.5), with the result

$$J_\alpha = -\frac{1}{2} \gamma^\beta \gamma_\alpha \psi^\mu \partial_\beta X_\mu, \tag{14.50}$$

where the normalisation is conventional and has been chosen to be the same as in Green et al. (1987) and Blumenhagen et al. (2013). Taking into account the suppressed spinor index, this current is a world-sheet vector-spinor. Its partner under supersymmetry is the energy-momentum tensor

$$T_{\alpha\beta} = \partial_\alpha X^\mu \partial_\beta X_\mu - \frac{i}{4} \bar{\psi}^\mu \gamma_\alpha \partial_\beta \psi_\mu - \frac{i}{4} \bar{\psi}^\mu \gamma_\beta \partial_\alpha \psi_\mu - \text{Trace}, \tag{14.51}$$

where 'Trace' indicates further terms, such that $T_{\alpha\beta}$ is traceless. The resulting world-sheet theory is conformally invariant.

Exercise 14.3.5 Show that the supercurrent (14.50) is conserved on-shell and satisfies the identity $\gamma^\alpha J_\alpha = 0$.

Exercise 14.3.6 Use the Noether method to derive the conserved supercurrent (14.50). Perform a variation of the RNS action, allowing the supersymmetry parameter ϵ to be space-time dependent. Write the variation in the form $\delta S \propto \int d^2\sigma \, (i(\partial_\alpha \bar{\epsilon}) J^\alpha + \cdots)$, where the omitted terms are total derivatives or vanish by the equations of motion. Briefly recall the argument why J^α is conserved on-shell if we restrict to rigid transformations, and show that J^α agrees, up to normalisation, with (14.50).

Exercise 14.3.7 Expressions such as the action, the supersymmetry variations and the supercurrent should be real. For expressions containing more than one anti-commuting variable, we need to specify what 'real' means, since exchanging two such variables gives rise to a minus sign. Our convention for a scalar built out of anticommuting spinors is that we take 'real' to mean 'invariant under Hermitian conjugation'.[11] Show that in this sense the action is real-valued and that the supersymmetry variations are consistent with the reality conditions we have imposed on the fields.

14.4 The RNS String

So far, we have extended the gauge-fixed Polyakov action to a two-dimensional field theory with rigid supersymmetry. For this to define a string theory we also need to generalise the constraints $T_{\alpha\beta} = 0$, and to work out which boundary conditions are admissible if we consider the field theory on a strip (open strings) or cylinder (closed strings). The systematic way of doing this is to supersymmetrise the Polyakov action before gauge fixing. The resulting action is invariant under

---

[11] Another convention used in the literature is 'invariance under complex conjugation'. Both conventions differ by signs, since Hermitian conjugation includes transposition, which reverses the order of terms.

local supersymmetry transformation in addition to reparametrisations. Theories with local supersymmetry are called *supergravity theories*. This reflects that the 'gauge fields' associated to the conserved currents $T_{\alpha\beta}$ and J_α are the metric or graviton $h_{\alpha\beta}$ and its supersymmetry partner, the gravitino ψ_α, which is a vector-spinor field. In two dimensions these fields do not carry local propagating degrees of freedom. The supersymmetrised Polyakov action is invariant under local Weyl transformations and their supersymmetry partners, called super-Weyl transformations, which makes it a theory of conformal supergravity. We will not introduce the supergravity formalism which is needed to formulate the locally supersymmetric extension of the Polyakov action. For our purposes it is sufficient to know that upon imposing the *superconformal gauge* $h_{\alpha\beta} = \eta_{\alpha\beta}, \psi_\alpha = 0$, the locally supersymmetric action reduces to the RNS action, supplemented with the constraints

$$T_{\alpha\beta} = 0, \quad J_\alpha = 0. \tag{14.52}$$

Besides the conservation equations

$$\partial^\alpha T_{\alpha\beta} = 0, \quad \partial^\alpha J_\alpha = 0, \tag{14.53}$$

the gauge-fixed theory satisfies

$$\eta^{\alpha\beta} T_{\alpha\beta} = 0, \quad \gamma^\alpha J_\alpha = 0. \tag{14.54}$$

This implies that the gauge-fixed theory is a superconformal field theory. The generators of the superconformal algebra are displayed below (see Equation (14.60)).

As for the bosonic string, the possible boundary conditions for the fields ψ^μ can be determined by imposing the absence of boundary terms under variation of the action. We will state the results without derivation. The first case is closed strings where the string coordinates X^μ satisfy periodic boundary conditions and the world-sheet is locally a cylinder. Since the ψ^μ are world-sheet spinors, and thus transform in a double-valued representation of the two-dimensional Lorentz group, one can choose either periodic or anti-periodic boundary conditions for them. The action is quadratic in the ψ^μ and thus periodic for both periodic and anti-periodic boundary conditions. Since the left- and right-moving components of the ψ^μ decouple by the equations of motion, one can choose periodic or anti-periodic boundary conditions independently for left- and right-movers, though to preserve Lorentz invariance the boundary conditions must then be the same for all values of μ. This leaves us with four combinations of boundary conditions:

$$\psi_+^\mu(\sigma^0, \sigma^1 + \pi) = \pm\psi_+^\mu(\sigma^0, \sigma^1), \quad \psi_-^\mu(\sigma^0, \sigma^1 + \pi) = \pm\psi_-^\mu(\sigma^0, \sigma^1). \tag{14.55}$$

Periodic boundary conditions are also called *Ramond boundary conditions* (R-boundary conditions), while anti-periodic boundary conditions are called *Neveu–Schwarz boundary conditions* (NS-boundary conditions).[12]

For open strings the left- and right-moving modes $\psi_\pm^\mu$ couple through the boundary conditions. If we preserve full Lorentz invariance and impose Neumann boundary

---

[12] Note that if one conformally maps the world-sheet cylinder to the complex plane, periodic boundary conditions are mapped to anti-periodic boundary conditions and vice versa. The computation is similar to the one relating the energy-momentum tensors on the cylinder and on the plane (see Section 6.1).

conditions on the string coordinates X^μ, then the admissible boundary conditions for $\psi_\pm^\mu$ are

$$\psi_+^\mu(\sigma^0, \sigma^1 = 0) = \psi_-^\mu(\sigma^0, \sigma^1 = 0), \quad \psi_+^\mu(\sigma^0, \sigma^1 = \pi) = \pm\psi_-^\mu(\sigma^0, \sigma^1 = \pi),$$

(14.56)

where the choice of a $(+)$ is called a *Ramond boundary condition* while the choice of a $(-)$ is called a *Neveu–Schwarz boundary* condition. As a generalisation, one can introduce D-branes and impose Dirichlet boundary conditions for the X^μ in some directions. This introduces additional $(-)$ signs into the boundary conditions for the $\psi_\pm^\mu$. Explicit expressions will not be needed in the following.[13]

Since we know the equations of motion and admissible boundary conditions, we can now write down the corresponding mode expansions. The mode expansions for the coordinates X^μ remain unchanged, and we only list those for the fermions $\psi_\pm^\mu$. For closed strings we have the following mode expansions:

$$\psi_-^\mu = \sum_{m \in \mathbb{Z}} d_m^\mu e^{-2im\sigma^-} \quad (R), \qquad \psi_+^\mu = \sum_{m \in \mathbb{Z}} \tilde{d}_m^\mu e^{-2im\sigma^+} \quad (R), \qquad (14.57)$$

$$\psi_-^\mu = \sum_{r \in \mathbb{Z}+\frac{1}{2}} b_m^\mu e^{-2ir\sigma^-} \quad (NS), \qquad \psi_+^\mu = \sum_{r \in \mathbb{Z}+\frac{1}{2}} \tilde{b}_m^\mu e^{-2ir\sigma^+} \quad (NS), \qquad (14.58)$$

which can be combined in four ways as R-R, R-NS, NS-R and NS-NS. The corresponding expressions for open strings are obtained by setting $d_m^\mu = \tilde{d}_m^\mu$, and $b_m^\mu = \tilde{b}_m^\mu$ as well as substituting $2\sigma^\mp \to \sigma^\mp$. In this case, we only have the choice between R boundary conditions and NS boundary conditions.

By generalising the procedure of covariant quantisation, one arrives at the following anticommutation relations for the fermionic modes:

$$\{d_m^\mu, d_n^\nu\} = \eta^{\mu\nu}\delta_{m+n,0} \ (R), \quad \{b_r^\mu, b_s^\nu\} = \eta^{\mu\nu}\delta_{r+s,0} \ (NS), \qquad (14.59)$$

where $m, n \in \mathbb{Z}, r, s \in \mathbb{Z} + \frac{1}{2}$. For closed strings there are analogous relations for the modes $\tilde{d}_m^\mu, \tilde{b}_r^\mu$. Observe that the R-modes d_m^μ together with the bosonic modes α_m^μ are the modes for a system of supersymmetric harmonic oscillators.

With these relations we can now construct, for each choice of boundary conditions, a corresponding Fock space. Physical states are defined by imposing the constraints $T_{\alpha\beta} = 0, J_\alpha = 0$. As for the bosonic string it is convenient to express these conditions in terms of Fourier modes. The corresponding expressions are

$$L_m = \frac{1}{2} \sum_{n=-\infty}^{\infty} :\alpha_{-n} \cdot \alpha_{m+n}: + \frac{1}{2} \left\{ \begin{array}{ll} \sum_{n=-\infty}^{\infty} \left(n + \frac{m}{2}\right) : d_{-n} \cdot d_{m+n}: & (R), \\ \sum_{r \in \mathbb{Z}+\frac{1}{2}} \left(r + \frac{m}{2}\right) : b_r \cdot b_{m+r}: & (NS), \end{array} \right.$$

$$F_m = \sum_{n \in \mathbb{Z}} \alpha_{-m} \cdot d_{m+n} \quad (R), \qquad (14.60)$$

$$G_r = \sum_{n \in \mathbb{Z}} \alpha_{-m} \cdot b_{m+r} \quad (NS).$$

The Fourier modes F_m or G_r of the supercurrent J_α extend the Virasoro algebra

$$[L_m, L_n] = (m - n)L_{m+n} + \frac{D}{8}(m^3 - m)\delta_{m+n,0} \qquad (14.61)$$

---

[13] See, e.g., Blumenhagen et al. (2013).

to the *superconformal algebra* with the additional relations

$$[L_m, F_n] = \left(\frac{m}{2} - n\right) F_{m+n},$$ (14.62)

$$\{F_m, F_n\} = 2L_{m+n} + \frac{D}{2} m^2 \delta_{m+n,0},$$ (14.63)

for R-boundary conditions and

$$[L_m, G_r] = \left(\frac{m}{2} - r\right) G_{m+r},$$ (14.64)

$$\{G_r, G_s\} = 2L_{r+s} + \frac{D}{2}\left(r^2 - \frac{1}{4}\right)\delta_{r+s,0},$$ (14.65)

for NS-boundary conditions. For closed strings there is a second set of relations for $\tilde{L}_m, \tilde{F}_m, \tilde{G}_r$. In all cases one obtains $\mathbb{Z}$-graded Lie superalgebras which contain the conformal (Virasoro) algebra, and the supersymmetry algebra as subalgebras. Due to normal ordering effects, one also has central terms. Observe that in the Virasoro algebra the coefficient $D/24$ has been replaced by $D/8$. This is due to a partial cancellation between contributions from X^μ and ψ^μ. One important consequence is that the critical dimension, where the theory becomes unitary after imposing the constraints and dividing out residual gauge symmetries is now $D = 10$ (8 transverse dimensions) instead of $D = 26$ (24 transverse dimensions).

The contributions of the ψ^μ also change the shift a of the ground state energy due to normal ordering effects, which enters into the physical state conditions

$$L_m|\phi\rangle = 0, \quad m > 0, \quad (L_0 - a)|\phi\rangle = 0, \quad a = \begin{cases} 0 & \text{(R)}, \\ \frac{1}{2} & \text{(NS)}, \end{cases}$$ (14.66)

$$F_m|\phi\rangle = 0, m \geq 0 \text{ (R)}, \quad G_r|\phi\rangle = 0, r > 0 \text{ (NS)}.$$ (14.67)

The computation which shows that $a = 0$ for R-boundary conditions is essentially the same as the one where we showed that for a supersymmetric harmonic oscillator the ground state energy is zero. For NS-boundary conditions the cancellation between world-sheet bosons and fermions is only partial.

We are now in position to construct the Fock spaces of the RNS-model. For open strings with NS boundary conditions, momentum eigenstates $|k\rangle$ satisfy

$$\alpha_m^\mu|k\rangle = 0, \quad b_r^\mu|k\rangle = 0, \quad m, r > 0.$$ (14.68)

The mass of a state is computed using the L_0-constraint:

$$\left(L_0 - \frac{1}{2}\right)|\phi\rangle = 0 \Rightarrow \alpha' M^2 = N - \frac{1}{2},$$ (14.69)

where

$$N = \sum_{m=1}^\infty \alpha_{-m} \cdot \alpha_m + \sum_{r=\frac{1}{2}}^\infty r b_{-r} \cdot b_r.$$ (14.70)

The lowest mass states are listed in Table 14.1 in the column 'Before GSO projection'.

We still have to impose the other constraints to obtain the physical states and then to divide out residual gauge equivalences. As for the bosonic string it turns out that one can obtain representatives for the physical states by restricting the Lorentz

Table 14.1. Lowest mass states of RNS strings with NS boundary conditions before and after GSO projection. For closed strings we list one chiral sector.

N	$\alpha' M^2_{\text{open}} = \frac{1}{4}\alpha' M^2_{\text{closed}}$	Before GSO projection	After GSO projection
0	$-\frac{1}{2}$	$\|k\rangle$	
$\frac{1}{2}$	0	$b^\mu_{-1/2}\|k\rangle$	$b^\mu_{-1/2}\|k\rangle$
1	$\frac{1}{2}$	$b^\mu_{-1/2}b^\nu_{-1/2}\|k\rangle, \alpha^\mu_{-1}\|k\rangle$	
$\frac{3}{2}$	1	$b^\mu_{-3/2}\|k\rangle, \alpha^\mu_{-1}b^\nu_{-1/2}\|k\rangle,$	$b^\mu_{-3/2}\|k\rangle, \alpha^\mu_{-1}b^\nu_{-1/2}\|k\rangle,$
		$b^\mu_{-1/2}b^\nu_{-1/2}b^\rho_{-1/2}\|k\rangle$	$b^\mu_{-1/2}b^\nu_{-1/2}b^\rho_{-1/2}\|k\rangle$

Table 14.2. Lowest mass states of RNS strings with R boundary conditions before and after GSO projection

N	$\alpha' M^2_{\text{open}} = \frac{1}{4}\alpha' M^2_{\text{closed}}$	Before GSO projection	After GSO projection
0	0	$\|k, A\rangle$	$\|k, A_+\rangle$
1	1	$\alpha^\mu_{-1}\|k, A\rangle, d^\mu_{-1}\|k, A\rangle$	$\alpha^\mu_{-1}\|k, A_+\rangle, d^\mu_{-1}\|k, A_-\rangle$

indices to indices with respect to the transverse rotation group $SO(8) \subset SO(1,9)$. The states with $N = 0, \frac{1}{2}$ are a scalar tachyon and a massless vector. Using the branching rules for $SO(8) \subset SO(9)$, one sees that the states with $N = 1$ form a massive symmetric rank 2 tensor, whereas the states at level $N = \frac{3}{2}$ correspond to a massive symmetric rank 3 tensor and a massive antisymmetric rank 2 tensor.

For R-boundary conditions, momentum eigenstates satisfy

$$\alpha^\mu_m|k\rangle = 0, \quad d^\mu_m|k\rangle = 0, \quad m > 0, \tag{14.71}$$

and the mass is computed from the L_0-constraint

$$L_0|\phi\rangle = 0 \Rightarrow \alpha'M^2 = N, \tag{14.72}$$

where

$$N = \sum_{m=1}^\infty \alpha_{-m} \cdot \alpha_m + \sum_{m=1}^\infty m b_{-m} \cdot b_m. \tag{14.73}$$

Note that

$$[N, d^\mu_0] = 0, \quad \{d^\mu_0, d^\nu_0\} = \eta^{\mu\nu}. \tag{14.74}$$

Therefore, the ground state is degenerate and carries a representation of the Clifford algebra $Cl(1,9)$,

$$d^{\mu_1}_0 \cdots d^{\mu_l}_0|k\rangle, \quad \mu_1 < \mu_2 < \cdots < \mu_k, \quad k = 0, 1, \ldots, D = 10. \tag{14.75}$$

Since the operators d^μ_0 are Hermitian, this is the representation by Majorana spinors, which has real dimension 32. We introduce the notation $|k, A\rangle$, where $A = 1, \ldots, 32$ for the ground state. The lowest mass states are listed in Table 14.2 in the column 'Before GSO projection'. This table simultaneously lists the states of an open string and of one chiral sector of a closed string, which take the same form, but have masses differing by a factor 4.

Physical states can be identified by restricting indices to the transverse rotation subgroup SO(8), which effectively imposes the light cone gauge. For spinor indices this means that we restrict to the Majorana spinor representation of SO(8), which has real dimension 16. Thus, the R ground state is a massless Majorna spinor. For the massive states one needs to decompose products of tensor and spinor representations into irreducible spinors. Here, it suffices to note that all states constructed using R boundary conditions are space-time spinors.

Up to accounting for a factor 4 in the mass formula, states take the same form for the open string and one chiral sector of the closed string. To obtain the full closed string spectrum we have to combine two chiral sectors. With NS-NS boundary conditions one obtains a tachyonic ground state and the same massless states states $G_{\mu\nu}, B_{\mu\nu}, \phi$ as for the bosonic state. States with NS-R and R-NS boundary conditions are space-time spinors. For states with R-R boundary conditions one needs to decompose the product of two spinor representations into irreducible representations. This gives antisymmetric Lorentz tensors of various ranks. States with R-R boundary conditions are bosonic.

14.5 Type-II Superstrings

So far, we only had partial success in generalising the Polyakov action. When using R, R-NS, and NS-R boundary conditions we have space-time fermions, and no tachyons. However we would like to use NS and NS-NS boundary conditions to also have photons and gravitons in the string spectrum, and with these boundary conditions we still have tachyons. As we have seen, when discussing compactifications and orbifolds, the choice of boundary conditions for closed strings is strongly restricted by modular invariance. It has been shown that there are only two modular invariant theories of closed strings based on the RNS model which contain space-time fermions, the Type-IIA, and Type-IIB superstring theories.[14] In both theories the Hilbert space has four sectors, which are labelled by the boundary conditions imposed on left- and right-moving world-sheet fermions as the NS-NS, NS-R, R-NS, and R-R sectors. Observe that all possible combinations of boundary conditions occur. We have seen before that modular invariant partition functions can be constructed by summing over all possible boundary conditions in world-sheet space and world-sheet time, and that the sum over boundary conditions in world-sheet time amounts to a projection of the spectrum in each sector, where sector refers to a choice of boundary conditions in world-sheet space. The projection used to define Type-II superstring theories is called the *GSO projection*. The same projection is used in the Type-I superstring theory, which is a theory of open and closed strings that we will discuss below.

In an NS sector, the GSO projection operator is

$$P_{\text{GSO}}^{\text{NS}} = -(-1)^{\sum_{r=\frac{1}{2}}^{\infty} b_{-r} \cdot b_r}. \tag{14.76}$$

---

[14] The modular invariant Type-0A and Type-0B theories only have bosonic states.

Since only states with $P_{GSO} = 1$ remain in the spectrum, the GSO projection eliminates all states with an even number of b-modes. Note that this eliminates all mass levels in an NS sector which do not match with an R sector, and in particular gets rid of the tachyon (see Table 14.1). In an R sector the GSO projector is

$$P_{GSO}^{R} = \bar{\Gamma}(-1)^{\sum_{m=1}^{\infty} d_{-m} \cdot d_{m}} \tag{14.77}$$

where we use the notation Γ^{μ} for space-time γ-matrices and $\bar{\Gamma}$ for the chirality operator. The GSO projection $P_{GSO} = 1$ eliminates all states which combine negative chirality with an odd number of d-operators or positive chirality with an even number of d-operators. The resulting spectrum of an R sector is displayed in table Table 14.2 in the column 'After GSO projection'. In this table Majorana–Weyl spinor indices are denoted by $A_{\pm}$. For low levels one can easily check that after GSO projection NS sectors and R sectors have an equal number of states. To extend this to all mass levels, one needs to compute the light-cone partition functions of NS and R sectors and show that they are equal. This is indeed true, and the equality of both partition functions follows from a remarkable identity found by Jacobi in 1829, and called *equatio identica satis abstrusa* (a rather strange formula). It is possible, within a formulation of the world-sheet theory which includes the Faddeev–Popov ghost fields, to explicitly construct a representation of the Poincaré Lie superalgebra on the Hilbert space, and thus to prove that the theory has space-time supersymmetry. As a quick check, we note that after GSO projection the NS and R sector massless ground states are a vector and a Majorana–Weyl spinor of the little group SO(8), with eight real states each. This matches with the dimension of the minimal massless representation of the ten-dimensional $(\mathcal{N}_{+}, \mathcal{N}_{-}) = (1, 0)$ Poincaré Lie superalgebra computed in Exercise 14.2.2. This representation is known as the massless vector supermultiplet.

Our discussion of the GSO projection for NS and R sectors applies to both open strings and the chiral sectors of closed strings. For closed strings, where we have to combine left- and right-moving NS and R sectors with one another, we have one additional choice, namely the relative chirality of the GSO projections in the R sectors. We can either choose the R sector ground states to have opposite chirality, which gives us a non-chiral theory called Type-IIA or we can choose them to have the same chirality, which gives us a chiral theory called Type-IIB. Both theories are supersymmetric, and the corresponding supersymmetry algebras are the $(\mathcal{N}_{+}, \mathcal{N}_{-}) = (1, 1)$ and $(\mathcal{N}_{+}, \mathcal{N}_{-}) = (2, 0)$ Poincaré Lie superalgebras. They have unique realisations as supergravity theories called Type-IIA/IIB supergravity, which are the effective field theories of the corresponding superstring theories. Both supergravity theories involve a single massless representation, the Type-IIA/IIB supergravity multiplet. To work out the massless spectrum, we take tensor products of the massless states listed in Tables 14.1 and 14.2, and decompose these into irreducible representations which we can associate with particles. We will now list the decomposition into irreducible representations for both the SO(1, 9) Lorentz representations and the representations of the little group SO(8).

The ground state of an NS-sector is a Lorentz vector, the [10] representation. The NS-NS sector is the same for Type-IIA and Type-IIB:

$$SO(1, 9) : [10] \times [10] = [54] + [45] + [1],$$
$$SO(8) : [8] \times [8] = [35] + [28] + [1].$$

These universal states are the graviton $G_{\mu\nu}$, the Kalb–Ramond field $B_{\mu\nu}$ and the dilaton ϕ. The ground state of an R-sector is a Majorana–Weyl spinor $[16]_\pm$ of the Lorentz group, with physical degrees of freedom given by a Majorana–Weyl spinor $[8]_\pm$ of the transversal rotation group SO(8). To obtain states in the NS-R and R-NS sector we need to decompose the product of a vector and spinor representation, which gives a vector-spinor and a spinor. Using the fact that products of vectors with spinors of chirality $\pm$ give spinors of chirality $\mp$,[15] we obtain

$$\text{SO}(1,9) \ : [10] \times [16]_\pm = [144]_\mp + [16]_\mp,$$
$$\text{SO}(8) \ : [8] \times [8]_\pm = [56]_\mp + [8]_\mp.$$

For Type-IIA, the NS-R and R-NS sectors contain two *gravitini* $\psi_\pm^\mu$ and two dilatini $\psi_\pm$ of opposite chirality. This reflects that we are dealing with an extended $(\mathcal{N}_L, \mathcal{N}_R) = (1,1)$ supersymmetry where by applying supertransformations to the graviton we obtain two rather than one vector-spinor superpartners, called gravitini. These are the gauge fields related to two independent supercurrents. Similarly, by applying supertransformations to the dilaton we obtain two spinors, called dilatini. For Type-IIB, this works in the same way, except that the two gravitini and the two dilatini have the same chirality. Finally, for the R-R sector we need to reduce the product of two spinor representations. The product of two Dirac or Majorana spinors reduces into the sum of all possible antisymmetric tensor representations. This reflects that the Clifford algebra over a vector space is isomorphic, as a vector space, to the exterior algebra. For products of Weyl spinors one obtains a subset of the possible antisymmetric tensors. One also needs to take into account that sometimes antisymmetric tensor representations further decompose into a self-dual and anti-self-dual part.[16] The decompositions that we list in the following are fixed by the dimensions and conjugacy classes of the irreducible representations.

For Type-IIA:

$$\text{SO}(1,9) \ : [16]_+ \times [16]_- = [1] + [45] + [210]. \tag{14.78}$$

The corresponding irreducible representations are antisymmetric tensors G_0, G_2, G_4 of rank $0, 2, 4$, respectively. The physical state conditions for these R-R states can be shown to be the Maxwell-like equations

$$\partial^{\mu_1} G_{\mu_1 \cdots \mu_{p+2}} = 0, \quad \partial_{[\mu} G_{\mu_1 \cdots \mu_{p+2}]} = 0. \tag{14.79}$$

This implies that these tensors are gauge-invariant field strengths, and not gauge potentials. Strings couple fundamentally to the B-field through the Wess–Zumino term in the Polyakov action, but they are neutral with respect to R-R gauge fields.[17] If we reduce the corresponding SO(8) representations we find

$$\text{SO}(8) \ : [8]_+ \times [8]_- = [8] + [56]. \tag{14.80}$$

These represent the gauge-inequivalent components of the gauge potentials C_1, C_3 of G_2, G_4. The zero-form field strength G_0 does not represent local degrees of

---

[15] This follows from the rules for conjugacy classes of representations, compare Appendix G.
[16] Here 'dual' refers to the Hodge dual, which in terms of components corresponds to full contraction with the ϵ-tensor.
[17] The sources for R-R gauge fields are D-branes, as discussed at the end of this section.

freedom. However, it is related to the existence of a *massive deformation* of Type-IIA supergravity and string theory. This massive deformation can be thought of as a cosmological constant that can only take discrete values.[18]

For Type-IIB the Lorentz representations decompose as

$$SO(1,9) \; : [16]_+ \times [16]_+ = [10] + [120] + [126] \tag{14.81}$$

and correspond to antisymmetric tensors G_1, G_3, G_5 of ranks 1,3,5. The rank 5 tensor G_5 is self-dual, $G_5 = *G_5$. The decomposition of $SO(8)$ representations is

$$[8]_+ \times [8]_+ = [1] + [28] + [35] \tag{14.82}$$

and corresponds to the gauge invariant degrees of freedom of the gauge potentials C_0, C_2, C_4. By adding up the contributions of all four sectors we see that in total we have 128 bosonic and 128 fermionic states, which agrees with the dimensions of the smallest massless representations of the Type-IIA and Type-IIB supersymmetry algebras computed in Exercise 14.2.2. These representations are the supergravity multiplets which provide the field content of the unique Type-IIA and Type-IIB supergravity theories, which therefore are the effective field theories of the Type-IIA and Type-IIB superstring theories.

Our observations about R-R gauge fields allow us to address the question which types of Dirichlet boundary conditions we should admit. Dirichlet boundary conditions break translation invariance, and consequently D-branes exchange momentum with open strings. Therefore, they are dynamical objects. The R-R sectors of Type-II string theories allow us to understand how D-branes arise as solitons. Strings are charged under the Kalb–Ramond field $B_{\mu\nu}$, since as we discussed the Wess–Zumino term $\propto \int_{\Sigma_2} d^2\sigma B_{\mu\nu} \partial_\alpha X^\mu \partial_\beta X^\nu \epsilon^{\alpha\beta}$ is the generalisation of the minimal coupling $\int_{\Sigma_1} d\sigma A_\mu x^\mu$ of a point particle to the electromagnetic field A_μ. Similarly, one can couple p-dimensional membranes with world-volume Σ_{1+p} to $(p+1)$-form gauge potentials $C_{\mu_1\cdots\mu_{p+1}}$ with field strengths $G_{\mu_1\cdots\mu_{p+2}} = (p+2)\partial_{[\mu_1} C_{\mu_2\cdots\mu_{p+2}]}$. Therefore, the R-R spectrum of Type-IIA/B string theory indicates the existence of Dirichlet-p-branes which act as sources for the field strengths G_{p+2}. We have also seen in Exercise 9.2.2 that when replacing a field strength G_{p+2} by its dual $\tilde{G}_{D-p-2}$ we also replace the original 'electric' gauge potential C_{p+1} by a dual, magnetic gauge potential $\tilde{C}_{D-p-3}$. This potential can be minimally coupled to a $D-p-4$ brane, which carries magnetic charge with respect to the field strength G_{p+2}. For example, the field strength G_4 of the IIA-string has a 2-brane as electrical and a 4-brane as magnetic source. These are D2 and D4 brane of the IIA-theory. The full spectrum of D-branes is summarised in Table 14.3.

The D(−1)-brane is an instanton ('D-instanton') which makes sense in Euclidean signature. The D3 brane is 'dyonic', that is, it carries both electric and magnetic charge due to the self-duality of G_5. The D8-brane can also be accommodated. It is sourced by the dual G_{10} of G_0, and related to the 'massive deformation' of the IIA theory. The D9 brane is simply ten-dimensional Minkowski space. It can further be shown that the presence of D-branes does not break space-time supersymmetry completely, but only partly. D-branes realise BPS states, which belong to special (short or BPS-) representations of the supersymmetry algebra. This often allows one

---

[18] See Bergshoeff et al. (1996) for details.

Table 14.3. D-branes in Type-IIA/IIB string theories					
Field strength	G_1	G_2	G_3	G_4	$G_5 = *G_5$
IIA/IIB	IIB	IIA	IIB	IIA	IIB
electric	D(−1)	D0	D1	D2	D3
magnetic	D7	D6	D5	D4	D3

to make exact statements, which hold without assuming that the coupling is small. In this way, D-branes (and other BPS solitons and BPS instantons) have opened a window into string theory beyond perturbation theory.

14.6 Type-I Superstrings

A systematic way of introducing open strings into Type-IIA or Type-IIB string theories is through adding D-branes. In this case, the open strings cannot move freely in ten-dimensional Minkowksi space. One can also add so-called *orientifold planes*, which implement the modding out by world-sheet parity. As discussed in Section 2.2.9, this brings us from oriented to non-oriented strings. A special case of this construction is the Type-I superstring theory, which can be obtained as an orientifold of the Type-IIB theory. The Type-I theory contains non-oriented closed and open strings, and it is the only superstring theory which allows open strings with Neumann boundary conditions in all directions in a ten-dimensional Minkowski space. Our analysis of the GSO projected open string spectrum has shown that open string states organise themselves into representations of minimal ten-dimensional $\mathcal{N} = 1$ supersymmetry algebra, while closed oriented superstrings have $\mathcal{N} = 2$ supersymmetry (IIA or IIB). To have a consistent theory of open and closed superstrings in ten-dimensional Minkowski space, it is necessary to reduce the supersymmetry of the closed string sector to $\mathcal{N} = 1$ through the orientifold projection.[19] There are further consistency conditions related to the absence of tadpoles and extistence of a consistent S-matrix. These imply that the open strings must carry Chan–Patton factors corresponding to the gauge group SO(32). The massless ground state of the open string sector is the ten-dimensional $\mathcal{N} = 1$ vector multiplet tensored with the adjoint representation of SO(32). The massless ground state of the closed string sector is the $\mathcal{N} = 1$ supergravity multiplet, which has $128 = 2^4 \times 8$ on-shell states. It arises by applying the four fermionic creation operators of the supersymmetry algebra to a Clifford vacuum carrying an eight-dimensional representation of the little group SO(8). The 64 bosonic states correspond to the graviton, B-field and dilaton, while the 64 fermionic states belong to a single gravitino and dilatino.

---

[19] Here we are considering strings in a fully supersymmetric background. Open strings localised on D-branes constitute additional solitonic sectors of the theory where half of the supersymmetry is broken and states organise into BPS-multiplets.

As we will see in Section 14.8, the Type-IIA and Type-IIB theories are related by T-duality, and thus belong to a single moduli space once we compactify. While Type-IIA, Type-IIB, and Type-I look like different theories if we insist on defining them on ten-dimensional Minkowski space, one can alternatively think of them as a single theory put into different backgrounds.

14.7 Heterotic Strings

The RNS model extends the gauge-fixed Polyakov action into a two-dimensional $(1, 1)$ supersymmetric conformal field theory. Since supersymmetry in two dimensions can be chiral, one can also consider the extension into a $(1, 0)$ supersymmetric theory, where only one chiral sector is supersymmetrised. The resulting theories are called *heterotic string theories*. Since the critical dimensions of both chiral sectors differ by $26 - 10 = 16$, one takes the 16 unpaired chiral world-sheet bosons $X^I(\sigma^+)$, $I = 1, \ldots, 16$ to be compactified, that is they satisfy boundary conditions

$$X^I(\sigma^0 + \sigma^1 + \pi) = X^I(\sigma^0 + \sigma^1) + w^I, \quad w = (w^I) \in \Gamma_{16}, \qquad (14.83)$$

where Γ_{16} is a sixteen-dimensional lattice. Modular invariance implies that this lattice must be even and self-dual with respect to the Euclidean scalar product. Up to rotations there are precisely two such lattices: the root lattice of the exceptional Lie group $E_8 \times E_8$ and the lattice generated by the root vectors and spinor weight vectors (of one chirality) of the group $SO(32)$. We denote these two theories by HE and HO. Their massless spectra comprise the $\mathcal{N} = 1$ supergravity multiplet together with vector multiplets in the adjoint representation of the gauge group. The gauge group is determined by the lattice Γ and is $E_8 \times E_8$ for the HE theory and a group locally isomorphic to $SO(32)$ for the HO theory.[20] Besides the HE and HO theory, there is a small number of modular invariant non-supersymmetric heterotic string theories. Only one of these is tachyon-free and has the gauge group $SO(16) \times SO(16)$.

14.8 Looking for the Big Picture

There are five consistent supersymmetric string theories in ten dimensions: IIA, IIB, I, HE, and HO. It can be shown that these are the only modular invariant supersymmetric string theories. There also exists a number of modular invariant, non-supersymmetric and tachyon-free theories, for example, a heterotic string theory with gauge group $SO(16) \times SO(16)$. These theories are less well understood than the supersymmetric theories, but seem to become the more interesting the longer we wait for supersymmetry to be discovered experimentally.

The five superstring theories are not independent. Some equivalences can be established within perturbation theory.

---

[20] See Appendix G for details.

- Type-IIA string theory on a circle of radius R is related by T-duality to Type-IIB string theory on a circle of radius $\frac{\sqrt{\alpha'}}{R}$. One says that Type-IIA/IIB are 'on the same moduli' after compactification. The ten-dimensional Type-IIB/IIA theory is the alternative decompactification limit $R \to 0$ of the Type-IIA/IIB theory compactified on a circle S_R^1. For bosonic strings this limit gave back the original theory. This is different for Type-II. T-duality acts as a 'chiral reflection', which acts non-trivially on the world-sheet fermions, with the result that the limits $R \to \infty$ and $R \to 0$ have different space-time chirality. In summary Type-IIA and Type-IIB 'are' T-dual.
- Similarly, the two heterotic theories HE and HO are T-dual. In this case T-duality acts non-trivially on the gauge sector.

Further equivalences have been discovered, and so far they have passed all consistency tests. They are non-perturbative and either relate strong to weak coupling (S-dualities) or map the coupling (dilaton) of one theory to geometric properties (moduli) of another theory (U-dualities).[21]

In this way, one can understand the strong coupling limits of all perturbative string theories.

- The Type-IIB theory has an exact $SL(2, \mathbb{Z})$ symmetry called S-duality which in particular maps strong to weak coupling.
- The Type-IIA theory has a strong coupling limit which is eleven-dimensional. The eleven-dimensional theory, which cannot be a string theory, has been dubbed eleven-dimensional M-theory. Its massless sector is given by eleven-dimensional supergravity, which is the unique supersymmetric theory in eleven dimensions.
- The HE theory has a strong coupling limit which is a version of M-theory where eleven-dimensional space-time takes the form $M_{10} \times I$, where I is an interval, and where the $E_8 \times E_8$ vector multiplets are located at the ends of this interval.
- The strong coupling limit of the HO theory is the Type-I string theory and vice versa.

Since in this way all five supersymmetric string theories are related to one another, they are believed to be different perturbative expansions of a single underlying theory, which also has an eleven-dimensional limit. This theory, whose precise definition remains to be found, is usually referred to as M-theory.

14.9 Final Remarks and Literature

There are many excellent textbooks on supersymmetry, which also cover supergravity, including Wess and Bagger (1992), West (1986), Freedman and Van Proeyen (2012). West (2012) develops supergravity along with string theory. Chapter II of Wess and Bagger (1992) gives a condensed treatment of the representation theory of four-dimensional supersymmetry, including BPS representations. For supersymmetry in arbitrary dimension and signature, see Van Proeyen (1999). For a

---

[21] The term U-duality is also used as a 'catch all' expression which includes T-duality and S-duality.

condensed mathematical treatment of Clifford algebras and Spin groups, see Lawson and Michelsohn (1989). Townsend (1997) discusses M-theory, its dualities, and its branes from the viewpoint of the BPS-extended supersymmetry algebra. The relation between supersymmetry and dualities across dimensions is reviewed in de Wit and Louis (1999). Books on the relation between supersymmetry, string theory, and particle physics include Binetruy (2008), Dine (2007), Ibáñez and Uranga (2012).

Our treatment of superstring theories in this chapter has been restricted to the very basics, and we have not included the alternative *GS formulation* where space-time supersymmetry is manifest while Lorentz covariance is not. More on this can be found in other textbooks, including Green et al. (1987), Polchinski (1998a), Polchinski (1998b), Becker et al. (2007), Kiritsis (2007), Blumenhagen et al. (2013), West (2012), Schomerus (2017), which also explore the network of string dualities. Our treatment of the RNS-model follows the one of Green et al. (1987) closely, though we have tweaked the conventions. The space-time supercharges can be constructed explicitly within the BRST approach to covariant quantisation, which extends the Fockspace by including the Faddeev–Popov ghost and superghost fields, Friedan et al. (1986). This formulation is also needed to construct Lorentz covariant fermionic vertex operators. A more recent approach to the construction superstring theories is the *pure spinor approach*, see Berkovits and Gomez (2017) for an introduction.

A
Appendix A Notation and Conventions

World-sheet: Σ.

World-sheet indices: $\alpha, \beta = 0, 1$.

World-sheet light-cone indices: $a, b = +, -$.

Holomorphic world-sheet indices: $z, \bar{z}$.

String world-sheet coordinates: $-\infty < \sigma^0_{(1)} \le \sigma^0 \le \sigma^0_{(2)} < \infty$ and $0 \le \sigma^1 \le \pi$.

Flat world-sheet metric: $\eta_{\alpha\beta}, \eta_{ab}$ (Lorentz signature, time-like direction has negative sign), $\delta_{\alpha\beta}$ (Euclidean signature).

Riemannian world-sheet metric: $h_{\alpha\beta}$.

Space-time: M, M_D. Minkowski space: $\mathbb{M}$.

Space-time vectors $x, y, \ldots$. Spatial part: $\vec{x}, \vec{y}, \ldots$. Transverse part: $\underline{x}, \underline{y}, \ldots$

Space-time indices: $\mu, \nu = 0, 1, \cdots D - 1$, or $M, N = 0, 1, \ldots D - 1$.

Spatial indices $i, j = 1, \ldots, D - 1$.

Indices for directions transverse to the world-sheet: $\underline{i}, \underline{j} = 1, \ldots, D - 2$.

Spinor indices (Dirac/Majorana): $a, b, \ldots, A, B, \ldots$

Spinor indices (Weyl, Majorana-Weyl): $A_\pm, B_\pm, \ldots$

Number of space-time dimensions: D.

Number of dimensions transverse to the world-sheet: $d = D - 2$.

Flat space-time metric: $\eta_{\mu\nu}$.

Riemannian space-time metric: $G_{\mu\nu}$ (string frame), $g_{\mu\nu}$ (Einstein frame).

We use the 'mostly plus convention' for Lorentz signature.

Toroidal compactifications: $D = d + n$.

Indices for internal directions: $i, j = 1, \ldots, n$, space-time indices: $\mu, \nu = 0, \ldots, d - 1$.

Planck mass/length M_P, L_P.

Newton's gravitational constant: G.

Gravitational coupling $\kappa = \sqrt{8\pi G/c}$.

Natural units: $\hbar = 1, c = 1$.

Planckian units: $\hbar = 1, c = 1, G = 1$.

Supergravity units: $\hbar = 1, c = 1, \kappa = 1$.

String tension T and Regge slope α': $T = \frac{1}{2\pi\alpha'}$.

String units: $\pi T = 1, \alpha' = 1/2$.

String length: $l_s = \sqrt{\alpha'}$. Alternative definition: $L_S = \sqrt{2\alpha'}$.

Dimensonless string coupling: g_S.

Appendix B Units, Constants, and Scales

Max Planck observed that, by using Newton's gravitational constant G, together with his constant h, and the speed of light c, one obtains a system of units where all physical quantities are measured using fundamental constants. When setting these constants to unity, all physical quantities become dimensionless. This is often considered as an indication that, by combining quantum theory and gravity, one can obtain a theory of nature which is in some sense 'complete'. Using $c, \hbar = \frac{h}{2\pi}$ and G we can define the Planck mass, Planck energy, Planck length, and Planck time:

$$M_P = \left(\frac{c\hbar}{G}\right)^{1/2} = 2.2 \cdot 10^{-5}\text{g}, \quad E_P = M_P c^2 = 1.2 \cdot 10^{19}\text{GeV},$$

$$L_P = \left(\frac{\hbar G}{c^3}\right)^{1/2} = 1.6 \cdot 10^{-35}\text{m}, \quad T_P = \left(\frac{\hbar G}{c^5}\right)^{1/2} = 5.4 \cdot 10^{-44}\text{sec}. \quad (B.1)$$

Planck units are defined by setting $\hbar = c = G = 1$. Note that these relations are specific to four dimensions, since the physical dimension of G varies with the space-time dimension. Also, there are variants of this system where G is set to a different numerical value. In particular, there is a version of the Planck units where G is replaced by $4\pi G$ everywhere, and in the supergravity literature one often sets the gravitational coupling $\kappa = \sqrt{8\pi G/c}$ to unity.

String theory offers an alternative way of defining a system of units where all quantities are dimensionless, since it contains a fundamental dimensionful constant, the string tension T, which in *natural units* ($c = \hbar = 1$) has dimension $[T] =$ Energy/Length = Mass/Length = Mass^2 = Length^{-2}. *String units* are defined by setting $T = 1/\pi$. Instead of T, one often uses the *Regge slope parameter*

$$\alpha' = \frac{1}{2\pi T},$$

which has dimension $[\alpha'] = \text{Length}^2$. In string units $\alpha' = \frac{1}{2}$. One definition of the *string length* is $l_S = \sqrt{\alpha'}$, so that $l_S = \frac{1}{2}\sqrt{2}$ in string units. None of these choices is canonical, and slightly different string units are obtained by changing the above definitions by multiplicative factors of order unity. For example, Green et al. (1987) define the string length as $L_S = \sqrt{2\alpha'}$. This has the advantage that $L_S = 1$ in string units. However, T-duality suggests that for closed strings there is a physical minimal length which is $\sqrt{\alpha'}$. When numerical factors are relevant, it is important to check which convention an author uses, while for qualitative arguments this is not important.

The relation between Planck and string units depends on the space-time dimension. We will use the gravitational coupling constant κ instead of G, because this is

the dimensionful coupling we associate with the three closed string vertex,

$$\kappa^2 = \frac{8\pi G}{c^2}.$$

Our version of Planckian units are the *supergravity units* $c = \hbar = \kappa = 1$. The dimension of κ is easily deduced from the Einstein–Hilbert action[1]

$$S \cong \frac{1}{\kappa^2} \int d^D x \sqrt{|g|} R.$$

We know that in natural units

$$[S] = 1, \quad [d^D x] = L^D, \quad [g_{\mu\nu}] = 1, \quad [R] = L^{-2},$$

where $L = $ Length. The last relation follows from $[\partial_\mu] = L^{-1}$ and that the Ricci scalar R contains two derivatives. Therefore

$$[\kappa] = L^{(D-2)/2}.$$

As a quick check note that $[\kappa] = 1$ in $D = 2$ (where the Einstein–Hilbert action is topological and computes the Euler number), and $[\kappa] = L$ in $D = 4$, which is consistent with our discussion of Planck units in four dimensions.

We will write κ_D when the space-time dimension is relevant. In D dimensions the ratio

$$\frac{\kappa_D}{(\alpha')^{(D-2)/4}} =: e^{\phi_{(D),0}} \tag{B.2}$$

is dimensionless and can be parametrised by the vacuum expectation value $\phi_{(D),0}$ of the D-dimensional dilaton. With the above conventional choice, $\phi_{(D),0} = 0$ corresponds to $\kappa_D = (\alpha')^{(D-2)/4}$. Depending on context it might be convenient to include a numerical factor in the relation (B.2), and many authors do. Neglecting such numerical factors, the exponential of the dilaton vacuum expectation value relates the Planck/supergravity scale and the string scale:

$$e^{\frac{2}{D-2}\phi_{(D),0}} \cong \frac{\kappa_D^{2/(D-2)}}{\sqrt{\alpha'}} \cong \left(\frac{L_{(D),P}}{L_S}\right) \cong \left(\frac{M_S}{M_{(D),P}}\right).$$

---

[1] Here, we leave open the overall normalisation of the Einstein–Hilbert action. The most common factors found in the literature are ± 1 and $\pm \frac{1}{2}$. The sign depends on the convention for the metric and the definitions of the Riemann and Ricci tensor. See Freedman and Van Proeyen (2012) for an overview of conventions and their relation.

Appendix C Fourier Series and Fourier Integrals

C.1 Fourier Series

Periodic functions with period 2π, $f(x + 2\pi) = f(x)$ can be represented by a Fourier series

$$f(x) = \frac{1}{\sqrt{2\pi}} \sum_{k=-\infty}^{\infty} c_k e^{ikx} \Leftrightarrow c_k = \frac{1}{\sqrt{2\pi}} \int_{-\pi}^{\pi} dx\, f(x) e^{-ikx}.$$

For a general period P, that is for $f(x + P) = f(x)$, these formulae are modified as follows:

$$f(x) = \frac{1}{\sqrt{P}} \sum_{k=-\infty}^{\infty} c_k e^{(2\pi ikx)/P} \Leftrightarrow c_k = \frac{1}{\sqrt{P}} \int_{x_0}^{x_0+P} f(x) e^{(-2\pi ikx)/P}.$$

Fourier series can be extended to generalised functions. One example is the periodic Dirac function or 'Dirac comb', $\delta_P(x) = \delta_P(x + P)$. By applying the inverse Fourier formula formally, we immediately obtain its representation as a Fourier sum:

$$\delta_P(x) = \frac{1}{P} \sum_{k=-\infty}^{\infty} e^{(2\pi ixk)/P}.$$

Sometimes, it is convenient to distribute the normalisation factors $\sqrt{2\pi}$ asymmetrically:

$$f(x) = \sum_{k=-\infty}^{\infty} \tilde{c}_k e^{ikx} \Leftrightarrow \tilde{c}_k = \frac{1}{2\pi} \int_{-\pi}^{\pi} dx\, f(x) e^{-ikx}.$$

We also need the extension of Fourier series to functions of several variables. For functions which have unit period in all variables, $f(x + y) = f(x)$, for all $y \in \mathbb{Z}^n$, the one-dimensional case generalises to

$$f(x) = \sum_{k \in \mathbb{Z}^n} \tilde{c}_k e^{2\pi ik\cdot x} \Leftrightarrow \tilde{c}_k = \int_{[0,1)^n} d^n x f(x) e^{-2\pi ik\cdot x}. \tag{C.1}$$

This can be further generalised to functions which are periodic modulo a rank n lattice $\Gamma \subset \mathbb{R}^n$. Then

$$f(x) = \sum_{k \in \Gamma^*} \tilde{c}_k e^{2\pi ik\cdot x}, \tag{C.2}$$

where the sum is over the dual lattice Γ^* (see Section 13.2.8) to ensure consistency of the expansion with the assumed periodicity. The Fourier coefficients are

$$\tilde{c}_k = \frac{1}{\text{vol}(\Gamma)} \int_{F.C.} d^n x f(x) e^{-2\pi ik\cdot x}. \tag{C.3}$$

where the integral is over a fundamental cell of the lattice Γ, which has volume $\mathrm{vol}(\Gamma) = \sqrt{|\det \Gamma|}$, where $\det \Gamma$ is the determinant of Gram matrix of the scalar product with respect to a lattice basis.

C.2 Fourier Integrals

Functions which decay sufficiently fast at infinity admit a representation as Fourier integrals, which can be viewed as 'expansions in plane waves', or 'expansions in irreducible representations of the translation group'.

The D-dimensional Fourier transformation[1]

$$f(x) = \frac{1}{(2\pi)^{D/2}} \int d^D k\, \tilde{f}(k) e^{ikx}$$

has the inverse

$$\tilde{f}(k) = \frac{1}{(2\pi)^{D/2}} \int d^D x\, f(x) e^{-ikx}.$$

The D-dimensional Dirac δ-function can be represented as the Fourier transform of the constant function $(2\pi)^{-D/2}$:

$$\delta^D(x) = \frac{1}{(2\pi)^D} \int d^D k\, e^{ikx}.$$

This follows immediately by formally applying the inverse Fourier transformation.

---

[1] Like for Fourier sums, there are conventions where the powers of $\sqrt{2\pi}$ are distributed asymmetrically.

Appendix D Modular Forms and Special Functions

D.1 The Dedekind η-Function

A holomorphic function $\phi(\tau)$ on the upper half plane $\mathcal{H} = \{\tau \in \mathbb{C} | \mathrm{Im}(\tau) > 0\}$ is called a *modular function* of weight h if it is meromorphic at infinity and transforms under the modular group $\mathrm{PSL}(2, \mathbb{Z})$ as

$$\phi\left(\frac{a\tau + b}{c\tau + d}\right) = (c\tau + d)^h \phi(\tau), \quad \text{where} \quad \begin{pmatrix} a & b \\ c & d \end{pmatrix} \in SL(2, \mathbb{Z}). \tag{D.1}$$

Modular functions which are holomorphic at infinity are called *modular forms*. Modular forms which vanish at infinity are called *cusp forms*. The upper half plane can be mapped conformally to the unit disk $\mathcal{D} = \{q \in \mathbb{C} | \, |q| < 1\}$ by $\tau \to q = e^{2\pi i \tau}$. The variables τ and q are used interchangeably.

The Dedekind η function has the product representation

$$\eta(q) = q^{1/24} \prod_{n=1}^{\infty} (1 - q^n).$$

Under the generators $\tau \to \tau + 1$ and $\tau \to -\frac{1}{\tau}$ of the modular group it transforms as:

$$\eta(\tau + 1) = e^{\frac{i\pi}{12}} \eta(\tau), \quad \eta\left(-\frac{1}{\tau}\right) = \sqrt{-i\tau}\, \eta(\tau).$$

The η-function is a modular form of weight $1/2$, up to phase factors, which form a so-called multiplier system. Its 24-th power is a cusp form proportional to the discriminant function $\Delta(q)$:

$$\Delta(q) = (2\pi)^{12} \eta^{24}(q).$$

In this book, we only touch upon the topic of modular forms and their generalisations, which play a big role in string theory. There are many books on the subject, see, for example, Serre (1973) for further reading.

D.2 ζ-Functions and ζ-Function Regularisation

D.2.1 The Riemann ζ-Function

The series

$$\zeta(s) = \sum_{n=1}^{\infty} \frac{1}{n^s}$$

converges for complex s with $\mathrm{Re}(s) > 1$. The Riemann ζ-function is defined by analytic continuation to the whole complex plane. The ζ-function is meromorphic with a simple pole at $s = 1$, with residue 1. Several special values are known, for example:

$$\zeta(-1) = -\frac{1}{12}, \quad \zeta(0) = -1, \quad \zeta(2) = \frac{\pi^2}{6}.$$

For $s > 1$ the ζ-function evaluates convergent sums, for example:

$$\zeta(2) = \sum_{n=1}^{\infty} \frac{1}{n^2} = \frac{\pi^2}{6}.$$

For $s < 1$ the ζ-function can be used to assign a finite value to divergent sums. This is called ζ-*function regularisation*. It is a mathematically meaningful procedure, because analytic continuation is unique. In physics ζ-function regularisation can be used to compute the quantum shift of the ground state energy, or Casimir energy, for quantum field theories in a finite volume, for example for the Maxwell field in a box. The experimental verification of the Casimir force predicted by the volume dependence of the Casimir effect is a classical test of quantum field theory.

Examples of ζ-regularised divergent sums:

$$\zeta(0) = \left(\sum_{n=1}^{\infty} 1\right)_{\zeta} = (1 + 1 + 1 + \cdots)_{\zeta} = -\frac{1}{2},$$

$$\zeta(-1) = \left(\sum_{n=1}^{\infty} n\right)_{\zeta} = (1 + 2 + 3 + \cdots)_{\zeta} = -\frac{1}{12}.$$

D.2.2 The Hurwitz ζ-Function

The Hurwitz-ζ-function is a generalised ζ-function defined by analytical continuation of the series

$$\zeta(s, a) = \sum_{n=0}^{\infty} \frac{1}{(n + a)^s}.$$

The series is absolutely convergent for $\mathrm{Re}(s) > 1$ and $\mathrm{Re}(a) > 0$, and can be extended to a meromorphic function for $s \neq 1$. The Riemann ζ-function is $\zeta(s) = \zeta(s, 1)$.

For $a = -n$, $n \in \mathbb{N}$, the values of the Hurwitz ζ-functions are related to the Bernoulli polynomials $B_n(s)$:

$$\zeta(s, -n) = -\frac{B_{n+1}(s)}{n + 1}.$$

The case $n = 1$ is relevant for computing the ground state energy in the twisted sector of the orbicircle CFT:

$$\zeta(s, -1) = -\frac{1}{2}B_2(s) = -\frac{1}{12}\left(1 - 6s + 6s^2\right).$$

D.3 The Gamma-Function and the Euler Beta-Function

The Γ-function is defined by its functional equation, together with its relation to the factorial for $u = 0, 1, 2, \ldots$:

$$\Gamma(u + 1) = u\Gamma(u) = u! \, . \tag{D.2}$$

It is holomorphic on $\mathbb{C}$ except for $u = 0, -1, -2, \ldots$. For $\mathrm{Re}(u) > 0$ it admits the integral representation

$$\Gamma(u) = \int_0^\infty dt \, t^{u-1} e^{-t}. \tag{D.3}$$

The Euler Beta-function is defined as

$$B(u, v) = \frac{\Gamma(u)\Gamma(v)}{\Gamma(u + v)}. \tag{D.4}$$

It is holomorphic for $u, v \in \mathbb{C}$ except at $u, v, u + v = 0, -1, -2, \ldots$. For $\mathrm{Re}(u), \mathrm{Re}(v) > 0$ it admits the integral representation

$$B(u, v) = \int_0^1 dt \, t^{u-1} (1 - t)^{v-1}. \tag{D.5}$$

Appendix E Young Tableaux

In this appendix, we summarise the rules for Young tableaux for the groups $SO(n)$, $SO(p,q)$ as far as they are needed to follow this book. Young tableaux are a graphical method for manipulating representations of groups. They can be applied to discrete groups, such as the permutation group in n elements S_n, as well as to Lie groups. While particle physics text books often discuss the method for unitary groups $SU(n)$, we are interested in rotation groups $SO(n)$ and Lorentz groups $SO(1, n-1)$, where the rules are slightly different.

In the Young tableaux formalism the fundamental representation, which for groups of the form $SO(p,q)$ is the vector representation, corresponds to a 'box' $\square$, while the trivial, one-dimensional representation corresponds to a 'bullet' •. New representations can be formed by taking tensor products of the fundamental representation with itself. This gives tensor representations, which in general are reducible. Irreducible representations are obtained by symmetrising or anti-symmetrising tensors with respect to all indices. This is encoded by arranging the boxes representing fundamental representations, or, more concretely, 'vector indices'. Horizontal arrangements correspond to symmetrisation, while vertical arrangements correspond to antisymmetrisation. While general representations can have mixed symmetry, we will only need representations which are completely symmetric or completely antisymmetric. Also, there is an additional feature for representations of pseudo-orthgonal groups. While for many groups including the general linear groups $GL(n, \mathbb{R})$ and the unitary groups $SU(n)$ tensor representations are irreducible if and only they have a definite symmetry, this does not apply for the groups $SO(p,q)$. For them antisymmetric representations are irreducible while symmetric representations can be decomposed by contraction of indices with the invariant metric. Symmetric traceless representations are irreducible, and denoted by a horizontal arrangement of boxes. The representations we frequently encounter in this book are listed in Table E.1. One of the main use of Young tableaux is to specify

Table E.1. Young tableaux for $SO(n)$ and $SO(p,q)$ representations

Representation	Young tableaux	Dimension
Scalar or trivial representation	•	1
Vector or fundamental representation	$\square$	$n = p + q$
Symmetric traceless tensor of rank 2	$\square\square$	$\frac{1}{2}n(n+1) - 1$
Antisymmetric tensor of rank 2	$\begin{matrix}\square\\\square\end{matrix}$	$\frac{1}{2}n(n-1)$

Table E.2. Branching rules for some $SO(n)$ and $SO(1, n-1)$ representations

$SO(n), SO(1, n-1)$	$\supset$	$SO(n-1)$
• 1	$\rightarrow$ $=$	• 1
$\square$ n	$\rightarrow$ $=$	$\square \oplus \bullet$ $(n-1) + 1$
$\begin{array}{c}\square\\\square\end{array}$ $\frac{1}{2}n(n-1)$	$\rightarrow$ $=$	$\begin{array}{c}\square\\\square\end{array} \oplus \square$ $\frac{1}{2}(n-1)(n-2) + (n-1)$
$\square\square$ $\frac{1}{2}n(n+1) - 1$	$\rightarrow$ $=$	$\square\square \oplus \square \oplus \bullet$ $\left(\frac{1}{2}(n-1)n - 1\right) + (n-1) + 1$

graphic rules for decomposing products of representations. We only need one such rule, the decomposition of the second rank tensor:

$$\square \otimes \square = \square\square \oplus \begin{array}{c}\square\\\square\end{array} \oplus \bullet$$
$$n \times n = \tfrac{1}{2}n(n+1) - 1 + \tfrac{1}{2}n(n-1) + 1.$$

We have added the dimensions underneath, as a check.

The second type of rules we need are branching rules, which tell us how to decompose an irreducible representation into irreducible representations of a subgroup. We use this to decompose representations of the Lorentz group $SO(1, n-1)$ into representations of a rotation subgroup $SO(n-1)$ and of the transverse rotation subgroup $SO(n-2)$, since these are the little groups which determine irreducible massive and massless representations of the Poincaré group. The branching rules needed in this book are listed in Table E.2.

We end with a few remarks, concerning further differences compared to SU-type groups and limitations of the method. First, pseudo-orthogonal groups $SO(p, q)$ also have spinor representations, which are relevant when we consider matter (fermions). These are not covered by the formalism described here. Second, as long as we restrict ourselves to the orientation preserving groups $SO(p, q)$, antisymmetric tensors of rank m are equivalent to antisymmetric tensors of rank $p + q - m$. For example, the fundamental representation of $SO(3)$ (the vector representation, considered as an representation by antisymmetric tensor of rank 1) is equivalent to representation by antisymmetric tensors of rank 2, with the equivalence transformation provided by the ϵ-tensor. This changes if we consider the full orthogonal group $O(3)$, because the ϵ-tensor is only a tensor if we fix an orientation. In general an antisymmetric tensor of rank two is not equivalent to a vector but to a pseudo-vector (axial vector), which transforms differently from a vector under reflections. Third, the fundamental representation of $SO(p, q)$ is real and therefore there is no need to distinguish between the fundamental and 'anti-fundamental' (complex conjugate) representation. This is different for SU-type groups, where the fundamental representation is complex.

E.1 Literature

Young tableaux, including their use for SO-type groups are explained in Hamermesh (1962). Due to the limitations of this method, it is important to be familiar with other methods of representation theory, see, for example, Cahn (1984), Cornwell (1997), and Fuchs and Schweigert (1997).

Appendix F Gaussian Integrals and Integral Exponential Function

F.1 Finite-Dimensional Gaussian Integrals

The basic Gaussian integral

$$I = \int_{-\infty}^{\infty} e^{-x^2} dx = \sqrt{\pi}$$

can be computed by evaluating the double integral I^2 in polar coordinates. The integral

$$\int_{-\infty}^{\infty} e^{-ax^2 + bx + c} dx = \sqrt{\frac{\pi}{a}} e^{\frac{b^2}{4a} + c}, \quad a > 0$$

can then be obtained by completing the square and substitution.

A higher-dimensional generalisation involving a symmetric positive definite matrix A and an n-component vector b is

$$\int d^n x \, e^{-\frac{1}{2}x^T A x + b^T x} = \sqrt{\frac{(2\pi)^n}{\det A}} e^{\frac{1}{2} b^T A^{-1} b}.$$

Here, we use that A is invertible and diagonalisable, and that $\det A$ is the product of the eigenvalues of A.

F.2 Infinite-Dimensional Gaussian Integrals

Infinite-dimensional Gaussian integrals can be defined rigorously,[1] but the expressions we need can be obtained by formally generalising finite-dimensional expressions. As an example, we consider the Euclidean version of the theory of a free massive scalar field, with action

$$S[\phi] = \int d^n x \frac{1}{2}\phi(-\Delta + m^2)\phi.$$

The functional $S[\phi]$ is quadratic in the field ϕ, and since the eigenvalues of the Laplace operator Δ are non-positive, $-\Delta + m^2$ is a symmetric positive differential operator, which we can view as an infinite-dimensional generalisation of a symmetric positive definite matrix A. Formal application of the Gaussian integral formula gives

$$Z = e^{-\Gamma} = \int D\phi \, e^{-S[\phi]} = \det^{-1/2}(-\Delta + m^2), \tag{F.1}$$

[1] See, e.g., Glimm and Jaffe (1981); Roepstorff (1994).

where we have fixed the overall normalisation by hand. The quantities Z and Γ are the partition function and effective action associated with the classical action S, respectively. We remark that in the massless case one has to take into account that $-\Delta$ has zero eigenvalues, which have to be treated separately.[2]

F.3 Integral Exponential Function and the Schwinger Proper Time Expression for the One-Loop Partition Function

In perturbative quantum field theory, the partition function Z, the generating functional $\log Z$ of connected graphs, and the effective action Γ correspond to Feynman graphs without external states ('legs'). In interacting theories, these quantities receive higher loop corrections. In Section 11.1, we evaluate the one-loop expression for $\log Z$ in the free theory, because this quantity is the field-theoretic analogue of the world-sheet one-loop partition function of string theory.[3] In this appendix, we fill in some details of the evaluation this quantity.

For the free theory of a massive real scalar (F.1) implies:

$$\log Z = \log \det{}^{-1/2}(-\Delta + m^2) = -\frac{1}{2}\log\det(-\Delta + m^2) = -\frac{1}{2}\operatorname{tr}\log(-\Delta + m^2).$$

The trace can be evaluated by switching to the momentum representation:

$$\begin{aligned}
\operatorname{tr}\log(-\Delta + m^2) &= \int d^n x \langle x| \log(-\Delta + m^2)|x\rangle \\
&= \int d^n x \int d^n k \int d^n k' \langle x|k\rangle\langle k| \log(-\Delta + m^2)|k'\rangle\langle k'|x\rangle \\
&= \int d^n x \int d^n k \int d^n k' \frac{e^{ikx}}{(2\pi)^{n/2}} \delta^n(k - k') \log(k^2 + m^2) \frac{e^{-ikx}}{(2\pi)^{n/2}} \\
&= V \int \frac{d^n k}{(2\pi)^n} \log(k^2 + m^2),
\end{aligned}$$

where

$$V = \int d^n x$$

is the volume of space-time, which we treat as a formal constant. The so-called Schwinger proper time expression for the one loop effective action $\Gamma = -\log Z$ arises through the replacement

$$-\log(k^2 + m^2) \to \lim_{\epsilon \to 0} \int_\epsilon^\infty \frac{dt}{t} e^{-(k^2 + m^2)t}. \tag{F.2}$$

Since the integral does not converge for $\epsilon \to 0$, this relation involves subtracting (ultraviolet) divergencies. For this one uses that the integral is related to the *integral exponential function*

$$\operatorname{Ei}(x) = \int_x^\infty \frac{dt}{t} e^{-t}, \quad x > 0,$$

---

[2] See, e.g., Weinberg (1987); Ginsparg (1988).
[3] The string theory partition functions is defined using connected world-sheets, therefore, it corresponds to $\log Z$ rather than Z itself.

which has the asymptotic expansion

$$\mathrm{Ei}(x) = -\gamma - \log(x) - \sum_{k=1}^{\infty} \frac{(-x)^k}{kk!},$$

where γ is the Euler–Mascheroni constant. Applying this expansion to the regularised integral

$$\int_{\epsilon}^{\infty} \frac{dt}{t} e^{-at} = \int_{a\epsilon}^{\infty} \frac{dt}{t} e^{-t} = \mathrm{Ei}(a\epsilon),$$

where $a = k^2 + m^2$, we obtain

$$\int_{\epsilon}^{\infty} \frac{dt}{t} e^{-at} = -\gamma - \log \epsilon - \log a + \cdots,$$

where the omitted terms vanish for $\epsilon \to 0$. The substitution (F.2) corresponds to a renormalisation of the divergent integral where the divergent term $-\log \epsilon$ is subtracted together with the constant $-\gamma$. Using this, one arrives at the expression (11.9) for $\log Z$.

Appendix G Lie Algebras, Lie Groups, and Symmetric Spaces

This appendix provides a brief summary of facts about Lie algebras, Lie groups, and symmetric spaces that are used in this book. A non-abelian Lie algebra is called *simple* if it does not have Lie subalgebras which are invariant under the adjoint action of the Lie algebra. Lie algebras which are direct sums of simple Lie algebras are called *semi-simple*, Lie algebras which are direct sums of simple and abelian Lie algebras are called *reductive*. In quantum theory gauge symmetries, more precisely internal finite-dimensional infinitesimal symmetries, are realised by unitary representations of reductive Lie algebras. The Poincaré Lie algebra, which is the algebra of space-time symmetries in Minkowski space, falls outside this class and its representation theory is reviewed in Section 8.1. In the following, we focus on simple Lie algebras, which are the building blocks of non-abelian internal symmetries.

The structure and representation theory of real simple Lie algebras is most easily understood by first studying their complexifications, that is complex simple Lie algebras. A real Lie algebra $\mathfrak{g}$ is called a *real form* of a complex Lie algebra $\mathfrak{g}_\mathbb{C}$, if $\mathfrak{g}_\mathbb{C}$ is the complexification of $\mathfrak{g}$. A real Lie algebra is called *compact* if all one-parameter groups obtained by exponentiation of generators are compact groups, that is, isomorphic to $U(1)$. Up to isomorphism, a complex simple Lie algebra has precisely one compact real form, though in general there are several non-isomorphic non-compact real forms. Only compact simple Lie algebras admit finite-dimensional unitary representations. These are the simple Lie algebras realising inner symmetries. It is quite common to freely move between a compact real Lie algebra and its complexification, for example, when using ladder operators for the Lie algebra of angular momentum.

A Lie algebra $\mathfrak{g}$ is, in particular, a vector space, so it has a dimension n. Out of the n linearly independent generators one can choose a maximal number $l < n$ of mutually commuting generators. These span a maximal abelian or *Cartan subalgebra* $\mathfrak{h}$, and l is called the *rank*. The remaining generators can be choosen to be eigenvectors with respect to the adjoint action of the Cartan subalgebra:

$$[H, E_\alpha] = \alpha(H)E_\alpha, \quad \forall H \in \mathfrak{h}, \tag{G.1}$$

where $\alpha \in \mathfrak{h}^*$ is a linear functional on $\mathfrak{h}$. The $n - l$ functionals α needed to complete specifying a basis of $\mathfrak{g}$ are called the roots of $\mathfrak{g}$. We can think of the roots $\alpha \in \Delta$ as a set of vectors in $\mathbb{K}^l \cong \mathfrak{h}^*$, where $\mathbb{K} \in \{\mathbb{R}, \mathbb{C}\}$. It can be shown that all roots can be written as non-negative integer linear combinations of a set of l simple roots α_i, $i = 1, \ldots, l$:

$$\alpha = \sum_{i=1}^{l} m_i \alpha_i, \quad m_i \in \mathbb{Z}, \ m_i \geq 0. \tag{G.2}$$

The Killing form

$$K : \mathfrak{g} \times \mathfrak{g} \to \mathbb{K}, \quad (a, b) \mapsto \mathrm{tr}(\mathrm{ad}(a), \mathrm{ad}(b)), \qquad (\mathrm{G}.3)$$

where $\mathrm{ad} : \mathfrak{g} \to \mathrm{End}(\mathfrak{g}), a \mapsto \mathrm{ad}_a, \mathrm{ad}_a(b) = [a, b]$ is the adjoint representation, is non-degenerate for (semi-)simple Lie algebras and negative definite for compact real forms. The Killing form can be used to define an isomorphism $\mathfrak{h}^* \to \mathfrak{h}$, $\alpha \mapsto H_\alpha$, by imposing $\alpha(H_\beta) = K(H_\alpha, H_\beta)$, which maps roots to elements of the Cartan subalgebra. The Killing form also induces a scalar product on $\mathfrak{h}^*$ and thus in particular between the roots, $\alpha \cdot \beta = K(H_\alpha, H_\beta)$.

Up to isomorphism, simple Lie algebras are classified by their *Cartan matrices*:[1]

$$C_{ij} = 2 \frac{\alpha_i \cdot \alpha_j}{\alpha_i \cdot \alpha_i}. \qquad (\mathrm{G}.4)$$

A Cartan matrix can be represented graphically by a *Dynkin diagram*.

For some simple Lie algebras all roots have the same length. In this case C_{ij} is symmetric and by normalising $\alpha_i \cdot \alpha_i = 2$ we have $C_{ij} = \alpha_i \cdot \alpha_j$. This normalisation is natural for toroidal compactifications in string theory, because non-abelian gauge symmetries appear when the Narain lattice contains vectors of the form $(\mathbf{k}_L, 0)$, $(0, \mathbf{k}_R)$, with $\mathbf{k}_{L/R}^2 = 2$. Simple Lie algebras where all roots have the same length are called *simply laced* Lie algebras, or ADE-Lie algebras, since the corresponding simple complex Lie algebras are $\mathfrak{a}_l, \mathfrak{d}_l, \mathfrak{e}_6, \mathfrak{e}_7, \mathfrak{e}_8$. The corresponding compact real Lie algebras are $\mathfrak{su}_{l-1}, \mathfrak{so}_{2l}, \mathfrak{e}_6, \mathfrak{e}_7, \mathfrak{e}_8$.[2] These are the simple Lie algebras that can be realised using toroidal compactifications in string theory.

A simple Lie algebra admits a so-called *Chevalley basis*, where all structure constants are integers. In terms of the roots $\alpha \in \Delta$, the non-vanishing commutation relations take the form

$$[H_\alpha, E_\beta] = 2 \frac{\alpha \cdot \beta}{\alpha \cdot \alpha} E_\beta, \quad [E_\alpha, E_\beta] = \begin{cases} H_\alpha, & \text{if } \beta = -\alpha, \\ \epsilon(\alpha, \beta) E_{\alpha+\beta}, & \text{if } \alpha + \beta \in \Delta, \\ 0, & \text{else.} \end{cases} \qquad (\mathrm{G}.5)$$

Note that the integer valued structure constants $\epsilon(\alpha, \beta)$ are not independent, since the commutator is antisymmetric and satisfies the Jacobi identity $[a, [b, c]] + \mathrm{cyclic} = 0$. The Cartan generators H_α are the images of the roots α under the isomorphism $\mathfrak{h}^* \to \mathfrak{h}$. Therefore, only l among the $H_\alpha, \alpha \in \Delta$ are linearly independent. One choice of independent Cartan generators is to take those associated with the simple roots, resulting in a basis is H_{α_i}, E_α for $\mathfrak{g}$. Another choice for a basis is H_i, E_α, where the Cartan generators H_i form an orthonormal basis with respect to the scalar product $(\cdot, \cdot) = -K(\cdot, \cdot)$, which is positive definite for compact real forms.

As an example, a Chevalley basis for $\mathfrak{su}_2$ is $H_\alpha, E_\alpha, E_{-\alpha}$, where

$$[H_\alpha, E_{\pm\alpha}] = \pm 2 E_{\pm\alpha}, \quad [E_\alpha, E_{-\alpha}] = H_\alpha. \qquad (\mathrm{G}.6)$$

As simple root for $\mathfrak{su}_2$, we have chosen the one-component vector $\alpha = (\sqrt{2})$.

Finite-dimensional representations $\Phi : \mathfrak{g} \to V$ of complex simple Lie algebras admit a decomposition into eigenspaces under the action of the Cartan subalgebra.

---

[1] To be precise, this classifies complex simple Lie algebras and their compact real forms. In general complex simple Lie algebra has several non-isomorphic non-compact reals forms, and their classification requires extra work. See, e.g., Gilmore (1974).

[2] Here we use the same symbol for the complex simple Lie algebras $\mathfrak{e}_l$ and their compact real forms. ·

These eigenspaces are labelled by *weight vectors*, whose components are the eigenvalues of this action. The weight vectors of representations generate the *weight lattice* Λ_W, which contains the *root lattice* Λ_R, which is the lattice generated by the root vectors, as a sublattice, $\Lambda_W \supset \Lambda_R$. Since any two weights belonging to the same representation differ by a root vector, representations belong to fixed equivalence classes (conjugacy classes) in Λ_W/Λ_R. The rank-l lattices Λ_W and Λ_R are abelian groups isomorphic to $\mathbb{Z}^l$, and their quotient Λ_W/Λ_R is a finite abelian group. When taking tensor products of representations, weights behave additively, and thus the decomposition into irreducible representations must respect addition in Λ_W/Λ_R. Therefore the conjugacy classes of the factors determine the conjugacy classes of the irreducible representations appearing in the product. This is useful, for example, when working out the massless spectrum of superstring theories, for which we stated the result in Section 14.5. Representations of compact real simple Lie algebras are obtained by restricting representations of their complex form.

To each simple Lie algebra $\mathfrak{g}$ one can associate a unique connected, simply connected Lie group G, but there are in general other, non-simply connected Lie groups with the same Lie algebra. Lie groups with the same Lie algebra are said to be *locally isomorphic*. For compact real simple Lie algebras the simply connected Lie group G has a finite centre which is isomorphic to the group Λ_W/Λ_R of conjugacy classes of representations. The other locally isomorphic Lie groups are obtained by taking quotients of G with respect to subgroups of the centre. As an example, relevant for the heterotic string, consider the Lie algebra $\mathfrak{so}_{32}$. The corresponding simply connected Lie group is Spin(32) and has centre $\mathbb{Z}_2 \times \mathbb{Z}_2'$. There are three non-isomorphic quotient groups: $SO(32) \cong \text{Spin}(32)/\mathbb{Z}_2^{\text{diag}}$, where $\mathbb{Z}_2^{\text{diag}} \subset \mathbb{Z}_2 \times \mathbb{Z}_2'$ is the diagonal subgroup, $PSO(32) \cong \text{Spin}(32)/(\mathbb{Z}_2 \times \mathbb{Z}_2')$ and $\text{Spin}(32)/\mathbb{Z}_2 \cong \text{Spin}(32)/\mathbb{Z}_2'$. Due to the relation between the centre of G and Λ_W/Λ_R, one can decide which group is realised globally as the gauge group of a physical theory by the conjugacy classes of representations which appear in the theory. The four conjugacy classes of Spin(32) representations are the class (0) of the adjoint representation, (v) of the fundamental or vector representation, and (s) and (c) of the two Majorana–Weyl representations. Using that (v), (s) and (c) correspond to the generators of the central subgroups $\mathbb{Z}_2^{\text{diag}}$, $\mathbb{Z}_2$ and $\mathbb{Z}_2'$, respectively, it follows that representations in the classes (0) and (v) are representations of SO(32), while Spin(32) representations in the classes (s) and (c) are only projective representations of SO(32). Similarly, Spin(32) representations in the classes (0), (s) are representations of Spin(32)/$\mathbb{Z}_2$. Modular invariance requires that the lattice Γ_{16} used to define heterotic string theories is an even self-dual lattice with respect to a Euclidean metric. Up to isometries, there are two such lattices, the root lattice of $E_8 \times E_8$[3] and the lattice generated by the weights of the classes (0) and (s) of Spin/$\mathbb{Z}_2$.[4] Therefore, the possible gauge groups for heterotic strings are $E_8 \times E_8$ and Spin(32)/$\mathbb{Z}_2$, though the latter is often referred to as SO(32).

If G is a group, then a *G-homogeneous space* is a space X on which G acts transitively. The isotropy or stabiliser group of a point $x \in X$ is the subgroup $H_x \subset G$ which leaves x invariant. The space X can be identified with the coset G/H_x. Since the

---

[3] E_8 only has one conjugacy class of representations, therefore, the root and weight lattices agree.
[4] We could, alternatively, choose the classes (0) and (c). The resulting gauge groups are isomorphic.

stabiliser groups of different points in X are conjugate in G, one can use the notation H. If H is a closed subgroup of a Lie group G, then G/H is a smooth manifold. The spaces $SL(2, \mathbb{R})/SO(2)$ and $O(d, d)/(O(d) \times O(d))$ – which appear at various points in this book – are homogeneous spaces. In fact they belong to the more special class of Riemannian globally symmetric spaces. A *Riemannian locally symmetric* space is a Riemannian manifold where the Riemann curvature tensor is parallel (covariantly constant). A *Riemannian globally symmetric space* is a Riemannian manifold where every point is the fixed point of an involutive isometry. Riemannian globally symmetric spaces are in particular Riemannian locally symmetric spaces. Riemannian symmetric spaces are, locally or globally, homogeneous spaces G/H, with an isometric action of G. The Riemannian metric of the symmetric space G/H is determined by the Killing form of the Lie algebra $\mathfrak{g}$ of G and the choice of the subgroup H.

G.1 Literature

See Cahn (1984), Cornwell (1997), Gilmore (1974), Fuchs and Schweigert (1997), and Bump (2004) for Lie groups, Lie algebras, and their representation theory. Root lattices, weight lattices, and their use in string theory are explained in Lerche et al. (1989); see also Cornwell (1997) for the connection between lattices and the relation between locally isomorphic Lie groups through covering maps. For the construction and classification of real forms, see Gilmore (1974) who also discusses symmetric spaces. For a mathematical treatment of Lie groups and symmetric spaces from the viewpoint of Riemannian differential geometry, see Helgason (1978).

References

Aldazabal, G., Marques, D., and Nunez, C. 2013. Double Field Theory: A Pedagogical Review. *Class. Quant. Grav.*, **30**, 163001.

Ammon, M. and Erdmenger, J. 2015. *Gauge/Gravity Duality*. Cambridge: Cambridge University Press.

Antoniadis, Ignatios, Bachas, C. P., and Kounnas, C. 1987. Four-Dimensional Superstrings. *Nucl. Phys. B*, **289**, 87.

Aspinwall, P. S., Bridgeland, T., Craw, A. et al. (eds.). 2009. *Dirichlet Branes and Mirror Symmetry*. Clay Mathematics Monographs, vol. 5. Providence, RI: American Mathematical Society.

Astashkevich, A. and Belopolsky, A. 1997. String Center-of-Mass Operator and Its Effect on BRST Cohomology. *Commun. Math. Phys.*, **186**, 109–36.

Athanasopoulos, P., Faraggi, A. E., Groot Nibbelink, S., and Mehta, V. M. 2016. Heterotic Free Fermionic and Symmetric Toroidal Orbifold Models. *JHEP*, **04**, 038.

Becker, K., Becker, M., and J. H., Schwarz. 2007. *String Theory and M Theory*. Cambridge: Cambridge University Press.

Belopolsky, A. and Zwiebach, B. 1996. Who Changes the String Coupling? *Nucl. Phys. B*, **472**, 109–38.

Bergshoeff, E., de Roo, M., Green, M. B., Papadopoulos, G., and Townsend, P. K. 1996. Duality of Type II 7 Branes and 8 Branes. *Nucl. Phys. B*, **470**, 113–35.

Berkovits, N. and Gomez, H. 2017 (11). An Introduction to Pure Spinor Superstring Theory. In: *Quantization, Geometry and Noncommunicative Structures in Mathematics and Physics*. Mathematical Physics Studies. Cham: Springer.

Berkovits, N., Sen, A., and Zwiebach, B. 2000. Tachyon Condensation in Superstring Field Theory. *Nucl. Phys. B*, **587**, 147–78.

Bern, Z., Dennen, T., Huang, Y.-T., and Kiermaier, M. 2010a. Gravity as the Square of Gauge Theory. *Phys. Rev. D*, **82**, 065003.

Bern, Z., Carrasco, J. J. M., and Johansson, H. 2010b. Perturbative Quantum Gravity as a Double Copy of Gauge Theory. *Phys. Rev. Lett.*, **105**, 061602.

Binetruy, P. 2008. *Supersymmetry*. Oxford: Oxford University Press.

Blumenhagen, R. and Plauschinn, E. 2009. *Introduction to Conformal Field Theory*. Lecture Notes in Physics, vol. 779. Berlin/Heidelberg: Springer.

Blumenhagen, R., Luest, D., and Theisen, S. 2013. *Basic Concepts of String Theory*. Berlin/Dordrecht/Heidelberg/London/New York: Springer.

Bump, D. 2004. *Lie Groups*. New York: Springer.

Cahn, R. N. 1984. *Semi-Simple Lie Algebras and Their Representations*. Benjamin and Cummings.

Callan, Jr, Curtis G., Martinec, E. J., Perry, M. J., and Friedan, D. 1985. Strings in Background Fields. *Nucl. Phys. B*, **262**, 593–609.

Candelas, P. 1987. Lectures on Complex Manifolds. In: Alvarez-Gaume, L. (ed.), *Superstrings '87*. Singapore: World Scientific.

Coleman, S. 1985. *Aspects of Symmetry. Selected Erice Lectures*. Cambridge: Cambridge University Press.

Cornwell, R. 1997. *Group Theory in Physics: An Introduction (vols 1 & 2)*. New York: Academic Press.

de Azcarraga, J. A. and Izquierdo, J. M. 1995. *Lie Groups, Lie Algebras, Cohomology and Some Applications in Physics*. Cambridge: Cambridge University Press.

de Boer, J., Cheng, M. C. N., Dijkgraaf, R., Manschot, J., and Verlinde, E. 2006. A Farey Tail for Attractor Black Holes. *JHEP*, **11**, 024.

de Wit, B. and Louis, J. 1999. Supersymmetry and Dualities in Various Dimensions. *NATO Sci. Ser. C*, **520**, 33–101.

Deligne, P., Etingof, P., Freed, D. S. et al. (eds.). 1999. *Quantum Fields and Strings: A Course for Mathematicians*. Providence, RI: American Mathematical Society.

D'Hoker, E. and Phong, D. H. 1988. The Geometry of String Perturbation Theory. *Rev. Mod. Phys.*, **60**, 917.

Di Francesco, P., Mathieu, P., and Sénéchal, D. 1997. *Conformal Field Theory*. New York: Springer.

Dijkgraaf, R., Verlinde, E. P., and Verlinde, H. L. 1988. C = 1 Conformal Field Theories on Riemann Surfaces. *Commun. Math. Phys.*, **115**, 649–90.

Dijkgraaf, R., Maldacena, J. M., Moore, G., and Verlinde, E. P. 2000. *A Black Hole Farey Tail*. E-print arXiv:hep-th/0005003.

Dine, M. 2007. *Supersymmetry and String Theory*. Cambridge: Cambridge University Press.

Dirac, P. A. M. 1964. *Lectures on Quantum Mechanics*. Belfare Graduate School Monograph Series. New York: Academic Press.

Dixon, L. J., Friedan, D., Martinec, E. J., and Shenker, S. H. 1987. The Conformal Field Theory of Orbifolds. *Nucl. Phys. B*, **282**, 13–73.

Dixon, L. J., Harvey, J. A., Vafa, C., and Witten, E. 1985. Strings on Orbifolds. *Nucl. Phys. B*, **261**, 678–86.

Dixon, L. J., Harvey, J. A., Vafa, C., and Witten, E. 1986. Strings on Orbifolds. 2. *Nucl. Phys. B*, **274**, 285–314.

Doubek, M., Jurco, B., Markl, M., and Sachs, I. 2020. *Algebraic Structures of String Field Theory*. Lecture Notes in Physics. Berlin/Heidelberg: Springer.

Duff, M. J., Nilsson, B. E. W., and Pope, C. N. 1986. Kaluza–Klein Supergravity. *Phys. Rept.*, **130**, 1–142.

Duncan, A. 2012. *The Conceptual Framework of Quantum Field Theory*. Oxford: Oxford University Press.

Erbin, H. 2021. *String Field Theory*. Lecture Notes in Physics. Berlin/Heidelberg: Springer.

Erler, T. 2013. Analytic Solution for Tachyon Condensation in Berkovits' Open Superstring Field Theory. *JHEP*, **11**, 007.

Erler, T. 2019 (12). *Four Lectures on Analytic Solutions in Open String Field Theory*. E-print arXiv:1912.00521.

Erler, T. 2020. Four Lectures on Closed String Field Theory. *Phys. Rept.*, **851**, 1–36.

Frankel, T. 2004. *The Geometry of Physics*. Cambridge: Cambridge University Press.

Freedman, D. Z. and Van Proeyen, A. 2012. *Supergravity*. Cambridge: Cambridge University Press.

Frenkel, E. and Ben-Zvi, D. 2001. *Vertex Algebras and Algebraic Curves*. Providence, RI: American Mathematical Society.

Friedan, D., Martinec, E. J., and Shenker, S. H. 1986. Conformal Invariance, Supersymmetry and String Theory. *Nucl. Phys. B*, **271**, 93–165.

Fuchs, J. and Schweigert, C. 1997. *Symmetries, Lie Algebras and Representations*. Cambridge: Cambridge University Press.

Fulton, W. and Harris, J. 1991. *Representation Theory*. New York: Springer.

Gaberdiel, M. R. and Suchanek, P. 2012. Limits of Minimal Models and Continuous Orbifolds. *JHEP*, **03**, 104.

Gilmore, R. 1974. *Lie Groups, Lie Algebras and Some of Their Applications*. London/New York/Sydney/Toronto: John Wiley & Sons.

Ginsparg, P. H. 1988. *Applied Conformal Field Theory*. E-print hep-th/9108028.

Giveon, A., Porrati, M., and Rabinovici, E. 1994. Target Space Duality in String Theory. *Phys.Rept.*, **244**, 77–202.

Glimm, J. and Jaffe, A. 1981. *Quantum Physics. A Functional Integral Point of View*. New York: Springer.

Goddard, P. and Olive, D. I. 1986. Kac–Moody and Virasoro Algebras in Relation to Quantum Physics. *Int. J. Mod. Phys. A*, **1**, 303.

Green, M. B., H., Schwarz J., and Witten, E. 1987. *Superstring Theory (2 vols)*. Cambridge: Cambridge University Press.

Greene, B. R. 1996. String Theory on Calabi–Yau Manifolds. Pages 543–726 of: *Fields, Strings and Duality. Proceedings, Summer School, Theoretical Advanced Study Institute in Elementary Particle Physics, TASI'96, Boulder, USA, June 2-28, 1996.*

Gutperle, M. and Strominger, A. 2002. Spacelike Branes. *JHEP*, **04**, 018.

Haag, R. 1996. *Local Quantum Physics*. New York: Springer.

Hamermesh, M. 1962. *Group Theory and its Application to Physical Problems*. Reading, MA: Addison Wesley.

Hamidi, S. and Vafa, C. 1987. Interactions on Orbifolds. *Nucl. Phys. B*, **279**, 465–513.

Helgason, S. 1978. *Differential Geometry, Lie Groups, and Symmetric Spaces*. Academic Press.

Henneaux, M. and Teitelboim, C. 1992. *Quantization of Gauge Systems*. Princeton, NJ: Princeton University Press.

Hohm, O. and Samtleben, H. 2019. *The Many Facets of Exceptional Field Theory*. E-print arXiv:1905.08312.

Hori, K., Katz, S., Klemm, A. et al. (eds). 2003. *Mirror Symmetry*. Clay Mathematics Monographs, vol. 1. Amer. Math. Soc.

Horowitz, G. T. and Polchinski, J. 1997. A Correspondence Principle for Black Holes and Strings. *Phys. Rev. D*, **55**, 6189–6197.

Hubsch, T. 1991. *Calabi–Yau Manifolds*. Singapore: World Scientific.

Hull, C. M. 1998. Timelike T-duality, de Sitter Space, Large N Gauge Theories and Topological Field Theory. *JHEP*, **07**, 021.

Hull, C. M. 2001. De Sitter Space in Supergravity and M Theory. *JHEP*, **11**, 012.

Humphreys, J. E. 1972. *Introduction to Lie Algebras and Representation Theory*. New York: Springer.

Ibáñez, L. E. and Uranga, A. M. 2012. *String Theory and Particle Physics*. Cambridge: Cambridge University Press.

Johnson, C. V. 2003. *D-Branes*. Cambridge: Cambridge University Press.

Kac, V. G. 1990. *Infinite Dimensional Lie Algebras*. Cambridge: Cambridge University Press.

Kaku, M. 1988. *Introduction to Superstrings*. New York: Springer.

Kaku, M. 1991. *Strings, Conformal Fields and Topology*. New York: Springer.

Kaplunovsky, V. and Louis, J. 1994. Field Dependent Gauge Couplings in Locally Supersymmetric Effective Quantum Field Theories. *Nucl. Phys. B*, **422**, 57–124.

Kaplunovsky, V. and Louis, J. 1995. On Gauge Couplings in String Theory. *Nucl. Phys. B*, **444**, 191–244.

Kawai, H., Lewellen, D. C., and Tye, S.-H. H. 1986. A Relation Between Tree Amplitudes of Closed and Open Strings. *Nucl. Phys. B*, **269**, 1.

Kawai, H., Lewellen, D. C., and Tye, S.-H. H. 1987. Construction of Fermionic String Models in Four-Dimensions. *Nucl. Phys. B*, **288**, 1.

Kiritsis, E. 2007. *String Theory in a Nutshell*. Princeton, NJ: Princeton University Press.

Lawson, H. B. and Michelsohn, M.-L. 1989. *Spin Geometry*. Princeton, NJ: Princeton University Press.

Lerche, W., Schellekens, A. N., and Warner, N. P. 1989. Lattices and Strings. *Phys. Rept.*, **177**, 1.

Maharana, J. and Schwarz, J. H. 1993. Noncompact Symmetries in String Theory. *Nucl. Phys.*, **B390**, 3–32.

Maldacena, J. M. 1996. *Black Holes in String Theory*. Ph thesis, Princeton University.

Marnelius, R. 1982. Introduction to the Quantization of General Gauge Theories. *Acta Phys. Polon. B*, **13**, 669.

Narain, K. S. 1986. New Heterotic String Theories in Uncompactified Dimensions < 10. *Phys. Lett.*, **B169**, 41.

Narain, K. S., Sarmadi, M. H., and Witten, E. 1987. A Note on Toroidal Compactification of Heterotic String Theory. *Nucl. Phys.*, **B279**, 369.

Nieto, J. A. 2001. Remarks on Weyl Invariant P-Branes and Dp-Branes. *Mod. Phys. Lett. A*, **16**, 2567–78.

Ooguri, H. and Vafa, C. 2007. On the Geometry of the String Landscape and the Swampland. *Nucl. Phys. B*, **766**, 21–33.

Ortin, T. 2004. *Gravity and Strings*. Cambridge: Cambridge University Press.

Peskin, M. E. and Schroeder, D. V. 1996. *An Introduction to Quantum Field Theory*. Reading, MA: Addison Wesley.

Plauschinn, E. 2019. Non-geometric Backgrounds in String Theory. *Phys. Rept.*, **798**, 1–122.

Polchinski, J. 1996 (11). Tasi Lectures on D-Branes. In: *Theoretical Advanced Study Institute in Elementary Particle Physics (TASI 96): Fields, Strings, and Duality*.

Polchinski, J. 1998a. *String Theory. Vol. 1: An Introduction to the Bosonic String*. Cambridge: Cambridge University Press.

Polchinski, J. 1998b. *String Theory. Vol. 2: Superstring Theory and Beyond.* Cambridge: Cambridge University Press.

Polyakov, A. M. 1981a. Quantum Geometry of Bosonic Strings. *Physics Letters B,* **103**(3), 207–10.

Polyakov, A. M. 1981b. Quantum Geometry of Fermionic Strings. *Physics Letters B,* **103**(3), 211–13.

Polyakov, A. M. 1987. *Gauge Fields and Strings.* Chur: Harwood Academic Publishers.

Rajaraman, R. 1982. *Solitons and Instantons. An Introduction to Solitons and Instantons in Quantum Field Theory.* Amsterdam: North Holland.

Ramond, P. 2010. *Group Theory.* Cambridge: Cambridge University Press.

Rocek, M. and Verlinde, E. P. 1992. Duality, Quotients, and Currents. *Nucl. Phys. B,* **373**, 630–46.

Roepstorff, G. 1994. *Path Integral Approach to Quantum Physics.* Berlin/Heidelberg: Springer.

Runkel, I. and Watts, G. M. T. 2002. A Nonrational CFT with Central Charge 1. *Fortsch. Phys.,* **50**, 959–65.

Sakamoto, M. 1989. A Physical Interpretation of Cocycle Factors in Vertex Operator Representations. *Phys. Lett. B,* **231**(3), 258.

Scherk, J. 1975. An Introduction to the Theory of Dual Models and Strings. *Rev. Mod. Phys.,* **47**, 123–64.

Scherk, J. and Schwarz, J. H. 1979. How to Get Masses from Extra Dimensions. *Nucl. Phys. B,* **153**, 61–88.

Schomerus, V. 2017. *A Primer on String Theory.* Cambridge: Cambridge University Press.

Schottenloher, M. 1997. *A Mathematical Introduction to Conformal Field Theory.* Berlin/Heidelberg: Springer.

Schubert, C. 2001. Perturbative Quantum Field Theory in the String Inspired Formalism. *Phys. Rept.,* **355**, 73–234.

Sen, A. 1998. Tachyon Condensation on the Brane Anti-Brane System. *JHEP,* **08**, 012.

Sen, A. 2018. Background Independence of Closed Superstring Field Theory. *JHEP,* **02**, 155.

Serre, J.-P. 1973. *A Course in Arithmetic.* New York: Springer.

Sexl, R. U., and Urbantke, H. K. 2001. *Relativity, Groups, Particles.* New York/ Vienna: Springer.

Siegel, W. 1988. *Introduction to String Field Theory.* Singapore: World Scientific.

Sonoda, H. 1988a. Sewing Conformal Field Theories I. *Nucl. Phys. B,* **311**(2), 401–16.

Sonoda, H. 1988b. Sewing Conformal Field Theories II. *Nucl. Phys. B,* **311**(2), 417–432.

Streater, R. F. and Wightman, A. S. 1964. *PCT, Spin and Statistics, and All That.* W.A. Benjamin, Inc.

Strominger, A., Yau, S.-T., and Zaslow, E. 1996. Mirror Symmetry is T-Duality. *Nucl. Phys. B,* **479**, 243–259.

Strominger, A. and Vafa, C. 1996. Microscopic Origin of the Bekenstein–Hawking Entropy. *Phys. Lett.*, **B379**, 99–104.

Sudarshan, E. C. G. and Mukunda, N. 1974. *Classcial Mechanics: A Modern Perspective*. Singpore: World Scientific.

Sundermeyer, K. 1982. *Constrained Dynamics*. Springer Lecture Notes in Physics 169. Berlin/Heidelberg/New York: Springer.

Townsend, P. K. 1997. *M-Theory from Its Superalgebra*. E-print hep-th/9712004.

Tseytlin, A. A. 1989. Sigma Model Approach to String Theory. *Int. J. Mod. Phys. A*, **4**, 1257.

van Beest, M., Calderón-Infante, J., Mirfendereski, D., and Valenzuela, I. 2021 (2). *Lectures on the Swampland Program in String Compactifications*. E-print arXiv:2102.01111.

Van Proeyen, A. 1999. *Tools for Supersymmetry*. E-print hep-th/9910030.

Wald, R. M. 1984. *General Relativity*. Chicago: University of Chicago Press.

Weinberg, E. 2012. *Classical Solutions in Quantum Field Theory*. Cambridge: Cambridge University Press.

Weinberg, S. 1987. *Covariant Path Integral Approach to String Theory*. 3rd Jerusalem Winter School in Theoretical Physics.

Weinberg, S. 1995. *The Quantum Theory of Fields*. Cambridge: Cambridge University Press.

Wess, J. and Bagger, J. 1992. *Supersymmetry and Supergravity*. Princeton, NJ: Princeton University Press.

West, P. 1986. *Introduction to Supersymmetry and Supergravity*. Singapore: World Scientific.

West, P. 2012. *Introduction to Strings and Branes*. Cambridge: Cambridge University Press.

Weyl, H. 1939. *The Classical Groups*. Princeton, NJ: Princeton University Press.

Woit, P. 2017. *Quantum Theory, Groups and Representations*. Cham: Springer.

Zwiebach, B. 2009. *A First Course in String Theory*. Cambridge: Cambridge University Press.

Index

Printed in the United States
by Baker & Taylor Publisher Services